Novel Aspects of Reproductive Physiology

Novel Aspects of Reproductive Physiology

Proceedings of the
Seventh Brook Lodge Workshop on
Problems of Reproductive Physiology

Edited by

Charles H. Spilman, Ph.D.
John W. Wilks, Ph.D.

Senior Research Scientists
Fertility Research
The Upjohn Company
Kalamazoo, Michigan

SP

SP MEDICAL & SCIENTIFIC BOOKS
a division of Spectrum Publications, Inc.
New York • London

Distributed by Halsted Press
A Division of John Wiley & Sons

New York Toronto London Sydney

SPECTRUM PUBLICATIONS, INC.
175-20 Wexford Terrace, Jamaica, N.Y. 11432

Library of Congress Cataloging in Publication Data

Brook Lodge Workshop on Problems of Reproductive Biology,
7th, 1977.
Novel aspects of reproductive physiology.

Includes index.
1. Reproduction--Congresses. I. Spilman, Charles
Hadley, 1942- II. Wilks, John W. III. Title.
[DNLM: 1. Reproduction--Congresses. W3 BR805 7th 1977
/ WQ205 B871 1977n]
QP251.B7296 1977 599'.01'6 77-28679
ISBN 0-89335-050-8

Distributed solely by the Halsted Press Division of John Wiley & Sons, Inc., New York, New York. ISBN 0-470-26368-7

Contributors

Terry G. Baker
Department of Obstetrics and Gynecology
University of Edinburgh
Edinburgh, Scotland

William H. Beers
The Rockefeller University
New York, New York

Henning M. Beier
Department of Anatomy
Med.-Thor. Institute
Aachen, West Germany

Hugh Bialecki
Cancer Research Institute
School of Medicine
University of California
San Francisco, California

Bent G. Boving
Departments of Obstetrics, Gynecology, and Anatomy
Wayne State University
Detroit, Michigan

Cyril Y. Bowers
Tulane University School of Medicine
New Orleans, Louisiana

Thomas A. Bramley
Department of Molecular Medicine
Mayo Clinic
Rochester, Minnesota

Kenneth D. Brown
Cancer Research Institute
School of Medicine
University of California
San Francisco, California

David W. Bullock
Department of Cell Biology
Baylor College of Medicine
Houston, Texas

Cornelia P. Channing
Department of Physiology
School of Medicine
University of Maryland
Baltimore, Maryland

Hao-Chia Chen
Reproduction Research Branch
NICHD
National Institutes of Health
Bethesda, Maryland

Hans W. Denker
Department of Anatomy
Med.-Theor. Institute
Aachen, West Germany

Allen C. Enders
Department of Human Anatomy
School of Medicine
University of California
Davis, California

Muriel Feigelson
Departments of Obstetrics,
Gynecology, and Biochemistry
Columbia University
New York, New York

Karl Folkers
Institute for Biomedical Research
The University of Texas
Austin, Texas

Stefan Fuchs
Institute for Biomedical Research
The University of Texas
Austin, Texas

John E. Gadsby
Agricultural Research Council
Institute of Animal Physiology
Babraham, Cambridge, England

Earl S. Gerard
Fertility Research
The Upjohn Company
Kalamazoo, Michigan

Ernest A. Gilling
Fertility Research
The Upjohn Company
Kalamazoo, Michigan

Denis Gospodarowicz
Cancer Research Institute
School of Medicine
University of California
San Francisco, California

Kenneth N. Gray
Section of Experimental Animals
The University of Texas
M.D. Anderson Hospital and Tumor Institute
Houston, Texas

R. Brian Heap
Agricultural Research Council
Institute of Animal Physiology
Babraham, Cambridge, England

Gary D. Hodgen
Reproduction Research Branch
NICHD
National Institutes of Health
Bethesda, Maryland

John Humphries
Institute for Biomedical Research
The University of Texas
Austin, Texas

John H. Jardin
Section of Experimental Animals
The University of Texas
M.D. Anderson Hospital and Tumor Institute
Houston, Texas

Frances A. Kimball
Fertility Research
The Upjohn Company
Kalamazoo, Michigan

Kenneth T. Kirton
Fertility Research
The Upjohn Company
Kalamazoo, Michigan

Alison Schwartz Kripner
Department of Physiology
School of Medicine
University of Maryland
Baltimore, Maryland

Thomas J. Lobl
Fertility Research
The Upjohn Company
Kalamazoo, Michigan

Janice R. Lorenzen
Department of Biological Sciences
Northwestern University
Evanston, Illinois

Harry J. Lynch
Department of Nutrition and Food Science
Massachusetts Institute of Technology
Cambridge, Massachusetts

Shuji Matsuura
Reproduction Research Branch
NICHD
National Institutes of Health
Bethesda, Maryland

William L. Miller
Fertility Research
The Upjohn Company
Kalamazoo, Michigan

Andrew V. Nalbandov
Animal Genetics Laboratory
University of Illinois
Urbana, Illinois

William D. Peckham
Department of Physiology
School of Medicine
University of Pittsburgh
Pittsburgh, Pennsylvania

Seymour H. Pomerantz
Department of Physiology
School of Medicine
University of Maryland
Baltimore, Maryland

Leo E. Reichert, Jr.
Department of Physiology
School of Medicine
University of Maryland
Baltimore, Maryland

Russel J. Reiter
Department of Anatomy
University of Texas
Health Science Center
San Antonio, Texas

Nancy D. Reichert
Reproduction Research Branch
NICHD
National Institutes of Health
Bethesda, Maryland

Griff T. Ross
Reproduction Research Branch
NICHD
National Institutes of Health
Bethesda, Maryland

Robert J. Ryan
Department of Molecular Medicine
Mayo Clinic
Rochester, Minnesota

Naguib A. Samaan
Department of Medicine
The University of Texas
M.D. Anderson Hospital and Tumor Institute
Houston, Texas

Paul C. Schwallie
Fertility Research
The Upjohn Company
Kalamazoo, Michigan

Neena B. Schwartz
Department of Biological Sciences
Northwestern University
Evanston, Illinois

Charles H. Spilman
Fertility Research
The Upjohn Company
Kalamazoo, Michigan

Sarah Lipford Stone
Department of Physiology
School of Medicine
University of Maryland
Baltimore, Maryland

Sidney Strickland
The Rockefeller University
New York, New York

Ari Van Tienhoven
Cornell University
Ithaca, New York

Israel Vlodavaky
Cancer Research Institute
School of Medicine
University of California
San Francisco, California

Yieh-Ping Wan
Institute for Biomedical Research
The University of Texas
Austin, Texas

Darrell N. Ward
Department of Biochemistry
The University of Texas
M.D. Anderson Hospital and Tumor Institute
Houston, Texas

John W. Wilks
Fertility Research
The Upjohn Company
Kalamazoo, Michigan

Kuo-Pao Paul Yang
Department of Medicine
M.D. Anderson Hospital and Tumor Institute
Houston, Texas

H.L. Nancy Yen
Department of Medicine
The University of Texas
M.D. Anderson Hospital and Tumor Institute
Houston, Texas

Preface

It has become strikingly apparent that the multiplicity of events in the reproductive process cannot be solely attributed to the gonadotropins or the gonadal steroids, nor can they be simply explained by the classical regulatory interrelationships of the hypothalamus and pituitary, or the gonad and its target organs. During recent years new biological activities have been observed in tissues and body fluids, or their extracts. New chemical entities have been observed to appear in the reproductive tissues at discrete times or specific cellular sites. New interactions between tissues and cells of the reproductive organs have been proposed. The purpose of the Seventh Brook Lodge Workshop on Problems of Reproductive Biology was to focus on the current status of knowledge regarding several of these novel regulatory molecules and concepts, and to provide an impetus for further advances in these areas.

The spiraling increase in the world's population continues to place constraints on available resources and the freedoms of mankind. The quality of life for the future will depend in part on the ability of man to control his fertility. New information and insights on the biology of reproduction will identify new regulatory mechanisms; hopefully, these novel sites of control will yield new and safe approaches to contraception. We dedicate this book to the achievement of that goal.

These proceedings represent the generous efforts of many individuals. We wish to thank the participants of the Workshop for preparing the manuscripts and for their valuable discussions. Drs. T. G. Baker M. Feigelson, D. Gospodarowicz, and W. D. Peckham deserve recognition for serving as session chairpersons. Many members of Fertility Research at The Upjohn Company assisted in the planning of the program. The staff of Brook Lodge extended their gracious hospitality and provided a comfortable environment in which to conduct the Workshop. The efforts of Diane Brush, Judy Woods, Lois Vande Giessen, and Jean Ericksen in transcribing the discussions are sincerely appreciated.

Charles H. Spilman
John W. Wilks

Kalamazoo, Michigan
November 1977

Contents

EMBRYO-UTERINE INTERACTIONS

PITUITARY AND GONADAL REGULATION

Novel Aspects of Reproductive Physiology

RETROSPECTS AND PROSPECTS

Retrospects and Prospects in Reproductive Physiology

A. V. NALBANDOV

There is a certain amount of satisfaction in looking back on roughly forty years of research activity, part of a period which included some of the most exciting episodes in sorting out the interrelations between the "master gland" and later the hypothalamus and of their target glands and tissues. The early parts of that era had the great advantage that the entire field of endocrinology could be followed by perusing two or three journals. It was a time for a philosophical approach to our discipline which, it seems to me, is largely absent among present-day practitioners in the field. It was a time when one felt that scientific contributions made would last and serve as a foundation for subsequent contributions of value.

In my capacity as a reviewer for at least three journals which publish endocrine materials, I am more and more leaning to the conclusion that present-day research seems to involve a preponderance of papers in which measurements of this or that hormone are reported. To be sure such background information must be available before intelligent speculation about hormone interrelations and control mechanisms involved is possible. However, it seems that in contrast to the earlier eras of endocrine research fewer hypotheses are being tested. I, myself, have the feeling as I read the literature that, in general, we are not breaking new ground but simply packing the old ground more solidly.

As chairman of the Publications Committee of the Endocrine Society, I was startled to learn shortly before relinquishing that office, that the average "life-span" of a paper appearing in the two journals of that Society is less than five years. This was measured by the frequency with which papers were cited. Even the "best" papers managed to be cited five to ten times during the first two years after publication while most papers were not cited at all. Five years after publication even the most quoted work completely disappears from the citation indeces. This, it seems to me, is in distinct contrast to the frequency of citation enjoyed by the papers of the old masters such as Evans, Hammond, Parkes, Simpson, Hisaw, Fevold, Koch, Price, Moore, Greep, Long, Everett, Sawyer, Corner, Hartman, and many others who laid the corner stones of our discipline. It is indicative that now-a-days few of the younger colleagues recognize any of these names and are rarely able to identify any of their major contributions.

The enormous explosion in the number of publications which has occurred in the last decade is probably responsible for the fact how often one finds old studies redone using different and frequently more refined techniques. One glaring example of this is the fact that the fairly recent "discovery" that estrogen causes LH release totally failed to acknowledge that this has been known since 1939 when Casida, at the University of Wisconsin, showed that the injection of estradiol causes ovulation in sheep. Even though the then available assay methods were unable to show a rise in plasma LH, the resulting paper stated that the induced ovulations were most probably caused by release of hypophysial LH. Several similar examples of failure to acknowledge earlier work could be given. This is probably the reason why I read so many recent papers with a "deja vu" feeling that I have known this for a long time and I frequently ask myself: "so what is new in this contribution?"

Some thirty years ago my earliest excursion into scientific endeavors involved attempts to correct what seemed to be a defect in normal reproductive phenomena and dealt with attempts to decrease embryonal mortality. It has been known for a long time that both laboratory and domestic animals show an embryonal mortality of about thirty percent during gestation. It seemed logical to associate embryo mortality with hormone deficiency and try to prevent it by the injection of various proportions of estrogen and progesterone as well as the latter hormone alone. Having invested five years of my life into this project, I came to the inevitable and reluctant conclusion that this approach was totally ineffective. Today, thirty years later, embryonal mortality is still about thirty percent in most animals and, as far as I know, nobody is near a solution to this problem. However, this excursion into so-called relevant research taught me a valuable lesson, namely, that before trying to solve a practical problem it is essential to study and to understand the basic causes of physiological events to be modified.

I think, as a generalization, it can be stated that it is exceedingly difficult to modify genetically controlled and thus fixed physiological events such as estrous cycles, parturition, and embryonal mortality. Although some phases of such chain events can be manipulated, e.g., shifted, but not without a more or less severe penalty in terms of lowered reproductive efficiency.

There are examples in which such man-controlled modifications are very successful. One prime example is artificial insemination in which, however, no real shifts in physiological events are involved but only the assumption of the duties of one partner of the sex act by man. This technique benefits animal agriculture and without it modern animal husbandry would be unthinkable. There are also examples in which abnormalities present in herds can be corrected much to the long-term deteriment of mankind. It has been recently found that both $PGF_{2\alpha}$ and/or GnRH cause cows with cystic ovaries to recover and to resume normal cycles. Because the success rate of these treatments is high and because cystic ovaries in both cattle and women are well known to be genetically controlled characteristics, I consider this finding as being detrimental to the long-term survival of herds in which these cyst-prone females will eventually predominate and require an ever increasing intervention by man to keep them reproducing.

For many years one of the major ambitions of the reproductive physiologist has been to acquire the ability to control cyclicity of domestic animals as well as the time of parturition. A variety of drugs has been used in these attempts but, alas, many of the results obtained were disappointing. It turned out that a variety of problems emerged such as species differences in degree of response and lowered fertilizability of ova after synchronization. In the case of controlling time of parturition with prostaglandins or synthetic glucocorticoids, the onset of labor can be reliably advanced by at most two days in swine and an unpredictable number of days in cows. In both species the use of prostaglandins entails all kinds of undesirable side effects.

To summarize my retrospective musings I might say that the shifting of genetically controlled reproductive phenomena turned out to be much more difficult than it seemed in the early days of our discipline.

I suppose that we will acquire greater success of imposing controls after we have gained a greater insight into the basic endocrine mechanisms underlying cyclicity and other reproductive phenomena typical of females.

For many years I have puzzled over the fact that, while most physiological systems such as digestion, respiration, blood circulation, etc., are relatively constant across the different species, reproductive physiology is extremely diverse. This diversity begins with anatomic differences of the reproductive tract of females in that in some the ovary is encapsulated in a bursa, in others it is engulfed by an open fimbria; some have two vaginal openings, and a corresponding modification of the penis in the male which is bifurcated in the opossum; several species have two cervixes, others only one; several species have two separate uterine horns, others have almost lost horn-like uteri and expanded the body of the uterus to accommodate the conceptus; some species have developed specialized areas on the uterine surface on which fetal attachment occurs, in others contact between fetal membranes and maternal tissue is over the entire uterine surface. This list could be expanded indefinitely.

I would like to mention briefly one of the most fascinating aberrations of sexual reproduction known to me. It involves the Asian shrew called *Suncus murinus*. The female of this species shows no cyclic changes in the vagina, uterus, or ovaries, the latter showing follicles of almost microscopic size over prolonged periods. This state of affairs persists until the female meets a male. The two start a fierce fight involving shrieks of anger and much biting. If the male has a low plasma androgen level he not infrequently is killed by the female. If the male is not subordinate the fighting persists for a few minutes or may last for as long as thirty to forty minutes depending on the dominance of the male. Suddenly, the shrieks of the female change into a bird-like chirp, apparently as a sign that she is ready for mating. Immediately after mating follicular growth begins and within 8 hours follicles reach ovulatory size. Ovulation, which is induced, occurs sixteen to eighteen hours after mating.

Curiously, ovariectomized or pregnant shrews also mate. No regression of tissues which are estrogen-dependent in other mammals is seen after ovariectomy. In addition, exogenous sex steroids have no measurable effect on steroid dependent tissues of intact or castrated females. In contrast, steroid dependent tissues of males degenerate after castration and do respond to exogenous steroids by growth. At present no explanation is available for this most bizarre reproductive pattern although an educated guess would be that reproduction in shrews may be controlled by the adrenal glands. Obviously the sex behavior of shrews bears no resemblance whatever to the sex control systems operating in most other mammals.

Differences also exist in hormonal control systems which could be extensively documented for the various species. One excellent example with which I am most familiar and which will suffice to illustrate the point is the mechanism of formation and maintenance of the corpus luteum. As you know, the luteotrophic hormone was discovered by Astwood and Greep in rats and identified to be prolactin. This finding was immediately generalized and interpreted to mean that prolactin must be luteotrophic in all mammals. As we now know this turned out to be untrue in that it was eventually found that LH is the most common luteotrophic hormone. In some animals (hamsters), both LH and FSH are necessary for corpus luteum formation and function. In still others like the rabbit neither gonadotrophin is luteotrophic but both corpus luteum formation and maintenance depend on estrogen. In several other species estrogen plays an auxiliary role in the life of the corpora lutea.

In some species such as the mare and the woman, conception results in the production of different specific gonadotrophic hormones while in most other species no such hormones have been found or convincingly demonstrated in spite of diligent search. In at least one species, the pig, the corpus luteum of the cycle is autonomous in that if a pig is hypophysectomized anytime during the cycle but after ovulation, the corpus luteum continues to grow and to synthesize progesterone apparantly without any trophic hormone. This, I consider a most fascinating aspect in that apparently the single ovulatory release of LH is sufficient to activate all the biochemical systems necessary for the hypertrophy and hyperplasia of lutein cells and for their ability to cause the synthesis of progesterone for the duration of the cycle of twenty-one days.

The corpus luteum of the pregnant pig is, however, like that of all other species studied, totally dependent on trophic hormones. In most pregnant animals the luteotrophin in pregnancy is LH. This is also true for part of the pregnancy for the rat. Thus, in this species both prolactin and LH participate in the maintenance and function of the corpus luteum but during different portions of gestation.

I think it is unfortunate that many of our colleagues are unaware of the tremendous diversity of control systems involved in the reproductive process and are thus willing to generalize from the rat to other species. It was fortunate for me that by force of circumstances I was able to compare reproductive processes of several species including both domestic and laboratory species. I have always maintained that much is to be gained both pedagogically as far as graduate education is concerned, as well as from the point of view of a more sophisticated insight into reproductive phenomena by being constantly reminded of the diversity of anatomic and neuroendocrine control systems governing reproductive phenomena.

Where are we going in our discipline? My thoughts on this subject are not intended to be exhaustive but merely glimpses of problems to which I would like to have the answers.

I have long been fascinated by the fact that females living in the wild are always either anestrous, pregnant, pseudopregnant, or lactating. This is in distinct contrast to domesticated animals and women in whom economic or moral factors dictate that pregnancies occur at times deemed desirable by their owners or by society. This, in turn, means that females in these categories are undergoing many repeated estrual or menstrual cycles which in women may last for years or a lifetime. Thus, these females are exposed to constant alternation of hormones to which they are exposed, estrogen followed by progesterone dominance, and this, in turn, succeeded by estrogen dominance, etc. I am intrigued by the fact that aberrations of the reproductive system are rarely found or at least rarely reported in wild-living females, are much more common in domestic animals, and most common in women. It is, for instance, a matter of record that breast, and perhaps other cancers of reproductive organs are much more prevalent in nulliparous women than they are in parous and especially multiparous ones. Whether there is a valid or only a spurious correlation between menstrual cycles uninterrupted by pregnancy for prolonged intervals and the incidence of genital cancers in women remains an intriguing question deserving further study. It certainly could be studied in laboratory animals.

Of many unsolved problems I find myself greatly attracted to the role pheromones play in reproductive sexuality especially in domestic animals. In fact, if I were starting out in the field, I would certainly want to concentrate my efforts in this area. This interest in pheromones is inspired by the well-known data obtained by R. R. Michael *et al*. They showed that recognition of sexually receptive Macaque females depends on the secretion of estrogen-controlled five volatile fatty acids by the vagina. Even nonconcupiscant castrated female monkeys, whose rumps were smeared with a mixture of the five volatile fatty acids, were recognized as being sexually receptive. That this recognition depends on the ability of the male to perceive the odors is shown by the fact that, if the nostrils of males are plugged up, they ignore sexually receptive females indefinitely. By the way, the same volatile fatty acids are found in vaginal secretions of women.

Based on these pioneering studies, one wonders whether these or similar or totally dissimilar pheromonal substances (called "copulins") are present in various domestic animals and are used as signalling mechanisms to identify sexually receptive females to males. Nothing is known about how a bull or a ram recognizes females in heat often over considerable distances. It is highly probable that some sort of airborne copulins are involved. It would certainly be of theoretical and practical importance to identify them and to synthesize them. One practical application could be to cause mating in females in which ovulation is induced with exogenous hormones and many of whom would not be in estrous. In dairy cattle heat is frequently very short and/or very weakly expressed and goes undetected. If, as it apparently does, this condition contributes to difficulties of conception, the availability of an appropriate copulin would find an important practical application. Even if it does not turn out to be of practical importance it would be certainly interesting to discover similarities or differences between signalling mechanisms between some monkeys and subprimate species. This is a very largely unexplored territory.

I find equally fascinating Martha K. McClintock's findings of synchronization of menstrual cycles in women living in close association over prolonged periods of time. Thus, college freshman girls moving into the same room in a college dormatory may have temporally widely separated menstrual periods at the beginning of the semester while toward the end they show a highly significant trend toward synchronization of periods. Such synchronization also occurs in mothers and daughters living in the same house. Similarly, synchronization of estrous cycles occurs in mice if a male is present. These so-called Whitten and Bruce effects are now known to be due to an androgen-controlled male odor present in the urine of intact males. No explanation is available for the synchronization of menstrual periods of roommates described by McClintock. It would be of great interest to know how these major shifts in the neuroendocrinologically controlled hormone system is accomplished and especially to know just what happens as far as such parameters as LH and FSH peaks, and hence steroid levels, are concerned.

Women, like mice, are apparently also subject to a male pheromone. This is inferred from the fact that girls who date frequently have significantly more regular and shorter menstrual cycles with narrowed standard errors of mean than do girls who date infrequently or not at all. Again, nothing is known whether this significant effect on the regularity of menstrual cycles is due to some male pheromone.

That pheromones can actually activate the hypothalamo-hypophysial-ovarian axis is shown in recent experiments in which urine of male mice was applied daily for eight days to the oronasal groove of one group of juvenile female mice while distilled water was used on a control group. On the ninth day of the experiment, both groups of mice were killed and their uteri weighed. The group treated with water had uteri weighing 45.8 ± 6.8 mg while those treated with male urine had uteri weighing 73.4 ± 7.8 mg -- a highly significant difference. The pheromone in urine of male mice has not been identified but it is known that it has a molecular weight of 860. The active fraction contains no steroids but gives a strong positive test for peptides.

Not many more details can be given on this, to me, most fascinating subject of pheromones. Our ignorance on the role of these substances in reproductive phenomena in domestic animals and man is great.

In this short overview of one of the most fascinating fields of research I tried to expose my own prejudices. This summary version of my thoughts is by no means intended to be exhaustive in the treatment of the field. If I have stimulated the thinking of some of the younger people, my task has been accomplished.

INTRA-OVARIAN REGULATION

1

Involvement of Plasminogen Activator in Ovulation

WILLIAM H. BEERS and SIDNEY STRICKLAND

Introduction

A schematic representation of a mature ovarian follicle is shown in Figure 1. This structure consists of a fluid filled antrum which is surrounded by basement membrane lined with several layers of granulosa cells. These components of the follicle exist in an

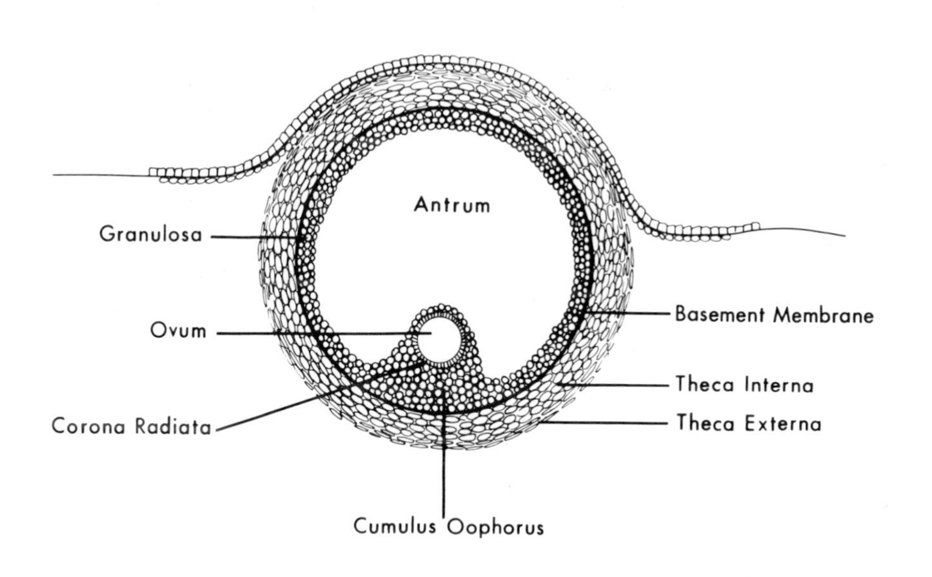

Figure 1. Schematic representation of an ovarian follicle.

avascular space. Thecal cell layers, which are vascularized, encompass the basement membrane, granulosa and antrum. At the time of ovulation all of these structures must be disrupted, and a tear must be made in the follicle wall in order for the ovum to escape from the ovary.

The biochemical mechanisms responsible for the breakdown of the follicle wall have not been established. It was proposed as early as 1916 that lytic enzymes may play a role in this process (Schochet, 1916), but efforts to demonstrate such substances proved to be difficult. It was shown, however, that certain proteolytic enzymes can weaken follicle wall strips *in vitro* (Espey, 1970) and that the injection of the serine protease trypsin, directly into the antrum of an intact follicle, will cause an event similar to ovulation (Espey, 1974).

Although the follicular compartment containing the granulosa cells and the antrum is avascular, the majority of the protein constituents of the follicular antral fluid are derived from serum. The barrier that separates the antrum from the circulation has properties that admit most blood proteins to the fluid at or near serum concentrations (Shalgi et al., 1973). This suggests that the zymogen of the serine protease plasmin, plasminogen, should be present in follicular fluid at concentrations between 0.1 and 1.0 mg/ml; levels that represent a substantial amount of potentially available protease. In addition, all of the known protease inhibitors of serum, with the exception of α_2 macroglobulin which is too large to be admitted, should gain access to follicular fluid. It is possible that these substances masked the activity of follicular proteases in earlier studies which failed to detect proteolytic activity in follicular specimens. Finally the enzyme that converts plasminogen to plasmin, plasminogen activator, has long been known to be present in the ovary (Albrechtsen, 1957). Little interest, however, has been shown in its potential as an enzyme that might have a role in ovulation.

When follicular fluid protease inhibitors are inactivated it is possible to detect plasminogen, plasminogen activator and their reaction product, the active protease plasmin, in follicular fluid specimens derived from mature mammalian follicles (Beers, 1975). Plasmin has many properties similar to trypsin and will weaken intact follicle wall strips (Beers, 1975). It has also been shown that plasmin will degrade basement membrane preparations (Bray et al., 1975). These properties, and the potentially high levels that plasmin can achieve within the follicle make it an attractive candidate as an enzyme instrumental in follicle wall degradation.

Correlation of Increased Plasminogen Activator with Ovulation

With the above observations in hand, studies were begun to determine whether the level of follicular plasmin changes in a way which would be suggestive of a role for it in ovulation. Plasmin activity could be modulated by changes in the level of any of the molecules under discussion: plasminogen, plasminogen activator or the protease inhibitors. Our studies suggest that the level of the zymogen and inhibitors remain constant in the follicle, presumably because these substances are serum borne and exchange freely in and out of the fluid. On the other hand, large increases in follicular plasminogen activator do occur near the time of ovulation (Beers et al., 1975; Strickland and Beers, 1976). Since, in principle, plasminogen activator could be derived from non-ovarian tissue and gain access to the follicle via the circulation, it was of immediate interest to determine whether the production of activator was ovarian cell associated. For our initial studies, granulosa cells from pre-ovulatory follicles were analyzed for plasminogen activator production using a sensitive fibrin agar overlay procedure described in Fig. 2. These determinations revealed that 25-30% of the cells in these preparations synthesize and secrete plasminogen activator. Moreover electron micrographic analysis of individual active cells, i.e. single cells associated with a lytic zone, established their identity as granulosa.

Having shown that the plasminogen activator which we had measured was granulosa cell associated, we then examined the extent to which its appearance was correlated with ovulation. In studies with immature rats induced to ovulate at a predictable time (see Table 1), granulosa cells from pre-ovulatory follicles were assayed using the procedure described in Fig. 3 which allows simple quantitation of the results. These cells were found to produce increasing amounts of activator during the hours preceding ovulation. The peak level of enzymatic activity, which represents an

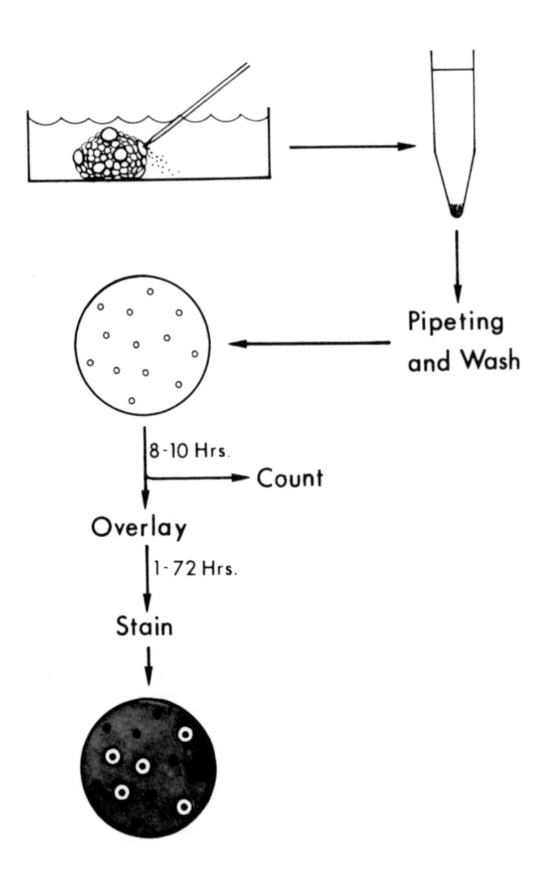

Figure 2. Procedure for detection of cell associated plasminogen activator production.

Cells were prepared from immature rats induced to ovulate as described in Table 1. At desired times after the second hormone injection, the ovaries were removed from the animals and the granulosa cells harvested by follicle puncture as previously described. After collection and washing by centrifugation the cells were allowed to attach to petri dishes under tissue culture conditions. After 10-12 hours the growth medium was removed and replaced with a mixture of fibrin, agar, thrombin, nutrient medium and serum. A fibrin clot then formed over the cell layer. The clot was dissolved locally in the vicinity of cells producing a fibrinolytic enzyme. Plasminogen activator is operationally defined as plasminogen dependent fibrinolysis. In these experiments, no fibrinolysis was observed if plasminogen was not included in the assay mixture. The lysis zones around active cells can be visualized as they develop, but are easier to score if the culture and fibrin agar overlay are stained with coomassie blue. This procedure is described in detail elsewhere (Beers et al., 1975).

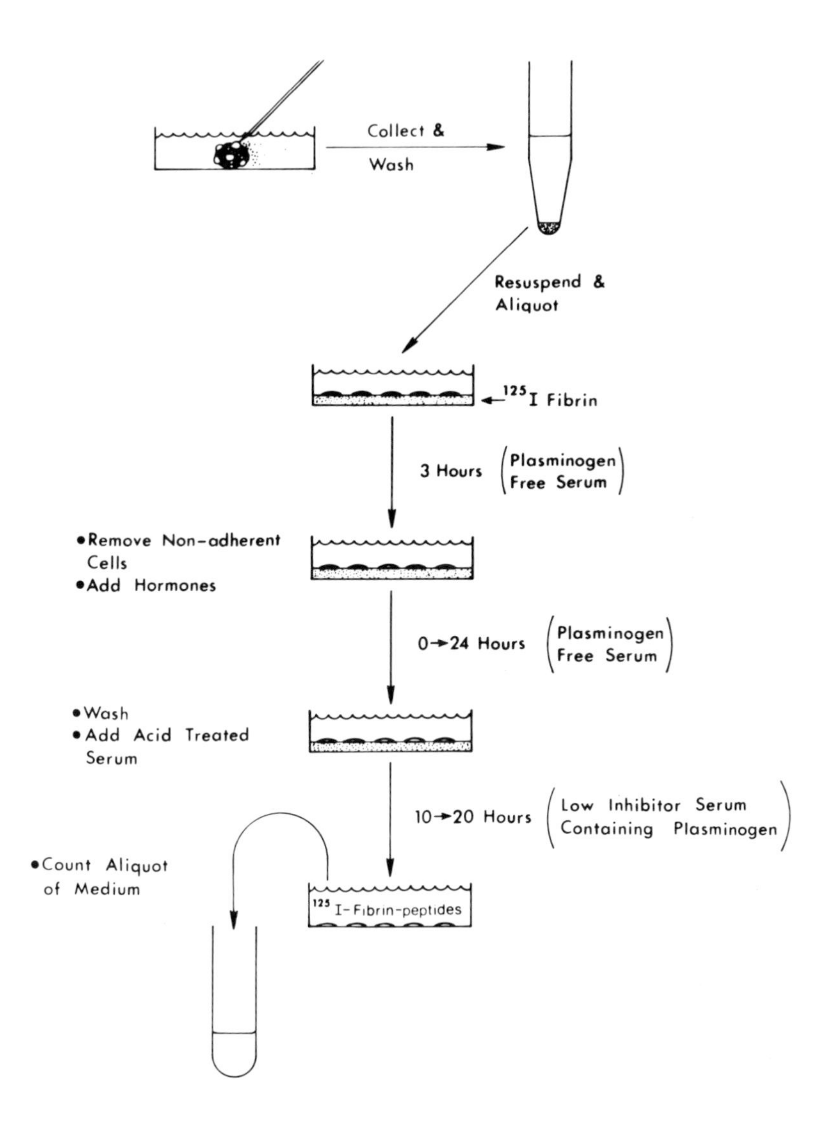
Collect &
Wash
Resuspend &
Aliquot
125I Fibrin
3 Hours
Plasminogen Free Serum
•Remove Non-adherent Cells
•Add Hormones
0→24 Hours
Plasminogen Free Serum
•Wash
•Add Acid Treated Serum
10→20 Hours
Low Inhibitor Serum Containing Plasminogen
•Count Aliquot of Medium
125I-Fibrin-peptides

Figure 3

Figure 3 Procedure for quantitation of plasminogen activator production by granulosa cells.

Cells were prepared as in figure 2 and then plated in growth medium containing plasminogen depleted serum at desired densities in tissue culture wells coated with ^{125}I-fibrin. After three hours the plated cells were washed, the medium replaced and the hormones added. After the desired period, the medium was again removed and replaced with medium containing serum which had been previously treated to reduce the level of protease inhibitors. If plasminogen activator is produced by the cells, it will convert the plasminogen, which is supplied by the serum supplement, to plasmin. This enzyme solubilizes the fibrin and fibrinolysis is monitored by analysis of the medium for soluble ^{125}I-labeled fibrin peptides. The procedure is linear with time and cell number and is described in detail elsewhere (Strickland and Beers, 1976).

increase of 50-100-fold, occurs near the time of ovulation; after this time there is a rapid decrease (Fig. 4). These changes are seen only in cells derived from follicles destined to ovulate. Granulosa cells from large, antral, but non-preovulatory follicles, prepared from the same ovaries at the same time as those from pre-ovulatory follicles showed no increase in enzymatic activity. Studies conducted with mature cycling rats yielded results consistent with those outlined above, i.e. increased levels of activity are only seen late on the night of proestrus, near the time of ovulation.

Analysis of hormonal regulation of plasminogen activator production by granulosa cells

Having established that the production of plasminogen activator by granulosa cells is closely correlated with ovulation, we then examined the endocrine control of this cellular function. Luteinizing hormone (LH) is generally regarded to be the gonadotropin that triggers ovulation and our initial studies were conducted with preparations of this hormone. These experiments were performed using cells harvested from large antral follicles obtained from immature rats as described in Table 1 and assayed as described in Figure 3. Although preparations of ovine LH caused a dose dependent increase in plasminogen activator production, ovine follicle stimulating hormone (FSH) and gonadotropins with predominantly FSH activity, e.g. pregnant mare serum gonadotropin (PMSG) were considerably more potent than LH (Table 2). Indeed all of the activity observed with the LH preparations could be accounted for on the basis of their bio-assayable FSH content. More recent, unpublished studies using highly purified rat FSH and LH reveal a 5-10,000 fold difference in their relative potencies. We feel that this result, which utilizes homologous rat gonadotropins to stimulate rat cells establishes that

Table 1. Gonadotropin injection regimens for studies of in vivo or in vitro stimulation of granulosa cells.

	1st injection		2nd injection	
	hormone	dose	hormone	dose
in vivo studies	PMSG	5 IU	hCG	25 IU
	PMSG	5 IU	ovine-LH	100 mg
	PMSG	5 IU	ovine-FSH	100 mg
in vitro studies	PMSG	5 IU	NONE	

Immature female rats, 26 days old, were administered 5 IU PMSG, subcutaneously. After 48 hours, they were either injected with a second gonadotropin, which caused ovulation 8-12 hours later, or decapitated and dissected for immediate preparation of granulosa cells.

PMSG: Pregnant mare serum gonadotropin
hCG: Human chorionic gonadotropin
FSH: Follicle stimulating hormone
LH: Luteinizing hormone

Table 2. Relative potencies of gonadotropins as effectors of plasminogen activator production in vitro.

Gonadotropin	Concentration providing half maximal stimulation
ovine-LH (S-19)	3 μg/ml
ovine-FSH (S-11)	0.1 μg/ml
hCG (pregnyl)	300 units/ml
PMSG (gestyl)	0.3 units/ml

The granulosa cells were prepared and analyzed as described in Figure 3. For each of the hormones listed, full dose-response curves were determined.

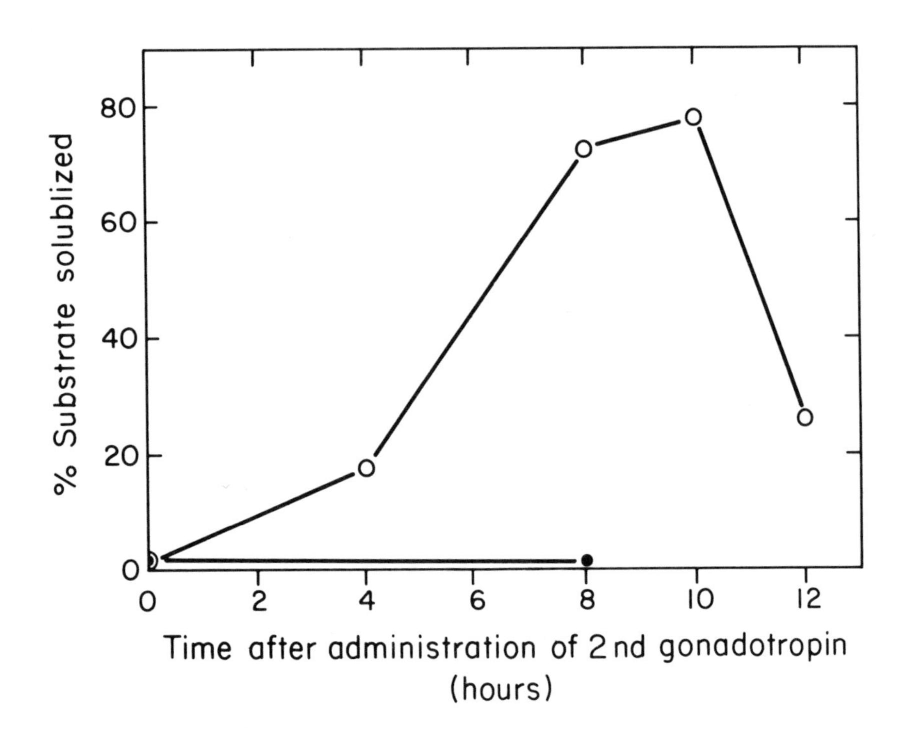

Figure 4. Correlation of the production of plasminogen activator by granulosa cells and ovulation.

Immature rats were administered the hormone regimen described in Table 1. After the second gonadotropin injection, in this case hCG, animals were sacrificed and the granulosa cells were prepared, separately, from pre-ovulatory and non-preovulatory follicles. The cells were assayed as in Figure 3 except that they were not exposed to any hormones.

The results are reported as the fraction of available ^{125}I-fibrin solubilized. Under the conditions employed, ovulation occurs 10-12 hours after the second injection. At 0 hours it is not possible to distinguish between pre-ovulatory and non-preovulatory follicles. At other times the open circles represent the values obtained with cells from pre-ovulatory follicles, the closed circles from non-preovulatory follicles.

the response is specific for FSH.

The effect of the gonadotropins is amplified if cyclic nucleotide phosphodiesterase inhibitors, such as theophylline or methylisobutyl xanthine, are included in the incubation media. This result suggests the involvement of cyclic nucleotides in the gonadotropin induced response.

Studies of the response of the cells to direct exposure to cyclic nucleotides support this contention: cyclic AMP and its derivatives cause a dose-dependent increase in plasminogen activator production by the cells; cyclic GMP and its derivatives, however, have no effect either as agonists or antagonists of the response. Finally, direct measurement of intracellular levels of cAMP in these cells, following FSH stimulation, demonstrate a dose dependent increase in this nucleotide that parallels the increase in plasminogen activator.* These results suggest that the FSH induced increase in plasminogen activator production by the granulosa cells is mediated by cAMP.

A number of other drugs and hormones have been tested on the *in vitro* granulosa cell system and these are tabulated in Table 3. Of the drugs tested, only the catecholamines and certain prostaglandins were effective. Although we had previously reported that norepinephrine and epinephrine were ineffective at concentrations up to 10^{-6}M, more recent studies with catecholamines at concentration between 10^{-6} and 10^{-4}M in the presence of cyclic nucleotide phosphodiesterase inhibitors reveal that the cells are capable of responding to these drugs. Furthermore, the β-adrenergic blocking agent propanolol effectively inhibits the response to the catecholamines.

Prostaglandins A, B and E are all effective in causing the granulosa cells to produce plasminogen activator. Many of the effects of these hormones are thought to be mediated by cyclic AMP and the response of the granulosa cells is amplified by cyclic nucleotide phosphodiesterase inhibitors. Prostaglandin $F_{1\alpha}$ and $F_{2\alpha}$ which generally are thought not to operate via cyclic AMP mediated mechanisms have no effect on the granulosa cell system. Although the response of the cells to saturating levels of FSH and the active

---

* Ginzberg, R., Lawrence, T., Gilula, N.B. and Beers, W.H., unpublished observations.

Table 3 Effect of substances other than gonadotropins on plasminogen activator production by granulosa cells.

Drug	Concentration range tested (M)	Approximate concentration producing half maximal stimulation (M)
Prostaglandin E_1	10^{-10} to 10^{-5}	5×10^{-7}
E_2	"	5×10^{-7}
A_1	"	2×10^{-6}
A_2	"	5×10^{-6}
B_1	"	$>10^{-5}$
B_2	"	1×10^{-5}
$F_{1\alpha}$	"	no response detected
$F_{2\alpha}$	"	"
Epinephrine	10^{-9} - 10^{-4}	$>10^{-4}$
Norepinephrine	10^{-9} - 10^{-4}	$>10^{-4}$
Dexamethasone	10^{-9} - 10^{-6}	no response detected
17β Estradiol	10^{-9} - 10^{-6}	"
Testosterone	10^{-9} - 10^{-6}	"
Progesterone	10^{-9} - 10^{-6}	"
17α-Hydroxyprogesterone	10^{-9} - 10^{-6}	"
20α-Dihydroprogesterone	10^{-9} - 10^{-6}	"

The granulosa cells were prepared and analyzed as described in Figure 3.

prostaglandins is not strictly additive, the magnitude of the response to both drugs is greater than to either one of them alone. In addition, the prostaglandin synthesis inhibitor indomethacin has no effect on the gonadotropin induced response. Thus it would appear that the prostaglandins and gonadotropins are operating via independent pathways which unite at some point, perhaps at the level of cyclic AMP. The prostaglandin antagonists, flufenamic acid and 7-oxa-13-prostynoic acid were without effect on the response of the cells.

Discussion

The studies described above demonstrate a close correlation between the production of plasminogen activator by granulosa cells and ovulation. This enzyme converts plasminogen, which is present in follicular fluid, to the active protease plasmin. Plasmin will digest follicle wall strips *in vitro* as well as preparations of basement membrane. These properties suggest the plasminogen activator/plasmin enzyme system could provide the mechanism necessary for follicle wall breakdown.

We have also examined the endocrine control of plasminogen activator production by these cells. The studies indicate that FSH, A, B and E prostaglandins and, at pharmacological doses, adrenergic catecholamines, will cause the cells to synthesize and secrete the enzyme.

The gonadotropin specificity of the response deserves some comment. LH is generally regarded to be the hormone that is responsible for ovulation (Schwartz and McCormack, 1972). If plasminogen activator is involved in events associated with ovulation, LH might be expected to effect its appearance within the follicle. Our experiments indicate that this is not the case. The gonadotropin induced response is apparently specific for FSH, a finding which may suggest a role for this gonadotropin in ovulation which has not been recognized previously. It should be noted that at the time of the pre-ovulating surge of LH there is a concommitant peak in circulating FSH (Gay et al., 1970). Furthermore, highly purified FSH, free of contaminating LH, will induce ovulation in the rat (Goldman and Mahesh, 1969; Harrington et al., 1970; Tsafriri et al., 1976). This is an area that needs to be more rigorously investigated, however it may be that LH and FSH are both involved in ovulation but that they are responsible for different biochemical events which are associated with that process.

The response of the cells to E-prostaglandins is of interest as well, since it has been reported that this substance reaches concentrations of 10^{-6}M within the follicle near the time of ovulation (LeMaire et al., 1975). According to our analyses, this level of prostaglandin is sufficient to cause greater than half maximal stimulation of plasminogen activator production by the granulosa cells. Although there are indications that adrenergic amines may influence certain events associated with ovulation (Kao and Nalbandov, 1972) the meaning of the catacholamine induced granulosa cell response is unclear. As seen above, the doses required are probably outside the physiologic range. It is possible that these drugs are structurally related to some other, more potent effector of the response and are acting as analogues of this hypothetical substance. This is a possibility we are currently investigating.

While our studies provide a novel and potentially meaningful insight into biochemical phenomena associated with ovulation, they also illustrate the need for further studies into this area of mammalian physiology. In particular, it is clear that many of our current concepts of the ovulatory process may have to be revised as more studies are conducted with highly purified hormones and defined cell types within the ovary.

Acknowledgements

We wish to express our appreciation to Dr. John Pike of Upjohn for his generous gift of prostaglandins and to Dr. Josef Fried of the University of Chicago for making available a sample of 7-oxa-13-prostynoic acid. The gonadotropins used in this study were provided by the National Institute of Arthritis, Metabolic and Digestive Diseases. This work was supported by grants from the Rockefeller Foundation (RF 70095), the National Institutes of Health (HD-09401), The American Cancer Society (BC-253) and the Jane Coffin Childs Memorial Fund for Medical Research. W.H.B. is a P.H.S. Research Career Development Awardee (HD-00189).

References

Albrechtsen, O. K. (1957) "The Fibrinolytic Activity of Human Tissues", British J. Haematol. 3, 284-291.

Beers, W. H., (1975) "Follicular Plasminogen and Plasminogen Activator and the Effect of Plasmin on Ovarian Follicle Wall", Cell 6, 379-386.

Beers, W. H., Strickland, S., and Reich, E., (1975) "Ovarian Plasminogen Activator: Relationship to Ovulation and Hormonal Regulation" Cell 6, 387-394.

Bray, B.A., Hsu, K. C., Wigger, H. J., and LeRoy, E. C., (1975) "Association of Fibrinogen and Microfibrils with Trophoblast Basement Membrane", Connect. Tissue Res. 3, 55-71.

Espey, L. L., (1970) "Effect of Various Substances on Tensile Strength of Sow Ovarian Follicles", Am. J. Physiol. 219, 230-233.

Espey, L. L., (1974) "Ovarian Proteolytic Enzymes and Ovulation", Biol. Reprod. 10, 216-235.

Gay, V. L., Midgley, A. R., Jr., and Niswender, G. D. (1970) "Patterns of Gonadotrophin Secretion Associated with Ovulation", Fed. Proc. 29, 1880-1887.

Goldman, B. D., and Mahesh, V. B., (1969) "A Possible Role of Acute FSH-Release in Ovulation in the Hamster, as Demonstrated by Utilization of Antibodies to LH and FSH", Endocrinol. 84, 236-243.

Harrington, F. E., Bex, F. J., Elton, R. L., and Roach, J. B. (1970) "The Ovulatory Effects of Follicle Stimulating Hormone Treated with Chymotrypsin in Chlorpromazine Blocked Rats", Acta. Endocr. 65, 222-228.

Kao, L. W. L., and Nalbandov, A. V. (1972) "The effect of antiadrenergic drugs on ovulation in hens", Endocrinology, 90: 1343-1349.

LeMaire, W. J., Leidner, R., and Marsh, J. M. (1975) "Pre and Post Ovulatory changes in the Concentration of Prostaglandins in Rat Graafian Follicles", Prostaglandins 9, 221-229.

Schochet, S. S., (1916) "A Suggestion as to the Process of Ovulation and Ovarian Cyst Formation", Anat. Rec. 10, 447-457.

Schwartz, N. B. and McCormack, C. E. (1972) "Reproduction: Gonadal Function and its Regulation", Ann. Rev. Physiol. 34, 425-472.

Shalgi, R., Kraicer, P., Rimon, A., Pinto, M., and Soferman, N., (1973) "Proteins of Human Follicular Fluid: The Blood-Follicle Barrier", Fertil. Steril. 24, 429-434.

Strickland, S., and Beers, W. H., (1976) "Studies on the Role of Plasminogen Activator in Ovulation", J. Biol. Chem. 251, 5694-5702.

Tsafriri, A., Lieberman, M. E., Koch, Y., Bauminger, S., Chobsieng, P., Zor, U., and Lindner, H. R., (1976) "Capacity of Immunologically Purified FSH to Stimulate Cyclic AMP Accumulation and Steroidogenesis in Graafian Follicles and to Induce Ovum Maturation and Ovulation in the Rat", Endocrinol. 98, 655-661.

DISCUSSION

Ryan: If one looks at the autoradiographic localization of labeled FSH, you don't find much binding by the large follicle cells. Also, at 48 hrs after PMSG or slightly after hCG, you find very little FSH stimulatable cyclase activity in rat ovaries, and if you take granulosa cells from large pre-ovulatory pig follicles, you find very little FSH stimulatable cyclase. If you don't have FSH receptors and if you don't have an FSH stimulatable cyclase in pre-ovulatory follicles, how is the FSH working?

Beers: First of all, I can't comment on the porcine studies. My feeling is that there probably are FSH receptors on the rat cells, but I'm not sure how to interpret the autoradiographic localization studies. For example, if you use radiolabeled LH and look for LH receptors it's very hard to demonstrate them; but if you use radiolabeled hCG you see a lot more. I can make a strong comment about the cyclic AMP levels, and that is that we see a roughly 20-fold increase in cyclic AMP following the exposure of granulosa cell cultures to FSH. Furthermore, by our calculations the intracellular concentration approaches 10^{-6} molar. So I don't know how to respond to the second part of your question, except that we find a large response. I can't comment on FSH receptors on these cells since we haven't done any receptor studies.

Channing: The rat and the pig may have different kinetics of follicular development in terms of FSH and LH receptors. A word of caution is that PMSG may alter the relative amounts of FSH receptors and LH receptors. In our studies in the monkey, PMSG causes an increase in both LH and FSH responsiveness. Have you done a study in normal proestrous rats in which the preovulatory FSH and LH surges are abolished by giving Nembutal? Under these conditions you could have follicles that won't ovulate and haven't been exposed to the surge of LH and FSH, but still can respond to gonadotropins. Have you taken these cells which are from mature follicles which should be reasonably physiological to examine, and examined their response to LH and FSH?

Beers: No, we haven't looked at those cells. We have done two other studies which bear on the question you raise. First, we've looked at fairly young rats, 15 to 26 days old, without any gonadotropin treatment and looked at the responsiveness of those cells to FSH. The amount of enzyme on a per cell basis and the dose response curves were identical to those I have shown. Second, we have used the cells that one obtains by the method that Griff Ross uses; hypophysectomized rats treated with large doses of DES. These cells also responded the same. We haven't found any gross evidence that the previous history of the animal is influencing the response.

Reichert: Some years ago in our studies on the purification of FSH and LH (Reichert and Parlow, Endocrinology 74:809, 1964) we found that very highly purified FSH preparations contained a hemoglobin splitting proteinase activity which could be separated from the FSH only with a great amount of difficulty, but which was separated from LH quite early in the purification scheme. Is there a possibility that the FSH preparation which was used in your *in vitro* system was contaminated with the proteinase activity which was carried through the purification?

Beers: It seems unlikely that the amount of FSH-associated protease would vary from preparation to preparation and from species to species exactly as does the Steelman-Pohley FSH activity of the preparations. I seriously doubt that we are looking at an activity that is independent of the cells. In fact not all of the studies involve *in vitro* stimulation of the cells. The studies of the cells from animals which received the hormones *in vivo* demonstrate an increase in plasminogen activator without *in vitro* exposure of the cells to FSH. In fact if you take out the ovaries or take out the follicles and homogenize and assay them you see essencially the same results.

Reichert: We also found that injections of FSH into hypophysectomized rats increased proteinase activity in extracts of rat ovary, whereas LH and hCG did not (Reichert, Endocrinology 70:697, 1962) and that the level of proteinase activity in saline extracts of ovaries from rats in estrus and diestrus were significantly higher than those from rats which were pseudopregnant or in late pregnancy (Reichert, Endocrinology 71:838, 1962).

Beers: We're not sure which hormone causes the increase that we see *in vivo* for the following reason. While FSH can cause ovulation in rats, so can LH which is reasonably free of FSH activity and hCG which has very little FSH activity. All of these gonadotropins will also elevate plasminogen activator levels when given in vivo. We're not sure what the solution to this apparent discrepancy is. I am tending to lean more and more towards the notion that the rise in activity that we see at ovulation is probably caused by both FSH and prostaglandins, and taking away the FSH won't make much difference as long as the prostaglandin levels increase.

Reiter: Considering that under certain circumstances prolactin is also released at the time of proestrus, it would be important to look at this hormone as well. Did you examine it?

Beers: Yes, and it had no effect.

Reiter: Did you say that FSH had equivalent activity to LH in terms of plasminogen activator?

Beers: I did say that, but the preparations of LH and FSH had the same amount of bioassayable FSH activity. The LH preparations contained something on the order of 1-3 percent FSH.

Folkers: What could be said about the purity of the rat FSH which you used?

Beers: The material is in the neighborhood of 50-85 percent pure.

Folkers: At 50 percent purity, it's not too high from a biological point of view.

Beers: If you look at it in the context of what else is available, it's very pure. If you're questioning whether something other than FSH is stimulating the cells, all we can say is that if it isn't FSH it's something that is purifying with FSH. The activity we are measuring goes right along with the increases in bioassayable FSH activity.

Gospodarowicz: I would like to know whether or not the real influence of plasminogen activator is a universal phenomenon which occurs in birds as well as in mammals. In mammals you get rupture of the follicles because of the possible presence of plasminogen activator and plasmin in the liquor folliculi. In the bird you don't have any liquor folliculi; you have only the ova lined with granulosa cells. However, when you look at the rupture of a chicken follicle, it is very impressive and takes place in a very particular region of the follicle, the stigma.

It occurred to me that the mammalian follicle might rupture through a similar mechanism. Do you have any comments on that?

Beers: I have some comments, not much data. First, one probably can't count on the same mechanism operating in any two species. We've done some work on other mammals and the situation appears to be similar to what I've shown for the rat. In birds it doesn't have to be the same. The follicle in the bird is very different from that in mammals and it's not clear that the same mechanism has to be working. The one time we looked for plasminogen activator in rupturing hen follicles we didn't see any activity. I'm not sure the chicken protease inhibitors are inactivated the same way as the mammalian ones, and we didn't do all the studies necessary to say whether or not there were inhibitors present. We didn't see an increase in activity, but it was not a definitive study.

Nalbandov: We just completed a study last year on prostaglandins in hen follicles. Basically, there is practically no prostaglandin in the largest follicle which is the one that's destined to ovulate. There is a tremendous increase in prostaglandin activity in the already ruptured follicle. This is predominantly a thousand-fold increase over the immediate post-ovulatory follicle as compared to one 24 hrs later. Of course, there the prostaglandin serves a totally different purpose in that it is involved in the expulsion of the finished egg from the uterus. For whatever it's worth, there is no increase in prostaglandin in the chicken pre-ovulatory follicle as compared to mammals.

Baker: I was fascinated with your result that FSH is more effective that LH. It might be interesting to relate this to other events in the follicle. In mouse ovaries *in vitro* FSH is much more effective than LH in inducing a resumption of meiosis in the oocyte. This is not due to LH activity because you can block the response to LH with antisera to the β-subunit of LH while this treatment has no effect on the FSH response. It is interesting that prostaglandin E_2 will mimic the effect of both hormones. So it looks as if FSH, at least in these rodents, may have an important role in the follicle.

I can suggest another model that you might consider using: that is the domestic pig, an animal with quite a long estrous cycle. Normally the LH surge will occur on about day 20 of the 21 day cycle. However, if you inject the pig with hCG on day 17 of the cycle, ovulation occurs at the right time interval afterwards (about 42 hrs), suggesting that the ovulatory mechanism of the follicle wall is competent. The intriguing thing is that the oocyte that is released is still at the germinal vesicle stage, and is totally immature in terms of both nucleus and cytoplasm. After artificial insemination (the female does not show estrus) the eggs become highly polyspermic and contain many hundreds of sperm. The cumulus cells do not mature and the granulosa cells can't luteinize. This provides a mechanism for looking at the process of ovulation and the effects of FSH, LH, or prostaglandins on it, without considering any other aspect of the maturation of the follicle overall. In other words, you can dissociate the ovulatory capacity of the follicle from the other maturational events. You might be interested to know that the estradiol concentration in these follicles is a third of normal. So they really are totally different from the normal *in vivo* situation, although ovulation can occur.

Schwartz: It seems to me that a lot of the comments which have been made reflect concern about the discrepancy of concentration between the FSH and LH. In virtually every *in vivo* ovarian system it seems to take more FSH than LH to cause an effect. Is it possible that the discrepancy in concentrations is because you are harvesting granulosa cells out of the follicle, and that harvested granulosa cells freed from the follicular wall layers are simply more susceptible to FSH than to LH?

Beers: Absolutely, but we haven't found a way to test that possibility yet. There's no question that these cells could have had LH receptors but lost them in culture. I think what we are describing in these studies is the *in vitro* system more than anything else.

Schwartz: It has been said that the rabbit does not have an FSH surge accompanying the LH surge. If that is correct, then maybe the rabbit *in vivo* system may be a really exciting one to look at. If there is FSH there and it is not detectable by the usual methods then ovulation may be intrinsic to LH in the rabbit. I think that might be an interesting species to study.

Beers: We have looked a little bit at rabbits and what we find are large increases in plasminogen activator that appear not to be granulosa cell associated. There are clear differences between the situation in the rat and rabbit, but we haven't defined them.

Beier: I would like to add another aspect to the discussion which is perhaps more peripheral, but it is more biological from the standpoint of the normal process of ovulation. You referred to the function of plasmin in the natural system as being the proteolytic agent which dissolves fibrin. My question is, if the plasminogen activator increases and, therefore, leads to higher concentrations of active plasminogen, what is the biological role of this? What does it mean for ovulation? You mentioned that the follicle wall is weakened. This is, of course, an assumption that doesn't tell too much about the substrate for plasmin. You also mentioned that the basement membrane is lysed or defective. Have you any idea if there is a particular activity of this plasmin for collagen? I think the most important thing happening in the weakening of the follicle wall, at the particular site of the stigma where an oocyte will escape from the follicle, is the change of the collagen in the theca. Is there any idea what the biological role of the plasminogen activator increase leading to activated plasminogen means in fact for the rupture at this site?

Beers: That, of course, is a question that is central to the issues I have raised. We don't know what the follicular substrates are for plasmin. The story with collagen is not entirely clear. What we know is that trypsin will solubilize insoluble collagen, but it does not attack the native collagen helix. However, once collagenase acts on collagen, and in some cases that means cleaving only one peptide bond, collagen becomes a substrate for trypsin. We haven't examined the action of plasmin on collagenase-treated collagen. It is known that in the rabbit, and I would be surprised if it weren't true in the rat, there are tonic levels of collagenase in ovarian tissue. There are no reported increases in collagenase activity that occur in the follicle, but if plasmin has tryptic-like activity on collagen, then the presence of a collagenase at a constant level would serve to promote increased collagenolysis as plasmin increases. It's a reasonable mechanism and one that remains to be tested.

Wilks: Have you looked for the synthesis of plasminogen activator in the thecal cells? Going back to prostaglandins, we know that in several species indomethacin can prevent ovulation. You mentioned that indomethacin did not prevent the increase of plasminogen activator synthesis with gonadotropins. What are your thoughts on the role prostaglandins play in ovulation?

Beers: With regard to production of plasminogen activator by thecal cells, I can only make a couple of comments. We have never successfully obtained a good population of thecal cells in culture. We have observed that a couple of percent of the cells in the granulosa preparations don't have granulosa-like morphology. Occassionally, using the fibrin-agar overlay assay, we have observed these cells making plasminogen activators. I don't know if these are thecal cells or not. We're trying to attack the problem in a different way, however. Of the cells that we assay, only 25-30% produce plasminogen activator. Cells of the cumulus appear not to produce any of the enzyme. Thus, the question becomes, are those 25-30% of cells localized in one region of the follicle wall? For instance, are they in the region of the stigma? In principle we can answer that question using sections of follicles and the overlay procedure. Moreover, if we can increase the resolution of the overlay, we might be able to say whether the thecal cells are also active. We're trying to do that now.

To answer your question on the role of prostaglandins in ovulation, it's my impression from the literature that the effects of indomethacin that result in inhibition of ovulation are probably on the pituitary. My feeling is that the increase in prostaglandins in the follicle at the time of ovulation influences the granulosa cells. I don't know where those prostaglandins are coming from. I would imagine they're coming from close by because the follicular concentration reaches such high levels. It could well be that inhibition of prostaglandin synthesis by indomethacin, in whatever cells are producing it, would influence ovulation. But, the evidence in the literature tends to point to the pituitary.

Ryan: Do you find a morphologic difference between cells that respond and those that don't respond? In particular, do you find microvilli on these cells? If you look with the scanning EM at granulosa cells in large follicles, you find many microvilli, but the photomicrograph that you showed lacked these structures.

Beers: The granulosa cells as a population, and at higher percentages than we observe making plasminogen activator, respond to FSH by rounding into a tight ball with long processes that make contact with other cells. It's a reversible change; the cells flatten back down. This phenomenon doesn't seem to be associated with mitosis, it doesn't last very long, it follows the same pattern that we see with plasminogen activator, it's cyclic AMP inducible, and its dose response curve is the same as that seen with the plasminogen activator response. In the scanning microscope we haven't seen changes in the microvilli, but upon rounding the intercellular processes become a lot more apparent. None of the obvious drugs, such as colchicine or cytochalasin B, have any effect on this response. We don't know the extent of which the rounding response correlates with the cells that are producing activator. Of course, when we can obtain enough highly purified FSH we will be in a position, with the single cell assay, to look at FSH binding, plasminogen activator production, and cell rounding on a cell by cell basis. In principle, then, we can establish those correlations.

Ryan: The reason I was asking this is that Winston Anderson and I have been looking at hCG localization on luteal cells following intravenous injection of tracers. We find that the ^{125}I labeled hCG localizes primarily on the microvilli of the luteal cell. The granulosa cells from large follicles have microvilli and a consistent response to LH and it is tempting to speculate that the LH receptors are also on the microvilli. However, if you go back to the small follicle granulosa cell, it's got a perfectly smooth surface and it's responsive to FSH but not LH. I wonder if your cell really isn't partially dedifferentiating when you put it in culture, at least during the early time periods when it may be making contact, and this may grossly change the cell surface and make it more responsive to FSH than LH.

Beers: It's entirely possible. Certainly we have no way of knowing, but it's something we tend to be leary about. We know that certain properties of the cells are changing in culture as we're working with them.

Gospodarowicz: We find working with bovine granulosa cells from large follicles that if you put them in culture for a very long period of time the cells start to luteinize, and they no longer respond to growth factor. Out of that population of cells, a population of cells emerge that do still respond to growth factor but show a completely different morphology. When you work with large follicles, you have a population of granulosa cells but you also have a small population of capillary epithelium which, after several generations, will overtake the granulosa cells. It's very well known that capillary epithelium can produce a large amount of plasminogen activator. Since you work with rather large follicles, I suspect that you may also have that population of cells.

Beers: I don't think so, at least not in large numbers. As I said, 85-90% of the cells all have a common morphology, and we're sure that those are granulosa cells. In addition, when we harvest cells we are very careful to puncture the follicle and then just give it a squeeze. I think that we're collecting the cells up to the basement membrane. This space is, of course, avascular. Furthermore, everything is done in such a short time period that even if there were contaminating cells in the population they don't have time to divide and take over.

Denker: I would like to go back to the question about the localization of the plasminogen activator activity inside the maturing follicle. If I understood you correctly, you'd imagine that plasminogen is being activated in the follicular fluid and so you would have to deal with a generalized plasmin activity in the whole follicular fluid which then had its action on the entire follicle wall. Wouldn't it be reasonable to assume that most activity should appear at the apex of the follicle where the stigma will develop? I think it would be most interesting to look at regional differences in plasminogen activator activity in the follicle wall, applying one of the very sensitive fibrin film techniques available which use pure fibrin-plasminogen films rather than agar support systems (e.g. Kwaan and Astrup: Lab. Invest. 17:140, 1967, or recent variants of it). It would be interesting to see whether the activity of the apex folliculi would prove to be higher than, for example, that of endothelia which you can always demonstrate easily with this technique. Have you tried this? On the other hand, we know that follicular fluid contains many enzyme inhibitors. Would you comment on this?

Beers: Plasminogen has an affinity for fibrin and may have an affinity for other substrates as well. It's thought that the zymogen may actually bind to its substrate, and is then converted to plasmin by plasminogen activator. This would create an active enzyme that's already bound to its substrate and would be uninfluenced by inhibitors until after it had acted. So, it may be that plasminization of any substrate is very much the result of local production of plasminogen activator. That's probably the only way to build up effective concentrations of plasmin. Our findings indicate that cumulus cells don't make the enzyme, and that tends to make us skeptical that plasminogen activator has anything to do with the break down of the structure of the cumulus near the time of ovulation. We feel that it's very likely you have to have local production of activator to overcome the effect of inhibitors.

Channing: This brings me to the point that we're used to examining steroid secretion by granulosa and thecal cells, and now we must reevaluate our thinking and include the concept that follicular cells secrete peptides and proteins. Perhaps we should think of at least four processes which occur in the final stages of antral follicle maturation. There might be a role of plasminogen activator for follicle development in general, rather than the process of follicular rupture. Other ovulatory processes, oocyte maturation and luteinization, may be under separate controls. We have to beware of using the general term ovulation and should specify that we're discussing follicular rupture, oocyte maturation, or luteinization. Perhaps the plasminogen activator mechanism is extremely important in follicular development, and perhaps other roles which are not well understood should be studied. Such a role for plasminogen activator is proposed because it is produced in response to FSH rather than LH, and FSH acts early rather than late in follicular development.

2

Studies on an Oocyte Maturation Inhibitor Present in Porcine Follicular Fluid

CORNELIA P. CHANNING, SARAH LIPFORD STONE,
ALISON SCHWARTZ KRIPNER, and SEYMOUR H. POMERANTZ

SUMMARY

The current status of studies on the purification and actions of an oocyte maturation inhibitor (OMI) obtained from porcine follicular fluid is summarized. Oocyte maturation inhibitor obtained from porcine follicular fluid is a polypeptide of less than 2000 daltons in weight. It has been isolated from follicular fluid using Amicon membrane filtration, Sephadex G-25 column chromatography and paper electrophoresis. A 2500 fold purification has been achieved using these procedures. Porcine OMI also inhibits rat oocyte maturation demonstrating that its action is not species specific. Follicular granulosa cells are probably the source of OMI and the inhibitory action of OMI upon porcine and rat oocyte maturation can be reversed by addition of luteinizing hormone to the cultures. The inhibitory action by both follicular fluid and purified OMI upon cumulus-enclosed porcine oocytes was shown to be reversible if the inhibitor was removed by 20-24 hours of incubation. A brief discussion of other follicular fluid inhibitors including ovarian "inhibin" and luteinization inhibitor also is included.

The follicular mechanisms which are responsible for maintaining the mammalian oocyte in the immature dictyate state of the first meiotic division from shortly before or after birth until the time of ovulation are unknown. Two basic mechanisms are possible. One is that a substance required for the completion of oocyte maturation is provided to the oocyte at the time of ovulation. The second mechanism is that an oocyte maturation inhibitor present within the follicle from the time of birth is either antagonized prior to ovulation by the preovulatory surge of gonadotropin or is diminished in the follicle by the action of the preovulatory gonadotropin surge. Earlier evidence by others as well as recent evidence from our laboratory indicates that the second mechanism is the most likely.

In 1935, Pincus and Enzmann observed that oocytes isolated from rabbit follicles matured spontaneously in culture. In 1955, Chang observed that addition of follicular fluid to cultured rabbit oocytes partially inhibited spontaneous meiosis. These pioneering studies and the observations of Foote and Thibault (1969) that coculture of porcine granulosa cells with porcine oocytes prevented spontaneous maturation of the cultured oocytes gave evidence that the granulosa cell may secrete an oocyte maturation inhibitor which accumulates in follicular fluid. In 1975, Tsafriri and Channing observed that addition of 50% porcine follicular fluid led to a 50% inhibition of maturation of cultured cumulus-enclosed oocytes whereas addition of 50% porcine serum did not have any inhibitory effect. In 1976, Gwatkin and Anderson made a similar observation using cultured hamster oocytes and bovine follicular fluid. Tsafriri and Channing (1975) went on to demonstrate that coculture of oocytes with 10^7 porcine granulosa cells from small, medium or large follicles led to complete inhibition of maturation of porcine oocytes. Granulosa cells obtained from small follicles were more "active" in inhibitory activity compared to cells from larger follicles: i.e. 10^5 cells from small follicles exerted significant inhibition of oocyte maturation whereas 10^5 cells from large follicles did not. Addition of a soluble extract of granulosa cells also led to inhibition of porcine oocyte maturation (Tsafriri, Pomerantz, and Channing, 1976a). Other studies which have been reviewed recently (Channing and Tsafriri, 1977) are also in support of the concept of follicular oocyte maturation inhibitor. An oocyte maturation inhibitor (OMI) has been isolated from porcine follicular fluid (Tsafriri, Pomerantz, and Channing, 1976b). In this communication we summarize recent progress on the partial purification of OMI and studies on the reversibility of the inhibitory action of OMI. A detailed description of these findings will be reported elsewhere (Stone _et al._, 1977).

Assay of Oocyte Maturation Inhibitor

OMI was assayed in cultures of cumulus-enclosed porcine oocytes as diagrammed in Figure 1. Oocytes were cultured for 43-46 hours in groups of 10-15/well in the presence of control medium or control medium plus inhibitor. The control medium consisted of medium 199 plus 15% porcine serum, 2.5 mM lactate, 0.03 mM pyruvate and 12.5 mU/ml insulin, as detailed previously (Tsafriri and Channing, 1975). At the end of the incubation period the oocytes were fixed and stained with aceto-orcein and classified as either immature (dictyate) or mature (including metaphase 1 and metaphase 2 with polar body).

FIGURE 1. Diagrammatic sketch of culture method for porcine oocytes and assay of oocyte maturation inhibitor (OMI) using cultures of cumulus-enclosed porcine oocytes.

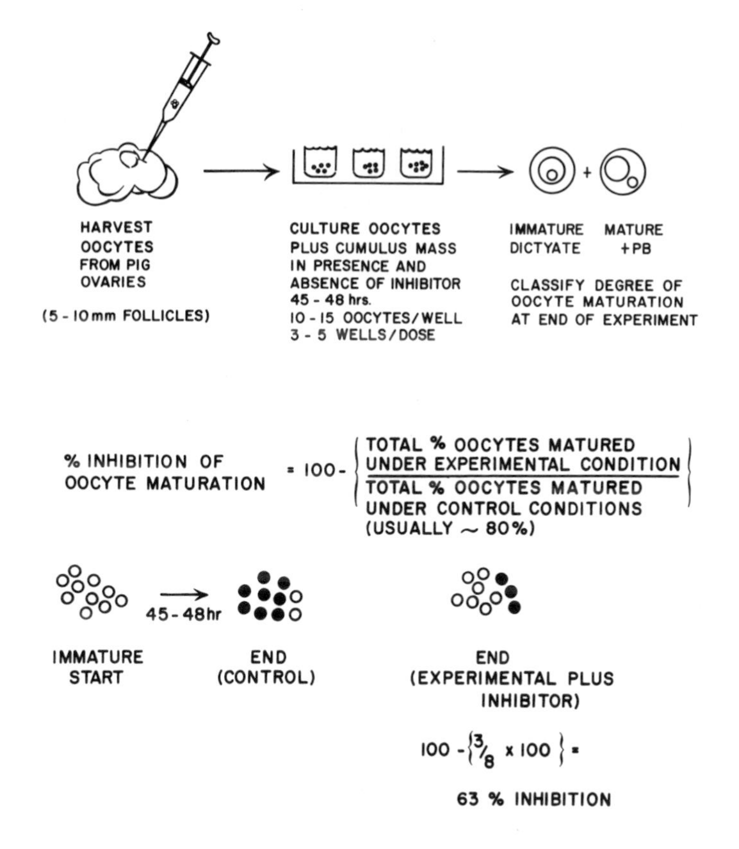

Throughout the purification of OMI the inhibitory activity of each fraction was expressed as percent inhibition of oocyte maturation which was calculated as follows:

$$\% \text{ Inhibition} = \frac{\% \text{ maturation control} - \% \text{ maturation experimental}}{\% \text{ maturation control}} \times 100$$

At each step of the purification procedure a dose-response curve was carried out using different dilutions of the inhibitory fraction. An example of such a dose response curve on a fraction derived from follicular fluid is shown in Figure 2. In this experiment 425 ml of porcine follicular fluid was filtered through an Amicon PM-10 membrane and the filtrate (less than 10,000 molecular weight) was collected and concentrated by lyophilization. The residue was dissolved in a small amount of water so that it represented 1/40 of the volume of the original follicular fluid. The dose-response curve is shown for various dilutions of this PM-10 concentrate.

Partial Purification of Porcine Follicular Fluid Oocyte Maturation Inhibitor

We have observed previously that addition of 50% porcine follicular fluid can exert 50% inhibition of cumulus-enclosed porcine oocyte maturation and that LH can overcome the inhibitory action of the follicular fluid or a low molecular weight fraction of it (Tsafriri, Pomerantz, and Channing, 1976a). If the porcine follicular fluid was filtered through an Amicon PM-10 membrane, with a molecular cut off weight of 10,000, the OMI passes through the membrane. Additional studies indicated that the inhibitory activity was retarded by an Amicon UM 2 filter with a molecular weight cut off of 1,000.

Sephadex G-25 Column Chromatography

Some of the material which passed through a PM-10 membrane and was retained by a UM 2 membrane was applied to a Sephadex G-25 column (1.5 x 40 cm) (Figure 3). The inhibitory activity was eluted off the column at a location corresponding to a molecular weight of about 2,000 (Tsafriri, Pomerantz, and Channing, 1976a). Sephadex G-25 chromatographic purification of the OMI was repeated using 3 different batches of 400-500 ml porcine follicular fluid obtained from medium and large sized follicles. In these latter instances the follicular fluid was

FIGURE 2. Ability of various doses of low molecular weight fraction of porcine follicular fluid (PM-10 concentrate) to inhibit porcine cumulus-enclosed oocyte maturation. Porcine follicular fluid was passed through a PM-10 membrane and the <10,000 molecular weight filtrate was concentrated, lyophilized and redissolved in a small volume of water so that it represented 1/40 the volume of the starting follicular fluid. At least 60 oocytes were used to estimate each point on the dose response curve. The material used for the assay contained 4.5 g/ml peptide material. Therefore, the amount of peptide added to the oocyte cultures in an assay volume of 0.2 ml at a dose of 1/400 was 2.2 mg.

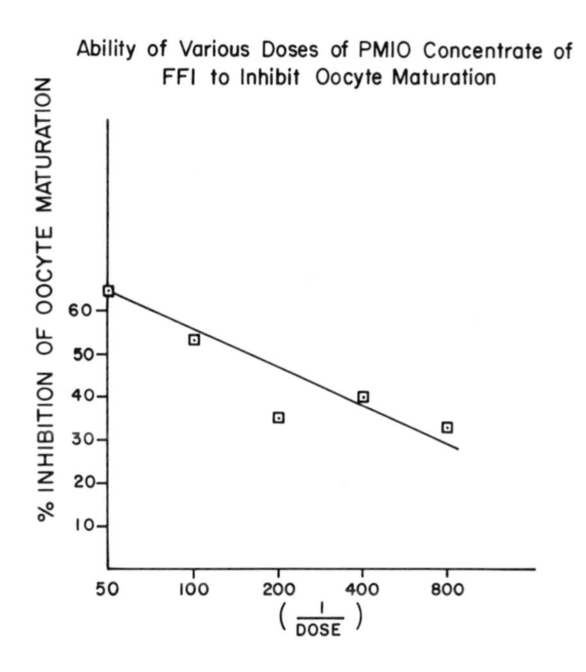

subjected to Amicon PM-10 membrane filtration and the filtrate was concentrated by lyophilization. The concentrated PM-10 filtrate was applied directly to the Sephadex G-25 column without an intermediate UM 2 filtration of the PM-10 concentrate since it was found that UM 2 filtration led to some loss of the OMI activity. In our later studies the size of the Sephadex column was increased to 5 x 70 cm and 0.01 M

NH_4HCO_3 buffer (pH 7.0) was used for elution of the OMI. The majority of the inhibitory activity came out in a region corresponding to a molecular weight of about 2000 daltons. This was designated as peak A and this fraction was used for further purification steps. A second peak of inhibitory activity was eluted later, which corresponded to a molecular weight of less than 1000. From the absorbancy at 280 nm of each column fraction it was observed that peak A was separated from the bulk of the peptide.

FIGURE 3. Purification of the follicular fluid inhibitor of oocyte meiosis by Sephadex G-25 chromatography. The sephadex G-25 column was operated as outlined in Tsafriri, Pomerantz, and Channing, 1976. Oocytes were cultured in 1:25 dilution of Sephadex G-25 eluates in culture medium 199. The maturation of oocytes in culture medium (TC-199), phosphate buffered saline (PBS) (in 1:25 dilution) and in UM 2 membrane concentrate (1:50 dilution) is indicated for comparison on the left side of the figure. The OD_{280} of fractions 26, 50, 53, 57, 59, 61, 65, and 70 were 0, 0.037, 0.031, 0.027, 0.020, 0.031, and 0.020, respectively (taken from Tsafriri, Pomerantz, and Channing, 1976, by permission from the Biology of Reproduction).

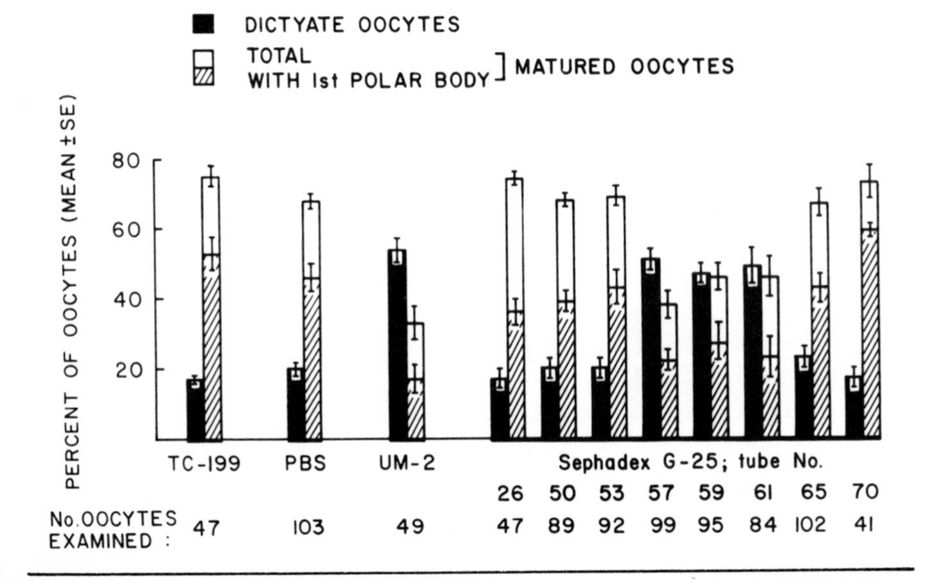

Paper Electrophoresis

In two trials, a 1 ml aliquot of Sephadex peak A (102.6 mg protein) was applied to Whatman 3MM paper for paper electrophoresis in formic-acetic acid buffer, pH 1.9 at 1500 V/3.5 hour. Tyrosine spotted on the paper served as an internal standard of migration. After the electrophoresis was completed the paper was air-dried and cut into twelve, 18 x 3 cm strips. Each strip was eluted with 15 ml H_2O for 2 hours, and then with 5 ml for 1.5 hours. The fractions were lyophilized and redissolved in 2 ml H_2O. The electrophoresis samples then were assayed for inhibitory activity in oocyte cultures at a dilution of 1:50 in Medium 199A. Because over 50% of the peptide content was found (by fluorescamine peptide analysis) to be in electrophoresis strips number 1 and 2, in a second electrophoresis run the 3 cm strips of regions number 1 and 2 were divided further into 1.5 cm x 18 cm strips and were designated as 1a, 1b, 2a, and 2b (Table 1). As indicated by the results, the strongest inhibitory activity was concentrated in electrophoresis samples 1b and 2a. In no other samples was the percent of oocyte maturation significantly different from that of controls. On the basis of peptide recovery using the fluorescamine assay at pH 9 (procedure in package insert of Roche Diagnostics) with β-MSH as a standard we estimate that we have achieved about a 2500 fold purification of OMI.

Reversibility of Oocyte Inhibition in Vitro

It is well known that porcine oocytes remain at the dictyate or immature stage of meiosis while bathed by the follicular fluid *in vivo* and are capable of resuming meiosis upon removal from the follicle and its fluid. A series of experiments was done to determine the ability of cumulus-enclosed oocytes to resume meiosis *in vitro* after various periods of incubation *in vitro* in 50% follicular fluid, PM-10 filtrate or in Sephadex peak A fraction. In preliminary trials the method for such reversal experiments was determined. After a period of oocyte incubation in inhibitory fraction, the medium around the oocytes was either replenished with fresh inhibitory medium (0.2 ml/ chamber of 10 oocytes) - "sham" reversal - or the medium was changed to non-inhibitory 50% serum for reversal. For each time period of incubation the "sham" reversal served as the inhibitory control for the reversal of inhibition or resumption of maturation. The oocytes

TABLE 1. Paper Electrophoresis of Sephadex Peak A Bioassay of Inhibitor in Oocyte Culture.

CULTURE MEDIUM	NUMBER OF OOCYTES	% OF MATURATION	PEPTIDE CONTENT OF STRIP (mg/ml)*	P
199A-Control	156	56.3		
Sephadex Peak A	94	30.9		
1a**	49	46.8		
1b	49	34.0	9.7	<0.01
2a	105	29.3	23.0	<0.01
2b	105	40.1	3.18	
3	90	53.2	2.45	
4	92	45.7	0.371	
5	101	45.9	0.233	
6	94	53.1	0.274	
7	96	56.6	0.137	
8	101	53.8	0.121	
9	103	51.6	0.106	
10	100	58.8	0.105	
11	102	52.0	0.082	
12	46	44.4	0.071	

*Based on peptide analysis using fluorescamine assay with B-MSH as standard peptide.

**Electrophoresis samples, assayed at dilutions of 1:50 in Control Medium 199A.

then were incubated further in the reversal or "sham" reversal medium for at least 24 hours so that oocyte maturation, if resumed, could be detected by chromosomal analysis. Further, preliminary trials indicated that oocytes remaining in inhibitory medium for as long as 36 hours later will not resume meiosis. Therefore, the oocytes cannot be maintained for the three-day culture periods that such long-term incubation in inhibitor would require. After 20 or 24 hr in 50% follicular fluid, the oocytes were able to resume meiosis when the culture medium was changed to 50% serum. With the Students' T-test for paired data, inhibitory and reversal media at 20 and 24 hours were significantly different, with $p < 0.003$ at 20 hours and $p < 0.01$ at 24 hours. After 28 hours in FF1, however, fewer oocytes were capable of resuming maturation upon medium change. Therefore, with the use of the *in vitro* culture system, the oocytes can be held in the dictyate stage of meiosis for up to 24 hours and still be able to resume meiosis after removal of the inhibitor. It is important that the inhibitor in FF1 does not appear to be toxic to the oocytes *in vitro* but merely "arrests" the oocytes at the immature stage. At the longer culture periods, 28-36 hours, the length of *in vitro* culture needed to detect reversibility may in itself be deleterious to the maintenance of healthy and maturing oocytes.

Similar reversibility studies were done using the partially purified inhibitor of FF1:the PM-10 filtrate (1:200 dilution with Medium 199 A and the inhibitory Sephadex peak A (1:200). Preliminary results shown in Figure 4 indicate that, after *in vitro* incubation in either PM-10 filtrate or in Sephadex peak A, upon changing the culture medium to a non-inhibitory medium the oocytes can undergo nuclear maturation.

Further tests using the partially purified fractions of FF1 are being carried out to substantiate the findings reported here.

Observations on Inhibition of Rat Oocyte Maturation Exerted by Porcine Follicular Fluid

The concept that porcine follicular fluid OMI is not species specific is supported by the observation that porcine follicular fluid as well as partially purified porcine follicular fluid OMI could inhibit maturation of cultured rat oocytes (Tsafriri, Pomerantz, Channing, and Lindner, 1977). Addition of 50% porcine follicular fluid or a low molecular weight fraction thereof to cumulus-enclosed oocytes harvested from immature rats 20 hours after treatment with pregnant mares' serum gonadotropin led to 75% and 53% ($P < 0.001$) inhibition of germinal vesicle breakdown. Follicular fluid also inhibited polar body formation in the cultures. As was the case in the porcine oocyte cultures, addition of 5 μg/ml ovine LH overcame the inhibitory action of either fraction. If oocytes were removed 44 hours after PMSG treatment, the inhibitory effect of the OMI was reduced indicating that the sensitiv-

FIGURE 4. Reversibility of inhibition of porcine oocyte maturation after 20 and 24 hours of incubation with two partially purified inhibitory fractions of porcine follicular fluid. Porcine cumulus-enclosed oocytes were incubated for 20 and 24 hours in either small molecular weight fraction of porcine follicular fluid (<10,000; designated PM-10 filtrate) at a 1:200 dilution or a Sephadex G-25 column eluate (Sephadex peak A 1:200 dilution) corresponding to a molecular weight of about 2000. After this pre-incubation in the medium containing the inhibitory fraction of FF1, the medium was replaced by control Medium 199 A (designated "reverse") and the oocytes cultured for an additional 24 hours.

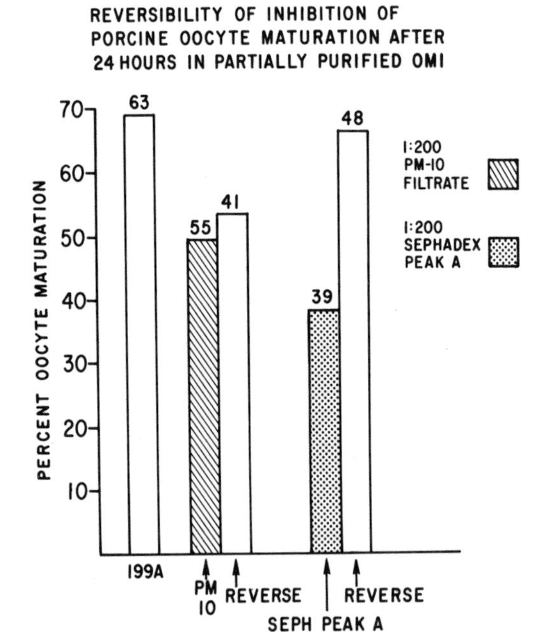

ity of rat oocytes to the inhibitor decreased during the course of follicular development. Since LH overcomes the action of OMI, it is possible that endogenous gonadotropins present between 20 and 44 hours after PMSG treatment had a longer time to act upon the oocytes to render them (or their cumuli) less sensitive to the actions of the OMI.

Relationship of Oocyte Maturation Inhibitor to Other Follicular Fluid Inhibitors

Follicular fluid contains other inhibitory substances besides OMI which are of interest as intraovarian regulators (Table 2).

TABLE 2. Ovarian Non-Steroidal Intrafollicular Inhibitors.

NATURE OF INHIBITOR	MOLECULAR SIZE (daltons)	FOLLICULAR FLUID SOURCE			CELL OF ORIGIN	ACTION
		Sm. FFl	Md. FFl	Lg. FFl		
Oocyte Maturation Inhibitor (OMI)	2,000	Yes	Yes	Yes	Granulosa Cells	Inhibit Oocyte Maturation.
Luteinization Inhibitor (LI)	>10,000	Yes	Yes	No	?	Inhibit Granulosa Cell Luteinization
Ovarian Inhibin	>10,000	Yes	Yes	Yes	?	Inhibit Serum FSH Level.
Luteinizing Hormone Binding Inhibitor (LHRBI)	~3,800	No	No	No	Corpus Luteum	Inhibit Binding of LH/hCG to Granulosa Cell and Corpus Luteum

Porcine follicular fluid contains an ovarian "inhibin" which inhibits blood levels of follicle stimulating hormone (FSH) in intact, as well as short-term (9 hr) and long-term, castrate male and female rats (Schwartz and Channing, 1977a, b, c; Marder, Schwartz, and Channing, 1977; Schwartz and Channing, unpublished observations). The ovarian inhibin differs from OMI in that it is >10,000 daltons in size (Schwartz and Channing, unpublished observations); whereas, the OMI is only about 2000 daltons in molecular weight. It is possible that the granulosa cells are the source of both inhibitors, although this has not been established in the case of inhibin. A dose response curve depicting the inhibitory effects of charcoal-treated porcine follicular fluid upon FSH levels in castrate female rats is shown in Figure 5.

FIGURE 5. Effect of charcoal-treated porcine follicular fluid or porcine serum on serum follicle stimulating hormone (FSH) in rats ovariectomized at 0800 in metestrus. The rats were given a single injection of follicular fluid or serum at 1130 hours and autopsied at 1700. Data shown are FSH levels measured in the serum obtained by decapitation at autopsy using a radioimmunoassay with NIH-FSH-RPI as a standard. Serum LH levels were not altered by treatment with porcine follicular fluid (taken from Marder, Channing, and Schwartz, 1977 with permission from Lippincott Co.).

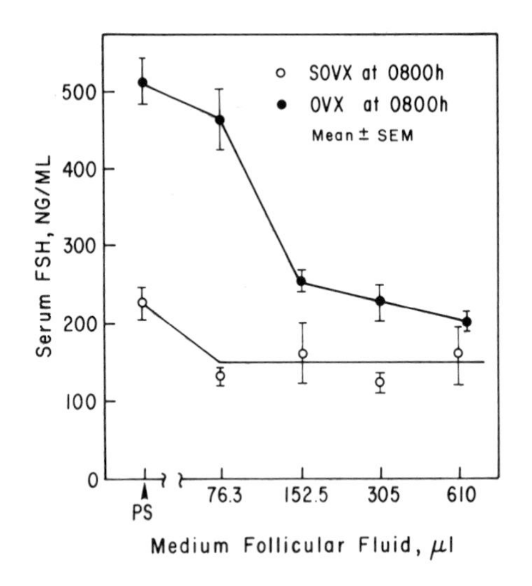

A third follicular fluid inhibitor, luteinization inhibitor, was demonstrated by Ledwitz-Rigby et al. to exist in porcine follicular fluid (Ledwitz-Rigby and Channing, 1972; Ledwitz-Rigby, 1973; and Ledwitz-Rigby et al., 1977). They observed that addition of 20-50% charcoal-treated follicular fluid obtained from small (1-2 mm) but not larger (6-12 mm) porcine follicles could inhibit spontaneous morphological luteinization and progesterone secretion by cultured granulosa cells from large follicles (Figure 6). The small follicular fluid could also inhibit LH stimulation of cyclic AMP accumulation. A difference between this inhibitor and OMI and ovarian inhibin is that the inhibin and OMI are present to some extent in fluid of all sizes;

<u>FIGURE 6</u>. Influence of 20% porcine follicular fluid and serum upon the morphology of porcine granulosa cell monolayer cultures. Porcine granulosa cells were cultured in Leighton tubes for 5 days in 20% porcine serum or 20% follicular fluid obtained from small (1-2 mm), medium (3-5 mm) or large (6-12 mm) follicles in a balance of medium 199. The cells were obtained from large (6-12 mm) follicles (H and E staining X450 magnification). (Taken from Ledwitz-Rigby <u>et</u> <u>al</u>., 1977, with permission).

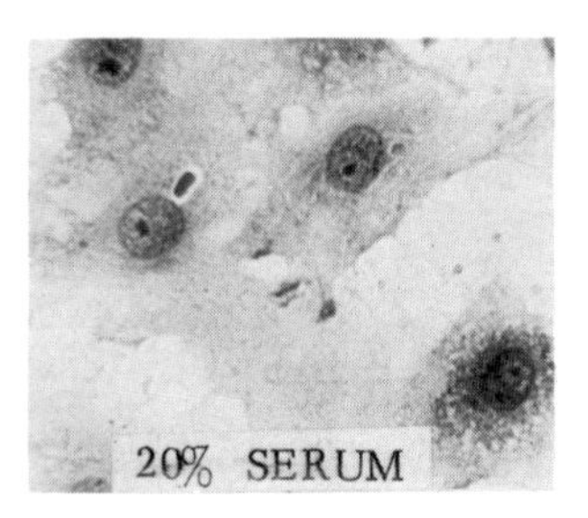

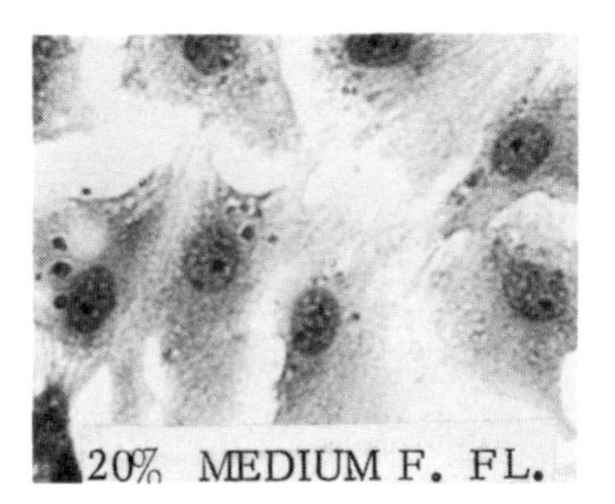

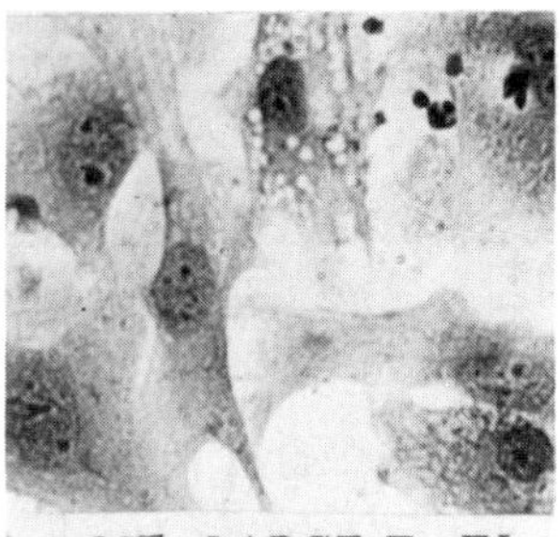

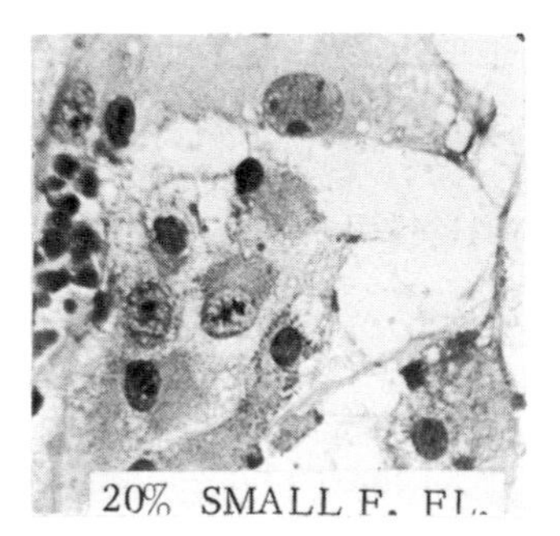

whereas, the luteinization inhibitor is present primarily in fluid of small follicles as well as to some extent in fluid obtained from medium follicles but absent in fluid from large follicles. It is possible that there is a luteinization stimulator present in fluid from large follicles which could act to override the luteinization inhibitor. Further studies on the chemical nature of these inhibitors will be required. More quantitative assays of the content of OMI in porcine follicular fluid must be done to determine differences in amount in different sized follicles. Since we have already shown that granulosa cells from small porcine follicles can exert a greater inhibitory effect upon oocyte maturation compared to cells obtained from large follicles, it could be predicted that the inhibitor should be present in greater amounts in fluid from small compared to large follicles.

Of interest, in 1967 Channing observed that addition of equine follicular fluid could inhibit luteinization of cultured equine cells. Bernard (1975) observed that bovine follicular fluid could inhibit luteinization of cultured rat granulosa cells. Lunenfeld and his colleagues observed that human follicular fluid obtained prior to but not after the LH surge could inhibit cyclic AMP accumulation in incubated rat ovaries (Kraiem and Lunenfeld, 1976).

A diagrammatic sketch of the relationship of intraovarian inhibitors is shown in Figure 7. There is no doubt that a better understanding of control of ovarian function will be gained by further investigation on these intrafollicular inhibitors. The use of antibodies to each of these will be useful as tools to observe their roles as well as to measure them under different physiological states.

The mechanism of action of these intraovarian inhibitors is not known and remains a fruitful area for future study. Oocyte maturation inhibitor may act directly upon the oocyte or may act indirectly via the cumulus cells. Studies are currently being carried out by Hillensjo, Kripner, and Channing in which the inhibitory action of partially purified porcine OMI is being observed on denuded and cumulus enclosed porcine oocytes. It is hoped that these sorts of studies should lead to an answer of whether or not the OMI acts directly upon the oocyte or has its effects mediated via the cumulus cells.

FIGURE 7. Diagrammatic representation of intraovarian inhibitors. Abbreviations used in the figure are as follows: OMI-oocyte maturation inhibitor, LI-luteinizing inhibitor, LHRBI-luteinizing hormone receptor binding inhibitor.

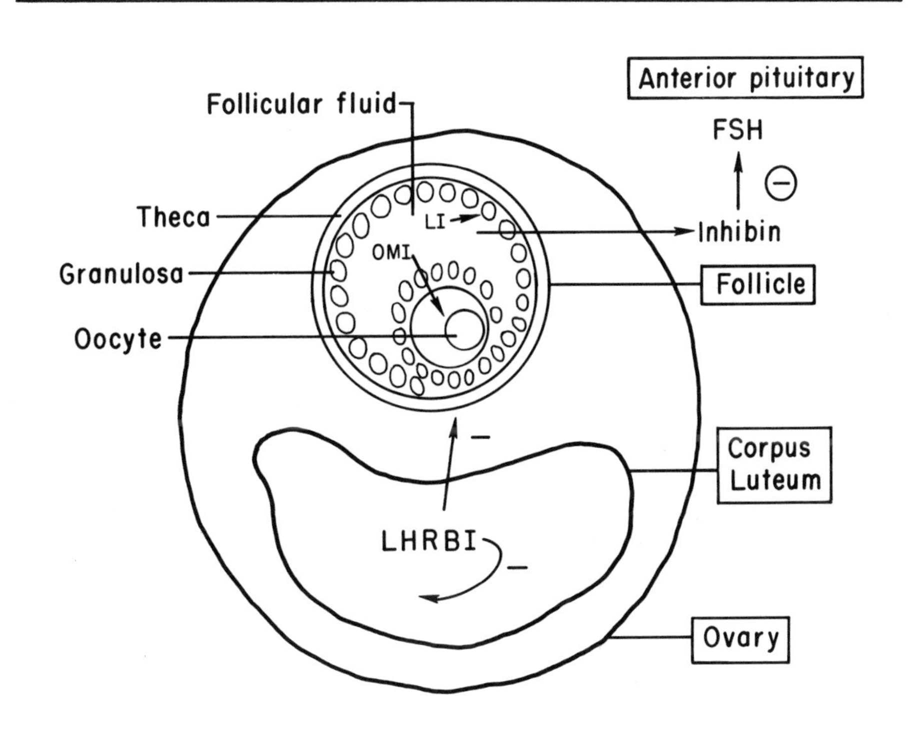

REFERENCES

Bernard, J. (1973). Effet du liquide folliculaire sur la luteinization in vitro des cellules granulosaires du rat. Compte Rendu Societe de Biologie 167:882.

Bernard, J. (1975). Effect of follicular fluid and oestradiol on the luteinization of rat granulosa cells in vitro. J. Reprod. Fert. 43:453

Channing, C.P. (1969). Studies on tissue culture of equine ovarian cell types: Culture methods and morphology. J. Endocrin. 43:381

Channing, C.P. and Tsafriri, A. (1977). Mechanism of action of luteinizing hormone and follicle-stimulating hormone on the ovary in vitro. Metabolism 26:413

Chang, M.C. (1955). The maturation of rabbit oocytes in culture and their maturation, activation, fertilization, and subsequent development in the fallopian tubes. J. Exp. Zool. 128:378

Foote, W.D. and Thibault, C. (1969). Recherches experimentales sur la maturation in vitro des oocytes de truie et de veau. Ann. Biol. Anim. Biochem. Biophys. 3:329

Gwatkin, R.B.L. and Anderson, O.F. (1976). Hamster oocyte maturation in vitro: inhibition by follicular components. Life Sciences 19:527

Kraiem, Z. and Lunenfeld, B. (1976). The cAMP accumulation inhibitor in follicular fluid of human origin. Abstract 169 in Program of the 58th Annual Meeting of the Endocrine Society held at San Francisco, June 23-25, 1976.

Ledwitz-Rigby, F. (1974). Possible site of inhibitory action of follicular fluid (FFl) upon porcine granulosa cell (GC) luteinization. In Program of the 7th Annual Meeting of the Society for the Study of Reproduction, Ottawa, Ontario, Canada, 1974 (Abstract 16).

Ledwitz-Rigby, F., Stetson, M., and Channing, C.P. (1973). Follicular fluid (FFl) inhibition of LH stimulation of cyclic AMP levels in porcine granulosa cells (GC). Biol. Reprod. 9:94

Marder, M.L., Channing, C.P., and Schwartz, N.B. (1977). Suppression of serum follicle stimulating hormone in intact and acutely ovariectomized rats by porcine follicular fluid. Endocrinology (in press).

Pincus, G. and Enzmann, E.V. (1935). The comparative behavior of mammalian eggs in vivo and in vitro. 1. The activation of ovarian eggs. J. Exp. Med. 62:665

Schwartz, N.B. and Channing, C.P. (1977). Suppression of the secondary rise in serum FSH at proestrus by porcine follicular fluid. Fed. Proc. 36:322 (Abstract 277)

Schwartz, N.B. and Channing, C.P. (1976). Evidence for ovarian "inhibin": suppression of the secondary rise in serum follicle stimulating hormone levels in proestrus rats by injection of porcine follicular fluid. Proc. Natl. Acad. Sci., U.S.A. (in press)

Schwartz, N.B. and Channing, C.P. (1977c). Suppression by porcine follicular fluid of the acute serum FSH rise following ovariectomy in the rat. Abstract 83 in Program of the 10th Annual Meeting of the Society for the Study of Reproduction held at Austin, Texas, August 15-17, 1977.

Stone, S.L., Pomerantz, S.H., Kripner, A.S., and Channing, C.P. (manuscript in preparation).

Tsafriri, A. and Channing, C.P. (1975a). An inhibitory influence of granulosa cells and follicular fluid upon porcine oocyte meiosis in vitro. Endocrinology 96:922

Tsafriri, A. and Channing, C.P. (1975b). Influence of follicular maturation and culture conditions upon porcine oocyte meiosis in vitro. J. Reprod. Fertil. 43:149

Tsafriri, A., Pomerantz, S.H., and Channing, C.P. (1976a). Follicular control of oocyte maturation. In Symposium on ovulation in the human. Frieburg, September 10-12, 1975. New York, Academic Press

Tsafriri, A., Channing, C.P., Pomerantz, S.H., and Lindner, H.R. (1977). Inhibition of maturation of isolated rat oocytes by porcine follicular fluid. J. Endocrinol. (in press)

Tsafriri, A., Pomerantz, S.H., and Channing, C.P. (1976b). Inhibition of oocyte maturation by porcine follicular fluid: partial characterization of the inhibitor. Biol. Reprod. 14:511

ACKNOWLEDGEMENTS

This research was supported by a grant from the Population Council of New York (M76.17), the Ford Foundation (760-0530), and also the National Institutes of Health (HD08834). The able assistance of Ms. Sandy Fowler is appreciated and the typing of Ms. Evelyn C. Wisowaty is gratefully acknowledged.

DISCUSSION

Reichert: In my experience factors which inhibit *in vitro* binding of a hormone to a receptor can be discouragingly ubiquitous. Have you looked at extracts of any other tissue to see whether or not such extracts would have effects similar to OMI in your system? In addition to the list of inhibitory factors which you presented, I would like to add another one. In studies in our laboratory, Dr. Nina Darga found that follicular fluid inhibits the binding of radio-labelled human FSH to bovine granulosa cells. Preliminary fractionation studies indicate the apparent molecular mass of this inhibitor to be in the range of 1000 daltons and that it is partially heat labile. So, this would be another example of a putative inhibitory factor present in follicular fluid.

Channing: I am very interested to hear about your nice study. The presence of an inhibitor of FSH action in follicular fluid is another mechanism to control granulosa cell maturation within the follicle and to explain why granulosa cells do not luteinize within the follicle prematurely in the presence of FSH. Whether or not a low molecular weight fraction of follicular fluid can be classified as an "extract" is an open question. We have observed that the low molecular weight fraction of porcine serum does not inhibit oocyte maturation to the same degree as does the low molecular fraction of follicular fluid. If a concentrated (1:10 dilution) low molecular weight fraction of serum is examined in the oocyte assay, it has inhibitory activity due to high amounts of ions and an adverse osmotic environment which are toxic to oocyte maturation. If the serum low molecular weight fraction is diluted 1:50, it loses its inhibitory activity; whereas 1:50 to 1:400 diluted low molecular weight fraction of porcine follicular fluid still has inhibitory activity.

Folkers: What are the difficulties or the practicalities of getting the oocyte maturation inhibitor to chemical purity with patience and time, and fractionation and assay?

Channing: Of course, we would like to get the oocyte maturation inhibitor to chemical purity as soon as possible. The rate limiting factor in the purification is the oocyte bioassay. The assay which requires culture of at least 40-50 oocytes per sample per dilution is sensitive to traces of impurities and certain buffers. Therefore, the choice of column and electrophoresis buffer to use in purification is restricted to those which can be removed by lyophilization or are not toxic to oocytes.

Folkers: How practical is the material that is actually extracted?

Channing: The source of OMI is practical being porcine ovaries obtained from the slaughterhouse. In about 3 weeks with plenty of hard-working students, fellows and myself, we can obtain a liter of follicular fluid from 6 batches of ovaries obtained at the slaughterhouse. The potential of OMI as a contraceptive is tremendous. If one inhibits oocyte maturation, no other event, then the ideal non-side effect contraceptive agent would be available.

Beers: A number of the effects on the oocyte that you ascribe to LH can be achieved with certain FSH preparations. Will FSH overcome the inhibition?

Channing: We have not done a rigorous series of studies on the ability of FSH to overcome the inhibitory action of OMI on oocyte maturation. In light of Dr. Reichert's finding of an FSH binding inhibitor in follicular fluid, we will go right back to the laboratory and examine the effect of FSH to overcome the actions of OMI.

Beers: Is it fair to say that resumption of meiosis is something that is restricted to oocytes that are going to be ovulated?

Channing: No, sometimes oocyte maturation resumes in the early stages of follicular atresia. It is possible that during atresia, the granulosa cells and/or cumulus cells loose their ability to make OMI. This is possible because in early stages of atresia granulosa cells show degeneration changes such as pynosis and karyorexesis.

Feigelson: Have you considered whether peak B of the oocyte maturation inhibitor might not be a subunit of peak A? Can you convert peak A to peak B by dissociation processes?

Channing: It certainly is a possibility that peak B of the oocyte maturation inhibitor is a subunit of peak A. Further studies will be required to find this out.

Ryan: Have you tried to see if any protease inhibitors, particularly small molecular weight protease inhibitors, mimic any of these activities? It has been known for a long time that trypsin will mimic the action of insulin. Our chapter in this volume presents data to indicate that a variety of serine proteases stimulate adenylate cyclase in the rat ovary and that protease inhibitors block the action of LH or hCG in activating this cyclase. There are data in Annelids that proteases will stimulate the resumption of meiosis (Plaucelliev, Exp. Cell. Res. 106:1, 1977).

Channing: Yes, we have. We have ordered some protease inhibitors and as soon as we have some time, we will add them to our cumulus-enclosed oocyte cultures.

Bullock: Can you tell us more about the timing of these events in the ovary and about the maturation inhibitor in particular? During the 40 hours after giving hCG that it takes a pig to ovulate, exactly how long does it take for the oocyte to mature and does the inhibitor level drop in follicular fluid in relation to maturation?

Channing: You have asked a very meaningful question. We hope to examine the changes of OMI in porcine follicular fluid during the 40 hours prior to ovulation. In order to do this optimally, we should have a radio-immunoassay for OMI which we don't have yet. Short of that, we intend to examine the OMI activity of the low molecular weight fraction of fluid of pigs obtained within the 40 hours prior to ovulation with the cooperation of Dr. Howard Brinkley at our College Park Campus. It will be necessary to examine only the low molecular weight fraction and not the whole follicular fluid since LH, which is present in pig follicular fluid immediately prior to ovulation, antagonizes the action of OMI. Since LH has a molecular weight greater than 10,000, a separation of it from OMI can be achieved by preparation of a low molecular weight fraction by passage of the fluid through an Amicon PM-10 membrane.

Bullock: You said in your talk that you think LH prevents the synthesis of the inhibitor. You now seem to be saying that you think LH might prevent the activity of the inhibitor, so that oocyte maturation would occur in the presence of LH even if inhibitor activity is present. What was the basis for your remark that LH inhibits the synthesis of the inhibitor and are you now suggesting that both things may be going on?

Channing: You have asked a very relevant question. It is quite possible that both events are going on; LH is antagonizing the action of the OMI on cumulus-enclosed oocytes and LH is bringing about a diminution in granulosa cell biosynthesis of OMI. Evidence for direct antagonism by LH is that if LH and OMI are added together, the inhibitory action of porcine follicular fluid as well as partially purified OMI upon porcine as well as rat cumulus-enclosed oocytes is overcome (Biol. Reprod. 14:511, 1976 and J. Endocrinol., in press). Recently A. Tsafriri observed that addition of ovine LH to cultures of rat granulosa cells led to a decrease in secretion of OMI into the culture medium (Tsafriri, A. In Vth International Congress of Physiology held at Paris, July, 1977). Additional studies are being carried out by us on the effect of LH upon OMI secretion by porcine granulosa cells.

Schwartz: Using antiserum to FSH or LH (Biol. Reprod. 10:236, 1974) in our laboratory and data of other laboratories using other techniques, it appears that, if there is an alteration in the FSH and LH ratio which is being "seen" by a matured follicle, there may be no follicular rupture. However, there may be patches of luteinization which appear in the granulosa cells; when these patches appear they are always at the opposite side from the oocyte. Do you have any explanation for the fact that luteinization might appear in patches anywhere and not uniformly throughout the follicle? If you have an explanation for that, do you have any explanation as to why luteinization is more likely to appear at some distance from the oocyte?

Channing: Your observation of patchy luteinization under conditions of altered FSH to LH ratio is most interesting and lends support for a follicular luteinization inhibitor. We do know that porcine cumulus-enclosed oocytes cannot make an inhibitor of luteinization in culture (Channing and Tsafriri, J. Reprod. Fert. 50:103, 1977). The fact that the cumuli region of the follicle is the last region to luteinize may be that the thecal cells in that region make less luteinization inhibitor. The follicular cellular source of luteinization inhibitor is not known.

Peckham: I have the impression that you feel that the oocyte maturation inhibitor is a peptide, but I am not quite sure what your basis for this feeling is.

Channing: Evidence that the oocyte maturation inhibitor is peptide in nature is that trypsin destroys the OMI activity in porcine follicular fluid. Additional studies are being carried out on the effects of trypsin on partially purified OMI.

Strickland: Do you know if the site of action of the inhibitor of FSH secretion is the pituitary or hypothalamus?

Channing: The site of inhibitory action of the ovarian follicular fluid FSH inhibitor is not known yet. Currently, we are examining whether or not it inhibits FSH secretion by pituitary monolayer cultures to see whether it has a direct effect on the anterior pituitary.

Beers: Another event that occurs in preovulatory follicles near the time of ovulation involves the cumulus-oocyte complex. During the same time period that plasminogen activator levels increase, there is a decrease from 100% to 0% in the intercellular communication between the cumulus and the oocyte. That means a decrease in the ability of these cells to transfer molecules up to about 1200 molecular weight units. You said that you haven't done studies on denuded oocytes which would shed some light on the possibility that communication between cumulus cells and the oocyte provides a mechanism of inhibiting meiosis. Do you feel that this kind of mechanism might be operative and that there is a likelihood that the inhibitory effects are being transmitted from the cumulus?

Channing: It is possible that the inhibitory action of OMI is mediated by its action upon the cumulus cells. Tests are being carried out on cumulus-enclosed and denuded oocytes to see whether or not the actions of the OMI are mediated by the cumulus cells.

Baker: There is one thing that still worries me slightly about this oocyte inhibitor: how can you explain the small antral follicle in which the oocyte can resume meiosis but then "block" at metaphase I? Do you think this is an effect of your inhibitor, or do you think it is something to do with the growth of the oocyte or a maturation process in the granulosa cells?

Channing: That is a fascinating question. As you know oocytes isolated from small porcine follicles cannot mature well in culture (J. Reprod. Fert. 42:149, 1975). Whether this is due to the fact that they are just immature or due to the fact that they are exposed to high level inhibitors cannot be resolved at the moment.

Folkers: Once you have collected the porcine ovaries from the slaughter house, what are the first general steps before Sephadex fractionation and electrophoresis?

Channing: Follicular fluid can either be aspirated from porcine ovaries at the slaughterhouse or at the laboratory. Under both conditions, we can obtain oocyte maturation inhibitor. The purification of the OMI has been published (Tsafriri, Pomerantz, and Channing, Biol. Reprod. 14:511, 1976). The procedure involves aspiration of the fluid from porcine follicles with a 20 gauge needle and syringe followed by centrifugation at 1000 x g to remove granulosa cells and freezing until further threatment. When 500 ml of fluid is accumulated it is passed through an Amicon PM-10 membrane and the low molecular weight fraction (<10,000 daltons) is collected, lyophilized and reconstituted in a small volume and applied to a Sephadex G-25 column.

Gospodarowicz: Don't you get an enormous fibrin clot when you remove the follicular fluid?

Channing: No, except with very large or cystic follicles.

Gospodarowicz: Does all the fluid you use come from small follicles?

Channing: Follicular fluid is harvested from small (1-2 mm), medium (3-5 mm) and large follicles (6-12 mm). Fluid from all follicle types contains OMI.

Gospodarowicz: Do you have any problems with the stickiness of the follicular fluid when placed on the Amicon filter?

Channing: We do not have severe problems with stickiness during the Amicon filtration of the fluid. Often times during filtration the large molecular weight fraction (PM-10 retentate) may form a precipitate which can be removed by centrifugation. The retenate is then added back and the filtration is continued. Some N_2 is used to assist the filtration.

3

LH-RBI — An Inhibitor of *In Vitro* Luteinizing Hormone Binding to Ovarian Receptors and LH-Stimulated Progesterone Synthesis by Ovary

KUO-PAO PAUL YANG, KENNETH N. GRAY, JOHN H. JARDINE,
H. L. NANCY YEN, NAGUIB A. SAMAAN, and DARRELL N. WARD

In most mammals including human beings, LH plays a predominant role in reproductive physiology. One of the important functions of LH in reproduction is to induce ovulation. An inhibitor of LH secretion or LH function would, therefore, have potential value as a contraceptive. In this respect, the existing steroid contraceptives which are combinations of progestogen and estrogen, produce their antifertility effect primarily through the mechanism of inhibition of LH secretion from the pituitary gland; while a contraceptive which acts through the mechanism of inhibition of LH function at the ovary level is not yet available. The current concept of hormone action implies that the initial step involves the binding of the peptide hormone to its receptor on target cell membranes. Therefore, the prevention of the binding of a particular hormone to its target receptors would provide a powerful control system to the function of the hormone involved. We have demonstrated that aqueous extracts of pseudopregnant or pregnant ovaries inhibit significantly the specific binding of LH to ovarian receptors in an in vitro system (Yang et al., 1976a,b). This inhibitor was named LH receptor binding inhibitor, or LH-RBI.

LH-RBI in Pseudopregnant or Pregnant Rat Ovary

LH-RBI was first found in the ovary of pseudopregnant and pregnant rats. In most of the experiments described in this paper (except those specified) ovaries were stored at -20°C for a period of 10 weeks or longer before being used for LH-RBI extraction. Figure 1 shows a schematic outline of the procedure for the extraction of LH-RBI from

Fig. 1. Scheme for Separation of LH-Receptors and LH-RBI.

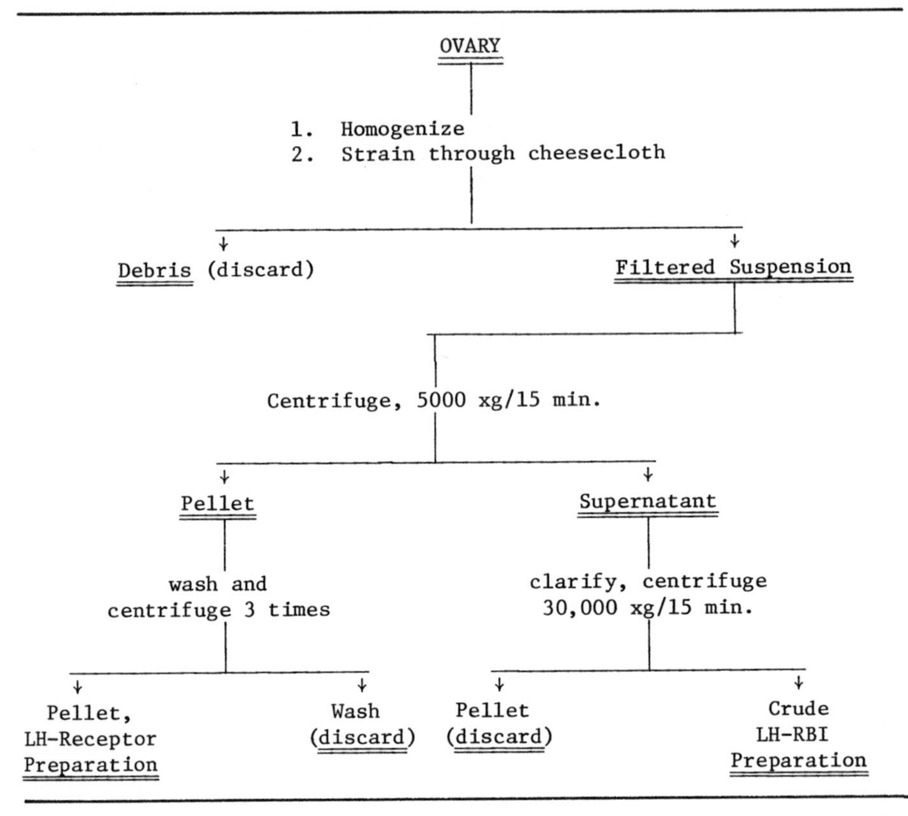

the pseudopregnant or pregnant ovaries. LH-RBI obtained from 10 mg of ovary by this procedure usually gave approximately 64% inhibition of specific binding of ^{125}I-labeled ovine LH in a radioligand receptor assay system which consisted of 10 mg ovary homogenate and 1 ng ^{125}I-LH in 0.2 ml Tris buffer (0.1 M) containing $MgCl_2$ (5 mM), sucrose (0.1 M), and bovine serum albumin (BSA) (0.1%), pH 7.5.

Freshly collected pseudopregnant ovaries, if without treatment, were found not to have significant LH-RBI activity. Upon storage at -20°C, LH-RBI activity extractable from the ovaries was found to increase gradually as the time of storage increased. This relationship between the LH-RBI activity and the time of storage is shown in Table 1.

Table 1. Effect of frozen-storage on the solubility of LH-RBI activity from pseudopregnant and pregnant rat ovaries.

Time in Storage (-20°C)		% Inhibition of [^{125}I] LH binding by ovarian extracts*	
Series 1.	Pseudopregnant ovaries[#]		
	None	0	(8)
	3 weeks	8	(2)
	5 weeks	15	(2)
	8 weeks	25	(2)
	9 weeks	48	(2)
	10 weeks	66	(12)
Series 2.	Pregnant ovaries[#]		
	None	15	(4)
	4.5 weeks	67	(2)
	6 weeks	69	(2)
Series 3.	Immature ovaries[#]		
	None	0	(5)
	10 weeks	0	(2)
Series 4.	Testes[#]		
	10 weeks	-9[@]	(4)
	29 weeks	-10[@]	(2)
Series 5.	Pseudopregnant livers[#]		
	10 weeks	25	(4)
	30 weeks	20	(2)

*Activity produced by 10 mg ovary-equivalent of indicated ovarian extract as a measure of "yield" for the LH-RBI in terms of response from extractable materials from the ovary on a unit weight basis. It is not convenient to define a "unit" for LH-RBI until more purified preparations are available. In the study applying testis or liver, 10 mg tissue-equivalent of testis or liver extract was applied in each test. Figures in parentheses indicate the number of determinations.

#Ovaries after 10 weeks in storage (-20°C) were used to prepare the receptors for measuring the LH binding. Frozen storage did not affect LH-binding appreciably, but studies in progress indicate an augmentation of the LH-RBI inhibitory sensitivity.

@The minus sign indicates that an apparent increase of binding was observed.

One should note that tissues other than luteinized ovary, such as testes, liver*, and non-luteinized ovary did not show LH-RBI activity upon storage at -20°C, indicating the tissue-specific nature of this phenomenon. Prolonged storage of the pseudopregnant ovary at -20°C for a period of 40 weeks (the longest storage period we have tested) did not reduce the maximal LH-RBI activity which was observed at 10 weeks of storage. The reason for the increased extractability of LH-RBI in the luteinized ovary upon freezing and storage is not clear (mainly for the reason that we have not enough information about the nature of the inhibitor). We postulate that LH-RBI is synthesized in the luteinized cells and then bound to a subcellular particulate fraction. Upon storage, LH-RBI is released (or becomes extractable) from the cellular site. Supporting evidence for this assumption is that LH-RBI is extractable from fresh pseudopregnant ovary by use of a non-ionic detergent, Triton X-100. Fresh ovaries were homogenized in water and the homogenate was then centrifuged to obtain the 30,000 xg supernatant fraction by the procedures described in Fig. 1. The 30,000 xg supernatant was then incubated with Triton X-100 (final concentration of Triton X-100 was 0.1%) at 4°C for 20 minutes. The incubated mixture was filtered through an Amicon ultrafiltration UM-2 membrane. After three washes of the retentate with distilled water (5 ml per wash), the retentate was recovered from the membrane and assayed for LH-RBI activity. Table 2 shows that 10 mg ovary equivalent extract of the <u>fresh</u> ovary after Triton X-100 treatment inhibited 70.7% of LH binding while the 0.1% Triton X-100 blank gave no significant inhibition. It should be mentioned that the UM-2 membrane retentate of the extract obtained from the Triton X-100 extraction of fresh pseudopregnant ovary did not bind appreciable amount of ^{125}I-LH. When the retenate was checked for possible contamination of soluble LH-receptors by an assay system

---

*After 10 weeks of storage, liver extracts showed a 25% inhibition of LH-binding; however, the inhibitor in the liver extract differed from LH-RBI in two aspects: first, a 10 mg tissue equivalent of LH-RBI inhibited more than 60% of original LH binding while an equal quantity of liver inhibitor inhibited only 25% of LH binding; secondly, LH-RBI selectively inhibited LH binding to ovarian receptors (Yang <u>et al</u>., 1976, b) while liver material inhibited LH binding to both ovarian receptors and testicular receptors.

Table 2. LH-RBI activity extracted from fresh pseudopregnant rat ovary with Triton X-100.

Systems	Fractions and treatment of fractions	% Specific binding of ^{125}I-LH	% Inhibition of binding
1. Control	No addition	19.01 (4)	
2. Experimental	Triton X-100 extracted LH-RBI	5.57 (4)	70.7
	Triton X-100 extracted LH-RBI & boiled 20 min.	5.65 (2)	70.3
	0.1% Triton X-100 blank*	18.50 (4)	2.7

*BSA (0.5%) solution was incubated with Triton X-100 (final concentration of Triton X-100 was 0.1%) at 4°C for 20 min. The mixture was filtered through the Amicon UM-2 membrane before being used for the assay.

described by Dufau _et al_. (1973, 1974), we found that 10 mg ovary equivalent of the UM-2 membrane retentate bound only 0.5% of the input 1 ng ^{125}I-LH at the end of 1 hour incubation at 37°C. Normally, 10 mg ovary equivalent of LH receptor (the 5,000 xg pellets of ovary homogenate in which the receptors have received no Triton X-100 treatment) bound about 20% of 1 ng ^{125}I-LH under similar experimental conditions.

LH-RBI in Other Species of Mammals

Although we first found LH-RBI activity in the pseudopregnant and pregnant rat ovaries, we have now found that LH-RBI activity is also present in the corpora lutea of some pregnant domestic animals such as cat, dog, sheep, and goat (Table 3). These LH-RBI activities extracted from different species of mammals also showed the phenomenon of increased extractability of LH-RBI after the tissues were frozen and stored.

Specificity of LH-RBI

LH-RBI appears to be present solely in the corpus luteum of a number of animals. In rats, it was found only in the ovary of pseudopregnant or pregnant rats, but not in the ovary of immature nonpregnant

Table 3. LH-RBI activity in corpora lutea of various species of mammals.

Mammals	Time in storage (-20°C)	% Inhibition of ^{125}I-LH binding of ovarian extracts*
Cat	None	31.1 (4)
	8.5 weeks	83.5 (2)
Dog	None	24.8 (4)
	12 weeks	54.7 (2)
Sheep	None	20.3 (2)
	14 weeks	50.2 (2)
Goat	None	39.8 (2)
	12 weeks	50.7 (2)

*Data are the per cent inhibition of ^{125}I-LH binding to rat ovarian receptor produced by 10 mg ovary equivalent of indicated ovarian extracts; numbers in parentheses represent the number of determinations on which the average % inhibition was calculated.

rats or adult nonpregnant rats (probably a mixture of rats at all stages of estrous cycle). Also, LH-RBI was not found in extra-ovarian tissues such as livers, oviducts and hearts of pseudopregnant rats or testes of male rats. In other species of mammals (cat, dog, sheep, and goat) LH-RBI activity was present only in corpora lutea but not in the non-corpora lutea parts of ovary. Dr. Cornelia Channing has also made comparable observations in the pig ovary, which we have confirmed using her extract preparations and our assay system. (See Sakai et al., 1977). As to the hormone specificity, LH-RBI did not inhibit the specific binding of ovine prolactin to prolactin receptor of pseudopregnant rat ovary in vitro, while the inhibition of LH binding by LH-RBI increased as the quantity of LH-RBI increased (Yang et al., 1976a). Approximately 90% of LH molecules which bound to ovarian receptors in the control was prevented from binding by the presence of LH-RBI extracted from 20 mg of pseudopregnant ovary.

Physico-Chemical Properties of LH-RBI

Heating the pseudopregnant or pregnant ovarian extracts at 100°C for 20 minutes or repeated freezing and thawing of these ovarian extracts destroyed only a small fraction (heating) or did not affect (freezing and thawing) the total inhibitory activities of LH-RBI. Treatment of LH-RBI with proteases (trypsin, pronase) destroyed more than 50% of LH-RBI activity. This study was done with an LH-RBI preparation partially purified by gel chromatography (Yang *et al.*, 1976a). It appears, therefore, that LH-RBI is at least partly peptide in nature.

LH-RBI does not bind LH. The inhibitory activity of LH-RBI is neither due to the binding of LH-RBI with LH nor due to damage of LH molecules by LH-RBI. Evidence supporting this conclusion was derived from the results of three separate experiments. In the first experiment, ^{125}I-LH was first incubated with LH-RBI for 1 hour at 37°C before a second incubation with the ovarian LH receptors; there was no greater inhibition of binding resulting from the preincubation of ^{125}I-LH with LH-RBI. In the second experiment, ovarian LH receptors were incubated with ^{125}I-LH in the presence or absence of LH-RBI; after incubation, the unbound or free radioactivity was applied to a Sephadex G-100 column, 1 x 100 cm, equilibrated with 0.1 M Tris buffer, pH 7.5. Unbound radioactivity of ^{125}I in the control (receptors + ^{125}I-LH only) was eluted from the Sephadex G-100 column in two major peaks (Yang *et al.*, 1976a) a front peak with Ve/Vo ratio of 1.69 (intact ^{125}I-LH) and a later peak with a Ve/Vo ratio of 2.06 (putative LH subunits). The profile of the unbound ^{125}I radioactivity obtained from the incubation of LH receptors with ^{125}I-LH in the presence of the ovarian extract showed another small peak eluted later than the intact LH peak and the LH subunit peak as seen in the profile of the control. Thus, this later peak, eluted in fraction of Ve/Vo ratio of 2.22, must represent degraded LH molecules smaller than LH subunits in size. However, the radioactivity represented by this degraded LH fragment peak was only 5% of the total unbound radioactivity, or 3% of the total input ^{125}I-LH for incubation. Therefore, the destruction of LH molecules by the LH-RBI as represented by this peak (Ve/Vo = 2.22) can only be considered as accounting for a very minor portion of the total inhibitory activity of LH-RBI. There was no significant ^{125}I radioactivity eluted before the intact ^{125}I-LH peak, as would be expected for a complex of solubilized LH receptors associated with ^{125}I-LH unless one assumes the receptor to have almost negligible molecular weight. Furthermore, 88% of the unbound ^{125}I-LH which was inhibited from binding to the receptors by LH-RBI appeared under the intact ^{125}I-LH peak. These results indicate that LH-RBI inhibits LH binding principally, if not entirely, by preventing the binding of intact ^{125}I-LH to LH receptors. In the third experiment, LH-RBI was incubated with ^{125}I-LH for 1 hour at 37°C. At the end of the incubation, polyethlene glycol was added according to the method of Dufau _et al_. (1973, 1974) for separation of soluble receptor-LH complex from unbound ^{125}I-LH. There was no significant radioactivity (^{125}I-LH) in the polyethlene glycol precipitate where LH-receptor complex would be expected. Ten mg ovary equivalent of LH-RBI bound only 0.2% of input 1 ng ^{125}I-LH while an equal amount of LH receptors (the 5000 xg pellets of ovary homogenate) bound >20% of 1 ng ^{125}I-LH in an experiment designed to test the system employed for this precipitation.

<u>Mechanism of LH-RBI Action on the Inhibition of LH Binding to Ovarian Receptors</u>

It is clear from these results that LH-RBI activity is not attributable to either degradative enzymes or soluble LH binding components (e.g. soluble receptors). Therefore, the inhibition seems to be due to one of two mechanisms: 1) competition between the LH-RBI and ^{125}I-LH for the same binding site on the receptors; 2) binding of LH-RBI and ^{125}I-LH to different sites on the receptors, with the binding of LH-RBI at the second site preventing the binding of ^{125}I-LH at the LH binding site. In either of the mechanisms, LH-RBI inhibits LH binding at the receptor level.

Experiments on the effect of LH-RBI on the binding of LH to the hormone receptors of ovary or testis were carried out for comparison. To our surprise, LH-RBI was found not to inhibit the binding of ^{125}I-LH to testicular LH receptors (Yang _et al_., 1976b). From this result, the possibility of the competition between LH-RBI and ^{125}I-LH for the same binding site becomes unlikely (Vide supra, mechanism 1). If LH-RBI inhibited the ^{125}I-LH binding through the mechanism of competition for the same binding site, LH-RBI would have inhibited the binding of ^{125}I-

LH to the testicular LH receptors. By elimination of alternatives, the data thus favors mechanism 2 (Vide supra). One may argue that the failure of LH-RBI to show the inhibition in the testicular system could also be due to low binding affinity of LH-RBI for the LH binding site in the testis. We examined this possibility using two criteria: 1) preincubating the testicular receptors with LH-RBI for 1 hour at 37°C before the addition of ^{125}I-LH and a second incubation (Yang _et al._, 1976b), and 2) increasing the quantity of LH-RBI for the incubation from 10 mg tissue equivalent up to 40 mg tissue equivalent. Again, however, no inhibition of the binding of ^{125}I-LH to testicular LH receptors was observed in either of these two conditions (Yang _et al._, 1976b).

Our suggestion that LH-RBI inhibits LH binding through the mechanism other than competitive inhibition was also supported by the results of a binding kinetics study. The ovarian LH receptors were incubated with increased concentrations of ^{125}I-LH in the absence or presence of a fixed concentration of LH-RBI. The results were used to construct the Scatchard plot (Scatchard, 1949). The dissociation constant, i.e. Kd for the ^{125}I-LH binding, was essentially the same in either the absence or presence of LH-RBI. Since the association constant (the reciprocal of Kd) of the ^{125}I-LH binding was not reduced by the presence of the inhibitor, this would suggest that LH-RBI did not compete with ^{125}I-LH for the same binding site on the receptor.

Our work suggests that the LH receptor of ovaries, but not testes, has a specific LH-RBI binding site in addition to the LH binding site; and that the binding of LH-RBI produced an "allosteric" type of inhibition to the binding of LH at the LH binding site (Fig. 2). In this respect, a demonstration of the specific binding of LH-RBI to the ovarian LH receptor would serve as direct evidence. This evidence, however, can only be available when a preparation of a homogeneous LH-RBI material suitable for radioactive labeling is achieved. Purification of LH-RBI is now in process in our laboratory.

Inhibition of LH-induced Progesterone Synthesis by LH-RBI

Table 4 shows the effects of LH and other hormones on the _in vitro_

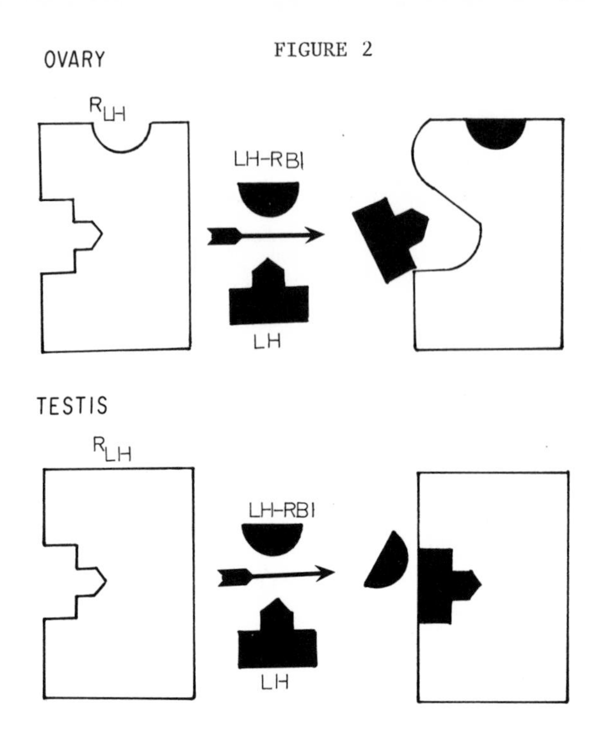

biosynthesis of progesterone in the ovarian tissue slices from pseudo-pregnant rat ovaries and the effects of LH-RBI on the basal and LH-stimulated biosynthesis of progesterone. There was a two-fold increase of progesterone level after a two-hour incubation at 37°C. LH caused a further increase of progesterone synthesis (>200% of the 2 hr-control). These results are comparable with those of Shemesh _et al_. (1975) who studied the _in vitro_ synthesis of progesterone by bovine corpus luteum slices. At the dose tested, hCG also stimulated progesterone biosynthesis by the ovary but to a smaller degree. Insulin had no effect on the ovarian steroidogenesis, indicating stimulation by LH and hCG was specific. When the LH-RBI extract was added along with LH or hCG to the incubate, the LH-stimulated progesterone biosynthesis was significantly inhibited and the progesterone level measured under this condition was in the range of that of the control in which neither LH nor

Table 4. Effect of crude LH-RBI on *in vitro* LH-stimulated and hCG-stimulated progesterone biosynthesis by pseudopregnant rat ovary slices.

Treatment	Progesterone* (μg/gm tissue)	
Unincubated control	11.0±0.4	(3)
Incubated control	24.0±1.9	(4)
LH (1.7 μg/ml)	55.2±1.0	(4)
LH + crude LH-RBI	24.0±1.5†	(4)
hCG (25 IU/mg)	36.4	(2)
hCG + crude LH-RBI	22.9†	(2)
Crude LH-RBI	24.3†	(2)
Insulin (1.7 μg/ml)	24.7±3.4	(3)
LH + ovarian extract (fresh ovary)	52.4†	(2)
Ovarian extract (fresh ovary)	23.1†	(2)

*Data indicated are mean ± Sem; number of determinations in parentheses.
†Progesterone levels were corrected for the amount of progesterone found in the LH-RBI extract or the extract of fresh ovaries.

the LH-RBI extract was added. However, the LH-RBI extract did not inhibit basal progesterone biosynthesis.

Since a crude preparation of LH-RBI was used in these experiments, one must consider whether the effect on steroidogenesis could be produced by components other than LH-RBI. The results presented in Table 4 suggest that the correlation of binding inhibition and inhibition of progesterone biosynthesis is attributable to a single substance. Note the failure of fresh ovarian extracts to inhibit, thus showing the inhibitor of LH-stimulated steroidogenesis has the same unusual requirement for freezing-storage seen in the LH-RBI as defined in our binding studies.

Fractionation of LH-RBI

As reported earlier (Yang *et al.*, 1976a) when an aqueous extract of pseudopregnant ovary was applied to a Sephadex G-200 column, 2.5 x 85 cm, equilibrated with 0.25 M $NH_4H\ CO_3$, LH-RBI activity was located in a broad fraction of effluent volume, suggesting adsorption effects, a multiplicity of active compounds, or both. Since the majority of LH-RBI activity was associated with the low molecular weight fractions, the aqueous extract of the pseudopregnant ovary was dialyzed against distilled water at 4°C for 15 hours. The dialysate and nondialyzable material obtained were tested for their effects on ^{125}I-LH binding. The LH-RBI activity was found both in the dialysate and to a lesser extent, in the nondialyzable portion of the extract (Yang *et al.*, 1976 a). Eighty-five percent of the LH-RBI activity in the dialysate remained after boiling for 20 minutes. However, only 20% of the LH-RBI activity remained in the nondialyzable portion of the extract after boiling under the same conditions. These data suggest the presence in the pseudopregnant rat ovary of two types of LH-RBI: a heat-labile LH-RBI which has a larger molecular weight (or is associated with a inert protein) and a heat-resistant LH-RBI which is a low molecular weight substance(s).

Purification of LH-RBI

The dialysate of the pseudopregnant ovarian extract was lyophilized and further fractionated on a Sephadex G-50 column developed with 0.005 N acetic acid. The eluates were separated into seven fractions (Yang *et al.*, 1976a) and tested for inhibition of ^{125}I-LH binding. The Sephadex G-50 column chromatography showed less trailing for the dialyzed fraction, but some inhibitory activity was still observed behind the major LH-RBI activity (Yang *et al.*, 1976a). Seventy percent of the total inhibitory activity of LH-RBI recovered from the Sephadex G-50 column was located in a single fraction, however. We believe that the inhibitory activity which eluted later than the major LH-RBI activity was due to the "trailing" of the LH-RBI in the Sephadex gels rather than to the presence of multiple inhibitors, although this point is not yet established unequivocally. From our calibration of the Sephadex G-50 column with proteins and peptides of known molecular weight it was estimated that the heat-stable LH-RBI has a molecular weight of about 3800 daltons.

Other Inhibitors of Gonadotropin Functions

Chang (1955) reported an inhibitory effect of follicular fluid on the spontaneous maturation of rabbit oocytes in culture. Liu and Davey (1974) described a trypsin-sensitive, heat-stable, low-molecular weight (about 1411) antigonadotropin from the ovaries of the insect *Rhodnius prolixus Stal*. Tsafriri and Channing (1975) reported an inhibitor for porcine oocyte meiosis, which has been further characterized (Tsafriri and Channing, 1976). The meiosis inhibitor was present in follicular fluid and had a molecular weight of less than 2000. It is clear from present information that these foregoing mentioned inhibitors and LH-RBI cannot be the same substance, although further research is required to establish whether these inhibitors may be chemically related.

Kent (1973) reported that a tetrapeptide isolated from hamster embryos inhibited ovulation *in vivo*. We tested whether this tetrapeptide, Thr-Pro-Arg-Lys, inhibited *in vitro* ^{125}I-LH binding to ovarian LH receptor, using concentrations ranging from 50 ng/ml to 500 μg/ml. However, no significant inhibition by this peptide on ^{125}I-LH binding was observed even at the highest concentration, i.e. 1×10^{-3} M (Yang *et al*., 1976a).

Bhalla and Reichert (1974) have described ethanol soluble factors from testes which inhibit FSH and LH binding, apparently by interaction with the hormones. That these factors are obtainable from testes, while LH-RBI is not, suggests the inhibitor we observed differs from that found by Bhalla and Reichert (1974).

Certain plant extracts have been shown to inhibit LH and FSH action *in vivo* (Braneman *et al*., 1976). In the instance cited, polyphenolic fractions from the plant were employed. The spectral absorption properties of partially purified LH-RBI preparations would not suggest a related composition, but of course, composition can only be evaluated on pure preparations.

Finally, the inhibitor of FSH action, designated "inhibin" by some, has received renewed research interest by several investigators (Steinberger and Steinberger, 1976; Schwarz and Channing, 1977; Franchimont *et al*., 1975; Nandini, *et al*., 1976 and Baker *et al*., 1976).

Inhibin is usually obtained from testicular extracts, which suggests it differs from LH-RBI. Moreover, its reported molecular weight range is higher than that of LH-RBI.

It is important for all the inhibitors of mammalian gonadotropin function to obtain a chemical characterization of the purified materials as early as possible. Only on this basis can we deal with differences or similarities on a meaningful basis. Therefore, purification and characterization of LH-RBI is our high priority goal of studies now in progress.

ACKNOWLEDGMENT

This study was supported in part by the following grants: American Cancer Society No. PDT-41E; U.S. Public Health Service, NIH, CA-05831, CA-16672, and AM-09801; and The Robert A. Welch Foundation G-147. Animals utilized in this study were maintained in facilities approved by the American Association for Accreditation of Laboratory Animal Care, and in accordance with current United States Department of Agriculture and Department of Health, Education and Welfare, National Institutes of Health regulations and standards.

REFERENCES

Baker, H. W. G., Bremmer, W. J., Burger, H. G., de Kretser, D. M., Dulmanis, A., Eddie, L. W., Hudson, B., Keogh, E. J., Lee, V. W. K. and Rennie, G. C. (1976). Testicular Control of Follicle-Stimulating Hormone Secretion. Rec. Prog. Hormone Res. 32, 429-476.

Bhalla, V. K. and Reichert, L. E., Jr. (1974). Gonadotropin Receptors in Rat Testes: Interaction of an Ethanol-Soluble Testicular Factor with Human Follicle Stimulating Hormone and Luteinizing Hormone. J. Biol. Chem. 249, 7996-8004.

Breneman, W. R., Zeller, F. J., Carmack, M. and Kelley, C. J. (1976). In Vivo Inhibition of Gonadotropins and Thyrotropin in the Chick by Extracts of Lithospermum ruderale. Gen. Compar. Endocr. 28, 24-32.

Chang, M. C. (1955). The Maturation of Rabbit Oocytes in Culture and Their Maturation, Activation, Fertilization and Subsequent Development in the Fallopian Tubes. J. Exptl. Zool. 128, 378.

Dufau, M. L., Charreau, E. H. and Catt, K. J. (1973). Characteristics of a Soluble Gonadotropin Receptor from the Rat Testis. J. Biol. Chem. 248, 6973-6982.

Dufau, M. L., Charreau, E. H., Ryan, D. and Catt, K. J. (1974). Soluble Gonadotropin Receptors of the Rat Ovary. FEBS Letters 39, 149-153.

Franchimont, P., Chari, S., Hagelstein, M. T. and Duraiswami, S. (1975). Existence of a Follicle Stimulating Hormone Inhibiting Factor "Inhibin" in Bull Seminal Plasma. Nature 257, 402-404.

Kent, H. A., Jr. (1975). Contraceptive Polypeptide from Hamster Embryos: Sequence of Amino Acids in the Compound. Biol. Reprod. 12, 504-507.

Liu, T. P. and Davey, K. G. (1974). Partial Characterization of a Proposed Antigonadotropin from the Ovaries of the Insect Rhodnius prolixus Stal. Gen. Compar. Endocr. 24, 405-408.

Nandini, S. G., Lipner, H. and Moudgal, N. R. (1976). A Model System for Studying Inhibin. Endocrinology 98, 1460-1465.

Sakai, C. N., Engel, B. and Channing, C. P. (1977). Ability of an Extract of Pig Corpus Luteum to Inhibit Binding of ^{125}I-Labeled Human Chorionic Gonadotropin to Porcine Granulosa Cells. Proc. Soc. Exptl.

Biol. Med. 155, 373-376.

Scatchard, G. (1949). The Attractions of Proteins for Small Molecules and Ions. Ann. N.Y. Acad. Sci. 51, 660.

Shemesh, M., Hansel, W. and Concannon, P. W. (1975). Testosterone Synthesis in the Bovine Corpus Luteum. Biol. Reprod. 13, 490-493.

Schwarz, N. B. and Channing, C. P. (1977). Suppression by Porcine Follicular Fluid of the Acute Serum FSH Rise Following Ovariectomy in the Rat. Abstract No. 37, 10th Annual Meeting, Soc. Study Reprod., p. 33.

Steinberger, A. and Steinberger, E. (1976). Secretion of an FSH-Inhibiting Factor by Cultured Sertoli Cells. Endocrinology 99, 918-921.

Tsafriri, A. and Channing, C. P. (1975). An Inhibitory Influence of Granulosa Cells and Follicular Fluid Upon Porcine Oocyte Meiosis *in vitro*. Endocrinology 96, 922-927.

Tsafriri, A., Pomerantz, S. H. and Channing, C. P. (1976). Inhibition of Oocyte Maturation by Porcine Follicular Fluid: Partial Characterization of the Inhibitor. Biol. Reprod. 14, 511-516.

Yang, K. P., Samaan, N. A. and Ward, D. N. (1976a). Characterization of an Inhibitor for Luteinizing Hormone Receptor Site Binding. Endocrinology 98, 233-241.

Yang, K. P., Samaan, N. A. and Ward, D. N. (1976b). Lutropin receptors from Male and Female Tissues: Different Responses to a Lutropin Receptor Binding Inhibitor. Proc. Soc. Exptl. Biol. Med. 152, 606-609.

DISCUSSION

Channing: We also found LH-RBI activity in pig corpus luteum extracts and also had to store them in a frozen state to get activity. If one obtains pig ovaries from the slaughterhouse, young, middle age, and old corpora lutea can be recognized from their color and the appearance of the follicles in that ovary. Extracts of old corpora lutea have higher amounts of LH-RBI activity than young corpora lutea, and the middle stage corpora lutea have intermediate amounts. It is important to observe differences in LH-RBI under different physiological states if the substance acts as a control mechanism. We use binding of LH or hCG to granulosa cells as a bioassay. It is quite possible that the corpus luteum could modulate follicular development by altering binding of LH either to itself, or to neighboring follicles.

Reichert: We have been studying the properties of a follicle stimulating hormone binding inhibitor in testes (Reichert and Abou-Issa, Biol. Reprod., In press). We find that buffer or water extracts of mature rat testes will inhibit the binding of radiolabelled human FSH to hormone specific receptors in the testes. The binding inhibitor is partially dialysable; approximately 60% will pass a Spectapor #1 membrane but will not pass a UM05 Amicon membrane, indicating a molecular weight in the range between 500 to 6,000. The binding inhibitor will promote dissociation of FSH from preformed hormone receptor complex. The binding inhibitor which does not pass through the membrane, the large molecular component, is heat stable; the material passing the membrane is heat labile. The problem is that similarly prepared extracts from liver, kidney, and brain, and pseudopregnant rat ovaries seem to have the same properties with respect to inhibition of FSH binding to receptor. We rationalize that only the testes have specific receptors for FSH, so that a binding inhibitor in target tissue may have some physiologic relevance despite the fact that it is also present in other tissues.

Baker: The concentrations of FSH and LH in human and ovine follicular fluid remain more or less constant throughout the life of the antral follicle and are generally quite a lot higher than in blood, so why doesn't the follicle undergo its various maturation processes long before the "normal" time? I wonder if the roles of these inhibitors isn't to prevent the action of the intra-follicular FSH and LH?

Channing: This is what prompted all of my studies in the beginning. If you have two follicles starting to grow, why does one grow and mature and why doesn't the other one? Luteinization inhibitor is found in higher levels in the smaller follicle, and it would be of interest to measure the FSH binding inhibitor in different size follicles. Have you done this Dr. Reichert?

Reichert: We have preliminary results which suggest that the amount of FSH binding inhibitor present per ml of follicular fluid may vary significantly among follicles of different sizes, although large follicles seem to contain more inhibitor by virtue of increased fluid volume. Additional studies will be required before definite conclusions can be reached about this, and these are currently under way in our laboratory.

Channing: Perhaps as the follicle matures you have less of the inhibitors synthesized; at least this seems to be the case for luteinization inhibitor and the levels of other inhibitors should be studied. We have studied LH-RBI in pig ovaries and it occurs in very low amounts in non-luteal tissues. It would be of great interest to measure LH-RBI in atretic follicles and follicles at different physiological states. Perhaps when a follicle matures the thecal layer synthesizes less LH-RBI in the presence of LH and FSH.

Folkers: Is it unreasonable to test frozen hypothalamic tissue for the presence of your inhibitor?

Ward: It is certainly not unreasonable; we have not done it, however.

Ryan: Several years ago, C.Y. Lee and I (Biochemistry 12:4609, 1973) noticed a phenomenon that may or may not be related to LH-RBI; namely, when labelled LH or hCG was incubated with ovarian tissue, a variable portion of the unbound fraction was no longer capable of binding to a fresh batch of receptor. The phenomenon was less pronounced with ovarian tissue from immature rats than with pseudopregnant rat ovaries, it was proportional to the amount of tissue used, it was related to the amount of gonadotropin used, and it also showed a difference between LH and hCG. Whatever was happening to the molecule happened to a greater extent and more rapidly with LH than hCG; the gonadotropin molecule was not grossly changed in structure as determined by gel-filtration or binding to antibody. Is the effectiveness of LH-RBI for hCG equivalent to its effectiveness for LH? Secondly, to address the question of whether the LH-RBI might be doing something in the tissue which then affects the gonadotropin molecule, is the capacity of the gonadotropin to bind to fresh mature or pseudo-pregnant ovary in any way altered after LH-RBI is added to receptor and immature ovary?

Ward: To answer the hCG question, hCG seems to be equally susceptible to LH-RBI within the framework of concentrations we have studied. We have not studied the binding properties of a hormone that has survived the binding process.

Beers: From your binding studies it looked like LH-RBI is acting via a non-competitive mechanism. The implication of that kind of behavior in a binding system is that the inhibitor is irreversibly bound to the receptor or is binding with an off rate that is orders of magnitude greater than the off rate for LH. Do you have any independent studies on the reversibility of receptor binding?

Ward: We haven't checked off rates for the LH in the presence of LH-RBI. The LH and hCG off rates are known to be slow. We have assumed they haven't been altered; Scatchard plots may or may not give us information of that type, but they do have the same slope in the presence or absence of LH-RBI.

Strickland: It is interesting that you get activation of the inhibitor upon storage since it doesn't seem to be a function of freezing and thawing. Activation might be a subtle chemical change which could be temperature dependent. Have you stored ovaries at either lower or higher temperatures and studied the appearance of LH-RBI?

Yang: When pseudopregnant ovaries were stored at higher temperatures than -20°C, LH-RBI activity was detected much sooner after storage of the tissues; for example, 5 hr at 23°C or 37°C and 2 days at 4°C.

Bullock: I would like to raise the question of the contraceptive implications of these various inhibitors which inhibit essential events in reproduction. It is a salutary experience to think through the contraceptive implications of these substances to the point where they could be used. How would such a new kind of contraceptive be used by women in India for example? Assuming one can overcome the problem of oral delivery of these inhibitors, which is an immense problem in itself, there remains the key question of how one times the ingestion of these compounds by the user. Unless one is going to administer them continuously, which is not regarded as desirable, there has to be some event that the women can use for taking the substance, and that has to be, in practical terms, menstruation or the absence of it. That consideration makes it difficult, at least for me, to envisage how something like an oocyte maturation inhibitor could be developed into a practical contraceptive. It is a little easier to envisage how LH-RBI could be used in these terms, because one could have a contraceptive that could be tied to a missed menstrual period and that would potentially prevent the rescue of the corpus luteum. I wonder if anyone working with these inhibitors would like to comment on how far their thinking has gone in terms of the ultimate contraceptive usage of these inhibitors?

Ward: I think germane to that will be an understanding of the chemical nature of the inhibitors. As far as the timing question, you have to know what sort of regimen you can put a person on to handle that particular substance, with whatever properties that it possesses.

Channing: Yes, we have thought about these problems. First of all we want to see if oocyte maturation inhibitor is active *in vivo*, or whether follicular fluid *per se* can alter the fertilizability of oocytes. If you would give follicular fluid, and the oocyte doesn't mature or isn't fertilizable, then would this be an approach? We don't know when you would have to give OMI in order to render the oocyte nonfertilizable. With the case of inhibin it is a little bit easier to comprehend, since it appears that FSH is required in the cycle prior to ovulation to cause the follicle or the oocyte to mature properly. Perhaps you could give the inhibin preparation at the end of the menstrual cycle so that the follicle would be exposed early in its life, and then later the egg wouldn't be fertilizable or some other function would be altered. One has to really know more about how these substances work, and then one could do timing studies and design the contraceptive. It may also be that early in life, exposure to inhibin at the proper time may render the oocytes nonfertilizable at a much later date. This also has to be considered.

Gospodarowicz: I think we should not get involved too much in biopolitics. I think that to find out what controls the maturation of the oocyte without even worrying what use it has in fertility control is important enough. I stress again that the studies are not that advanced, so I don't think we can even discuss what will be the role in fertility control.

4

Hormone Binding, Proteases and the Regulation of Adenylate Cyclase Activity

NANCY D. RICHERT, THOMAS A. BRAMLEY, and ROBERT J. RYAN

I. Bacterial Binding of hCG

Addition of porcine follicular fluid[1] to the growth medium of cultured rat luteal cells markedly increased the gonadotropin hormone binding capacity of these cultures. Subsequent studies (Richert and Ryan, 1977a) demonstrated that a bacterial contaminant in the follicular fluid was responsible for this effect. When the contaminant (identified as Pseudomonas maltophilia) was grown in 0.5 liter flasks of trypticase soy broth (TSB), we found that the bacteria would bind the labeled hormone ([^{125}I]hCG) with high affinity (K_d = 2.3 x 10^{-9}M) and that the bacterial binding sites could be saturated.

The capacity for hormone binding paralleled the growth of the bacterial culture (Fig. 1). The bacterial binding sites exhibited hormone specificity for hCG or LH. A variety of other hormones, sugars, and glycoproteins were unable to compete with [^{125}I]hCG for binding (Fig. 2). HCG binding sites were not observed on other Pseudomonads (Ps. testosteroni, Ps. aeruginosa), or on other gram-negative organisms (E. coli, E. cloacae); see Table 1.

It is unlikely that the hormone binding sites on the follicular fluid isolate of Ps. maltophilia resulted from a genetic recombination between the bacteria and the porcine granulosa cells because another strain of Ps. maltophilia (obtained from the American Type Culture Collection) also bound [^{125}I]hCG to the same extent (Table 1).

[1] The follicular fluid was aspirated from large, medium or small pig follicles and was not sterilized.

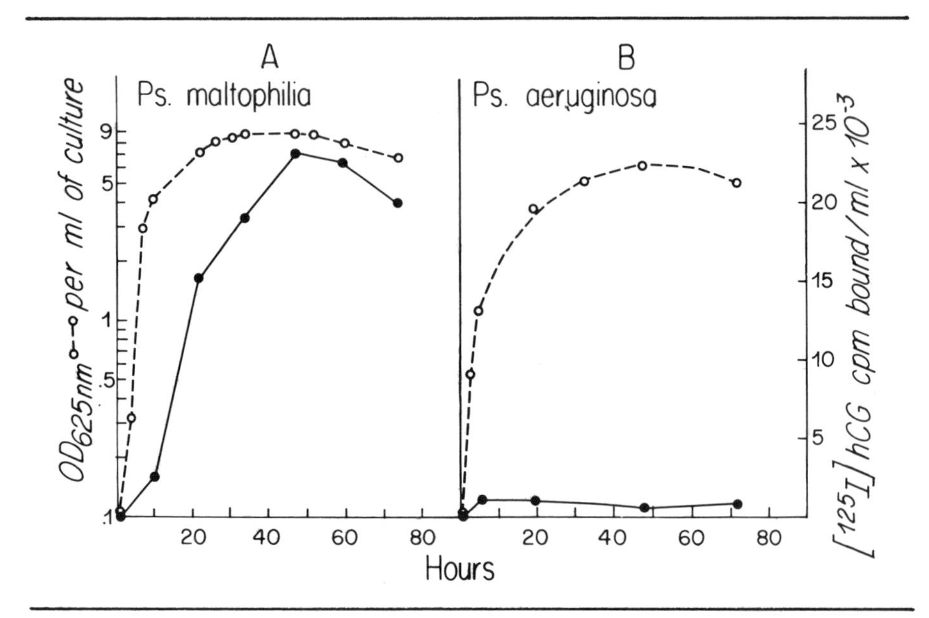

Fig. 1. Growth curve (0) and ^{125}I-hCG binding (•) in TSB at 30° with Ps. maltophilia (A) and Ps. aeruginosa (B). Batch cultures (500 ml) were inoculated at time 0. At various times, duplicate 10-ml aliquots were removed from the flask for measurement of optical density of the culture at 625 nm with TSB as a blank. For binding of [^{125}I]hCG, the aliquots were centrifuged at 20,000 x g, washed, and resuspended in 1.0 ml of 40 mM Tris/10% sucrose buffer. Binding assays contained 2 ng of [^{125}I]hCG and 100 µl of the final bacterial pellet in a volume of 1 ml of 40 mM Tris. Nonspecific binding was measured in the presence of unlabeled hCG (Richert and Ryan, 1977a).

II. Bacterial Production of hCG

To explain the existence of a hormone binding site on a procaryotic organism, we investigated the possibility that the bacteria might produce an hCG-like molecule. The ectopic production of hCG by human tumors (Weintraub and Rosen, 1971), and the fact that hCG shares a region of sequence homology with cholera toxin (Kurosky, Markel et al., 1977) suggested that this hormone might have a primitive origin. Furthermore, Wuerthele-Caspe Livingston and Livingston (1974) reported a bacterial protein which immunologically resembled hCG. Cohen and Stramp (1976) also identified an hCG-like material from bacterial cultures which stimulated in vitro testosterone secretion by Leydig cells of the rat testis.

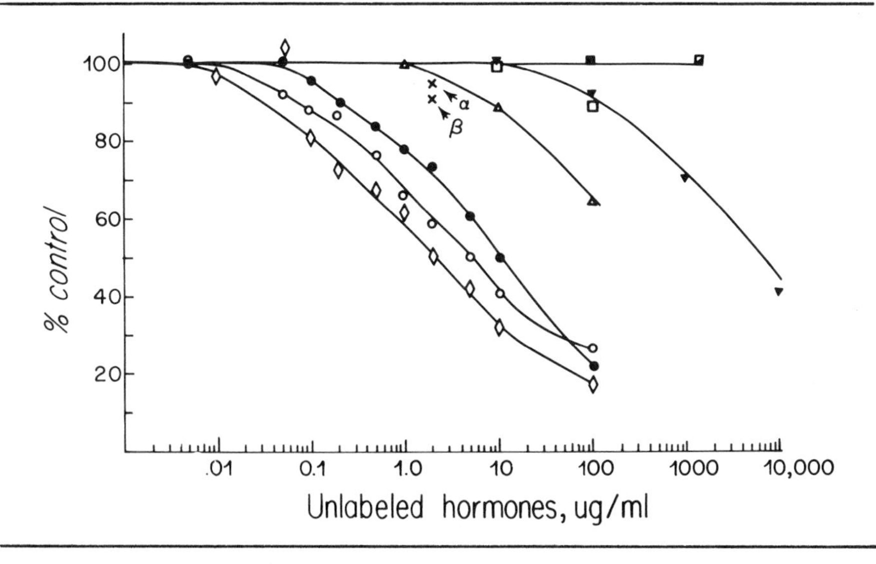

Fig. 2. Competition for binding of [^{125}I]hCG to Ps. maltophilia by various unlabeled hormones. Two ng of [^{125}I]hCG was incubated with 210 μg of bacterial protein in the presence of increasing concentrations of unlabeled hormones. Incubations were performed at 20° overnight and terminated with Carbowax: ◊, hLH; O, hCG; •, oLH; X, hCGα and β subunits: Δ, oFSH; □, oPRL; ▼, ovalbumin, ovomucoid, bovine gamma globulin; ■, sugars (glucose, galactose, mannose, maltose, D-glucosamine, N-acetylneuraminic acid); □, KI (Richert and Ryan, 1977a).

Table 1. Binding of [^{125}I]hCG by various procaryotic organisms.

Organism (source)	% [^{125}I]hCG bound per ml culture	% [^{125}I]hCG bound per mg protein
Ps. maltophilia (follicular fluid)	24.7	19.3
Ps. maltophilia (ATCC 13637)	15.7	13.5
Ps. testosteroni (ATCC 11996)	0.5	0.8
Ps. aeruginosa (ATCC 27853)	1.5	1.5
E. cloacae (follicular fluid)	1.8	N.D.
E. aerogenes (Mayo Clinic)	1.4	1.0
E. coli (ATCC 25922)	0.9	1.0
Bacillus sp. (follicular fluid)	1.2	3.5

Suspension cultures were sampled at various times for growth and binding determinations. Binding data represent maximal binding in all cultures at stationary phase. Data from Richert and Ryan (1977a).

To determine whether hCG was produced by <u>Ps. maltophilia</u>, the bacterial cultures were grown to stationary phase in 2-4 liter flasks of TSB, then centrifuged at 20,000 x g for 20 min. to pellet the bacteria. The cell-free supernatant was filtered through a 0.45 mμ Nalgene filter, lyophilized and chromatographed on a 2.5 x 90 cm column of Sephadex G-100 in 0.1 M ammonium bicarbonate buffer (pH 8.0).

The elution profile from the column is presented in Fig. 3. Each fraction was tested for its immunological similarity to hCG by radioimmunoassay (RIA) using antisera prepared against native hCG, or the hormone α and β subunits (Good et al., 1977). Fractions were also tested for their ability to stimulate adenylate cyclase in rat ovarian membrane preparations.

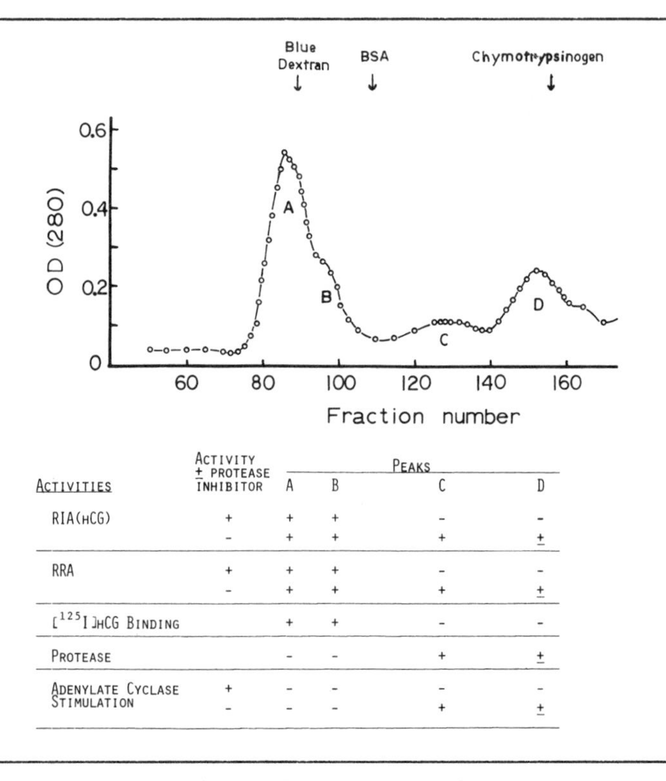

ACTIVITIES	ACTIVITY ± PROTEASE INHIBITOR	PEAKS A	B	C	D
RIA(HCG)	+	+	+	-	-
	-	+	+	+	±
RRA	+	+	+	-	-
	-	+	+	+	±
[^{125}I]HCG BINDING		+	+	-	-
PROTEASE		-	-	+	±
ADENYLATE CYCLASE STIMULATION	+	-	-	-	-
	-	-	-	+	±

Fig. 3. Sephadex G-100 elution profile of <u>Ps. maltophilia</u> culture media. Fractions were assayed for hCG-like activity in RIA, RRA and rat ovarian cyclase systems with and without a serine protease inhibitor (PMSF). Fractions were also assayed for [^{125}I]hCG binding and protease activity.

The results in Fig. 3 indicate that the material in Peaks A and B (> 90,000 daltons) immunologically resembles native hCG (by RIA) and also competes with hCG in a radioreceptor assay (RRA). This large molecular weight material, however, fails to stimulate adenylate cyclase activity in a membrane-enriched fraction of the rat ovary. Peak C (30,000 daltons) which does not immunologically cross-react with hCG antisera, will stimulate adenylate cyclase activity. Peak D is devoid of either of these activities. A small amount of adenylate cyclase stimulation by pooled fractions of Peak D is attributed to minor contamination with Peak C.

To maximize the production of the hCG-like material, the cultures were grown under various conditions and for different times. The results showed that the hCG-like material was maximal during the stationary phase of bacterial growth, but would disappear and then reappear. This fact, together with the quantitative variability in Peaks A and B suggested that a proteolytic enzyme released by the bacteria might be degrading the large molecular weight hCG-like material. The column fractions were, therefore, tested for proteolytic activity using a standard casein hydrolysis procedure (Laskowski, 1955). The results summarized in Fig. 3 showed that proteolytic activity was present in Peak C. The significance of this activity in stimulation of ovarian adenylate cyclase is discussed in Section III.

The presence of a bacterial protease in the culture media complicated the process of isolating and purifying the putative hCG-like molecule from these cultures. Furthermore, subsequent experiments demonstrated that the material in Peaks A and B was capable of binding [^{125}I]hCG in a standard receptor assay. This suggests that the 90,000 molecular weight material may be solubilized bacterial receptor which is shed during the growth of the culture. By binding the iodinated tracer in the RIA, this solubilized receptor would decrease the [^{125}I]-hCG cpm bound to antibody and thus appear to be an hCG-like molecule.

Other experiments have failed to demonstrate any functional change in the bacterial cultures after addition of exogenous hCG. There is no alteration in cell growth, or cell death, nor is there any stimulation of procaryotic adenylate cyclase in response to the hormone. For these reasons, we feel that it is highly unlikely that the bacteria themselves produce an hCG-like molecule. The presence of the hormone receptor on these organisms may simply represent another example of structural similarity between procaryotic and eucaryotic cell surfaces, analogous to the presence of ABO blood group antigens on Salmonella species (for review, see Markowitz, 1976).

III. Stimulation of Adenylate Cyclase by Pseudomonas maltophilia Protease

The stimulation of rat ovarian adenylate cyclase by Peak C (the 30,000 MW material) from the bacterial culture media was further examined. Figs. 4A and 4B compare the dose-response curves for the ovarian cyclase stimulation by hCG and by Peak C. HCG produces a sigmoidal increase in activity between the concentrations of 0.01 and 1 μg/ml and the maximal response (4- to 6-fold over basal vs. 10-fold for NaF) persists with doses of hCG up to 100 μg/ml. Maximum stimulation by Peak C causes an 8- to 10-fold increase in cyclase activity. At concentrations greater than 15 μg/ml, Peak C suppresses the response, and ultimately the cyclase activity is reduced below basal.

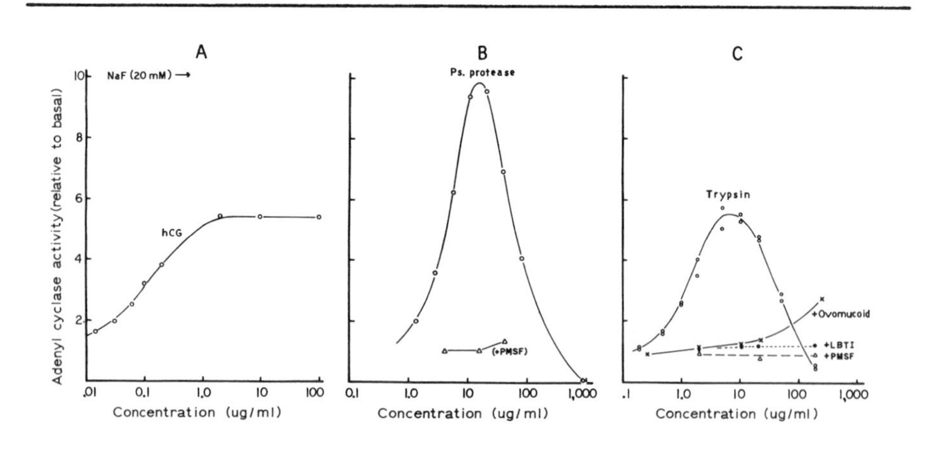

Fig. 4. Stimulation of Ovarian Adenylate Cyclase by hCG, Pseudomonas Protease and Trypsin. A membrane fraction of superovulated rat ovaries was incubated with increasing concentrations of hCG, Pseudomonas protease (Peak C) or trypsin to assay adenylate cyclase activity. Assays were conducted at 30° for 20 min. Activity is calculated as pmoles cAMP/20 min/mg protein and expressed as fold-increase relative to basal. The inhibitors (PMSG, ovomucoid, LBTI) were preincubated with the proteases for 15 min at 24° prior to dilution in the assay (Richert and Ryan, 1977b).

Because Peak C was found to have proteolytic activity (Fig. 3), attempts were made to separate the protease from the cyclase-stimulating factor by DEAE cellulose chromatography. Both activities, however, eluted in the void volume of the column. As an alternate approach, the Peak C material was treated with a serine protease inhibitor, phenylmethanesulfonyl fluoride (PMSF). This inhibitor blocked both proteolytic activity and the cyclase stimulating activity (Fig. 4B) and suggested that the cyclase stimulating factor was a serine protease. A serine protease having a molecular weight of 30,000 daltons has been isolated from Ps. maltophilia cultures (Boethling, 1975).

IV. Stimulation of Adenylate Cyclase by Other Serine Proteases

To determine whether other serine proteases have a similar effect on rat ovarian adenylate cyclase, we tested a variety of commercial proteases of bacterial and mammalian origins (Richert and Ryan, 1977b). Fig. 4C demonstrates that bovine pancreatic trypsin affects rat ovarian cyclase in nearly the same manner as the Pseudomonas protease. Pretreatment of trypsin with proteolytic enzyme inhibitors (ovomucoid, lima bean trypsin inhibitor [LBTI], or PMSF) abolished the cyclase stimulation. A time course of trypsin stimulation of cyclase showed a lag phase of less than 5 minutes, and the stimulation was linear for nearly 50 min (data not shown).

The cyclase stimulating effects of other serine proteases are indicated in Fig. 5. Trypsin, chymotrypsin, pronase, subtilisin, and protease (from Streptomyces griseus) all activate ovarian adenylate cyclase in nearly the same manner. Maximum stimulation occurs at protease concentrations ranging from 5-10 μg/ml. Higher protease concentrations suppress activity as shown for trypsin in Fig. 4C. Plasmin was relatively ineffective in cyclase stimulation and thrombin, plasminogen, and trypsinogen (not shown) were without effect at the doses tested. Chymotrypsinogen was active only at high doses where proteolytic activity could be demonstrated.

V. Serine Protease Stimulation of Progesterone Secretion

Further studies were performed to determine whether serine proteases could stimulate progesterone synthesis (a cAMP-mediated event) in slices of immature rat ovary in vitro (Richert and Ryan, 1977d). The results (Table 2) indicate that trypsin will increase ovarian

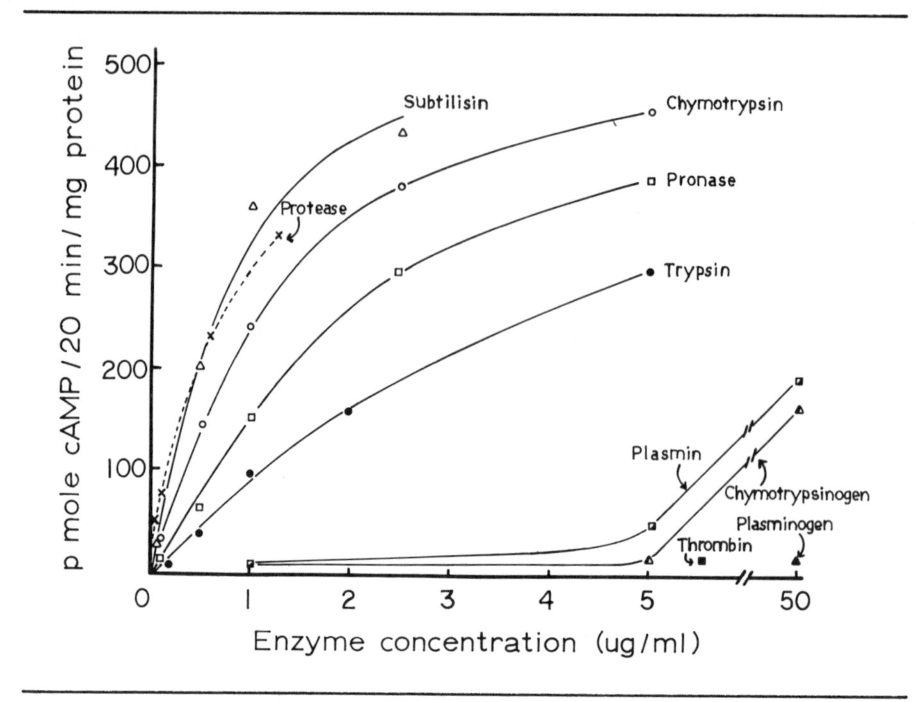

Fig. 5. Stimulation of rat ovarian adenylate cyclase by increasing doses of various bacterial and mammalian serine proteases and zymogens. From Richert and Ryan, 1977b.

Table 2. The effects of trypsin on stimulation of progesterone synthesis by slices of immature rat ovaries, with or without pretreatment of the slices with 50 μg/ml neuraminidase. From Richert and Ryan, 1977d.

Treatment	Dose (μg/ml)	Response (ng progesterone/mg wet wt/2 hrs)
Preincubated 30 min in buffer		
Control	—	5.3
Trypsin	500	4.5
Preincubated 30 min with 50 μg/ml neuraminidase		
Control	—	3.5
Trypsin	500	24.0

ovarian steroidogenesis; however, pretreatment of ovarian slices with neuraminidase is required for this effect. The reason for this neuraminidase requirement is not known, although it is interesting that the enzyme also markedly enhances progesterone synthesis by hCG. We speculate that neuraminidase unmasks surface proteins (e.g., the hCG receptor) or makes surface glycoproteins more susceptible to tryptic digestion.

VI. Mechanism of serine protease activation of adenylate cyclase

Serine proteases have an amino acid sequence CAGY (cys-ala-gly-tyr) which is also found in cholera toxin, and in the β subunits of the glycoprotein hormones TSH, FSH, LH and hCG (Kurosky, Markel et al., 1977). The CAGY sequence per se is not responsible for ovarian adenylate cyclase stimulation because the β subunits of LH and hCG are biologically inactive (Catt et al., 1973). Furthermore, the CAGY sequence isolated from thrombin is without effect on the cyclase (unpublished observation). We considered an alternate possibility that the serine proteases, by virtue of the CAGY sequence might bind to the hCG receptor, then modify the hormone receptor in such a way as to mimic the effect of the hormone itself. Proteolysis would be required because the zymogen precursors were inactive in stimulating cyclase, as was trypsin in the presence of trypsin inhibitors (Fig. 4c).

The results (Fig. 6) indicate that all of the proteases which actively stimulate cyclase in the ovary, also inhibit [^{125}I]hCG binding to the hormone receptors in this tissue. The loss of hormone binding results from a loss or change in the hormone receptor itself, rather than destruction of the labeled hormone. The same inhibition was observed if the tissue was pretreated with trypsin, and the trypsin was neutralized with excess LBTI before the addition of [^{125}I]hCG. The results (Fig. 7) show that the dose-response correlations for protease activation of cyclase and protease-induced loss of [^{125}I]hCG binding are nearly identical.

Despite this correlation, however, there are several experiments which suggest that serine proteases do not specifically interact with the hCG receptor to cause cyclase stimulation. First, in desensitized ovaries, which lack a functional hCG receptor, there is no hormonal (hCG) stimulation of adenylate cyclase, but trypsin, chymotrypsin, and

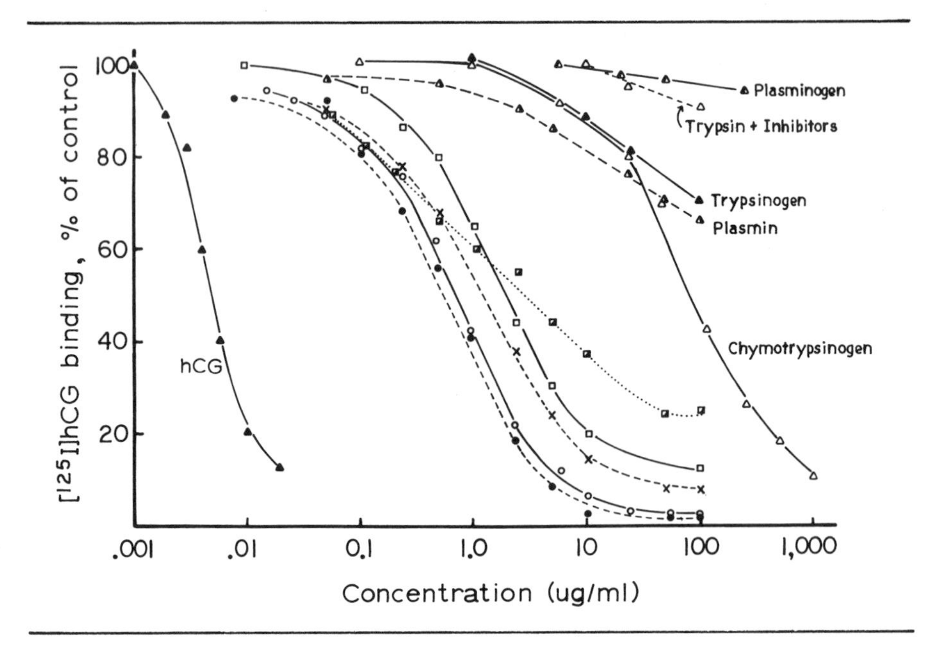

Fig. 6. Effect of Serine Proteases on [^{125}I]hCG Binding to the 2,000 x g Pellet Fraction of Superovulated Rat Ovaries. Increasing concentrations of serine proteases (or unlabeled hCG) were incubated with 100 μg ovarian membrane protein, and 2 ng [^{125}I]hCG in a final volume of 1 ml in 40 mM Tris buffer containing 0.1% BSA for 30 min at 37°. In control incubations (containing no proteases or unlabeled hCG), 49.1% of the total labeled hormone was bound. Data is expressed as % of control binding normalized to 100%. Symbols are •—•, pronase; 0—0, protease (Strep. griseus); X—X, chymotrypsin; □—□, trypsin; ◪—◪, subtilisin (Richert and Ryan, 1977b).

NaF stimulation is comparable to that observed in control ovaries (Table 3). Secondly, when maximal doses of hCG and trypsin are simultaneously added to adenylate cyclase assays, their combined effects are greater than when either is added alone (Fig. 8). The additive effects of trypsin and hCG suggest that they do not work through a common receptor on the cell surface.

It is interesting that trypsin will enhance the NaF stimulation of adenylate cyclase (Fig. 8). For this reason, we considered the possibility that the proteases might directly activate the catalytic subunit of the cyclase (or unmask new catalytic sites). If serine proteases directly activate the cyclase enzyme, then protease stimulation (like NaF stimulation) should persist in Lubrol solubilized

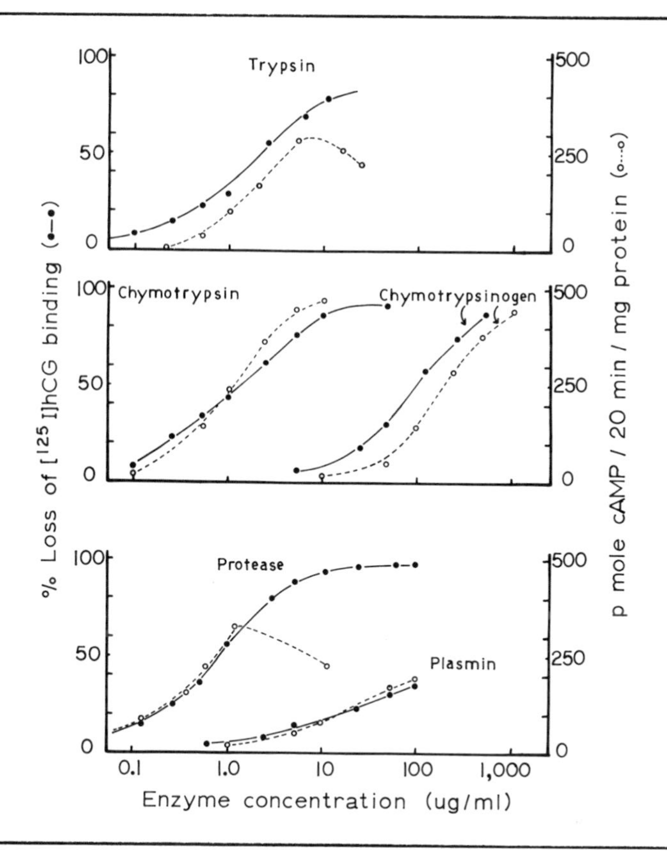

Fig. 7. Correlation Between Doses of Proteases required to Stimulate Rat Ovarian Adenylate Cyclase Activity and Diminish the Binding of [^{125}I]hCG to Rat Ovarian Receptors. From Richert and Ryan, 1977b).

Table 3. Stimulation of Adenylate Cyclase Activity in Membrane Fractions Prepared From Control and Desensitized Rat Ovaries. 2000 x g Pellet fractions were prepared from control and hCG-desensitized rat ovaries. The membrane preparations were incubated in the adenylate cyclase assay alone, or in the presence of hCG (2 μg/ml), trypsin (5 μg/ml), chymotrypsin (5 μg/ml) or NaF (20 mM). From Richert and Ryan (1977b).

	Adenylate Cyclase Activity (pmoles/20 min/mg protein)				
Rat Ovary	Basal	hCG	Trypsin	Chymotrypsin	NaF
Control	130	884	749	1085	1881
Desensitized	115	126	644	888	1045

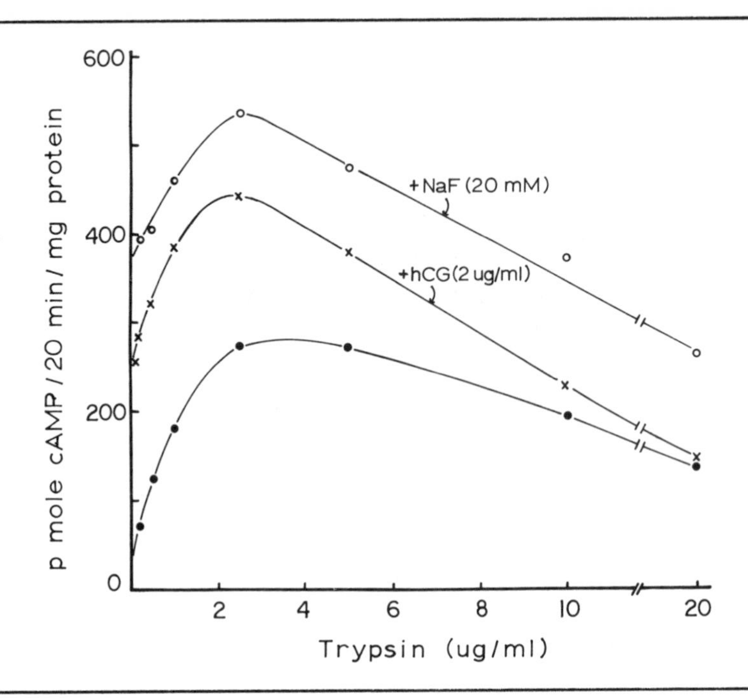

Fig. 8. The Effects of Trypsin on Rat Ovarian Adenylate Cyclase Activity, alone •—•, or in the Presence of Maximal Stimulating Doses of hCG X—X, or NaF 0—0. From Richert and Ryan, 1977b.

membranes. The results (Table 4), however, show that this is not the case. Lubrol abolishes both hormonal and serine protease stimulation of adenylate cyclase in the same manner, while fluoride-stimulated activity is enhanced by the detergent. The fact that Lubrol does not stimulate basal cyclase activity suggests that the protease stimulation cannot be explained simply as a change in membrane permeability. Lubrol does not inactivate the proteolytic activity of trypsin or chymotrypsin (data not shown).

VII. Effect of Protease Inhibitors on Activation of Rat Ovarian Adenylate Cyclase

To further investigate the mechanism of protease stimulation of cyclase, a variety of specific protease inhibitors was used (Richert and Ryan, 1977c). TPCK(L-1-tosylamide-2-phenylethyl chloromethyl ketone) is an active site titrant and irreversible inhibitor of chymotrypsin that does not affect trypsin (Schoellmann and Shaw, 1963).

Table 4. Effects of Lubrol Solubilization on Adenylate Cyclase Activity in Rat Ovarian Membrane Preparations. From Richert and Ryan, 1977b.

Additions		Adenyl Cyclase Activity (pmoles/20'/mg protein) Lubrol Concentration (%)					
		0	0.01	0.03	0.05	0.07	0.1
None		53	47	34	35	24	14
NaF	(20 mM)	390	378	550	560	512	368
hCG	(2 μg/ml)	288	106	35	34	33	20
Trypsin	(5 μg/ml)	147	98	43	24	21	13
Chymotrypsin	(5 μg/ml)	186	108	77	56	41	32

If serine proteases directly activate the catalytic subunit of adenylate cyclase, then TPCK should selectively inhibit chymotrypsin stimulation without affecting trypsin stimulation. However, the results (Fig. 9) demonstrate that TPCK blocks both trypsin and chymotrypsin stimulation in the same manner. Control experiments showed that trypsin remains fully active in a proteolytic assay in the presence of TPCK. More surprising is the fact that hormonal (hCG) stimulation of cyclase is also inhibited by TPCK while basal and NaF activities are unaffected.

These results suggest that TPCK might be inhibiting a membrane protease which is responsible for regulating cyclase activity in the ovary. The loss of hormonal stimulation of cyclase cannot be explained by inhibition of [^{125}I]hCG binding to the ovarian hormone receptors (Table 5). Furthermore, TPCK does not inactivate the hormone. If hCG is pretreated with TPCK, then serially diluted before the assay to remove the inhibitor, the hCG stimulation of cyclase is unaffected. Fig. 9 also demonstrates the effect of TAME(p-tosyl-L-arginine methyl ester), a synthetic substrate for trypsin that cannot be hydrolyzed by chymotrypsin (Schwert et al., 1948). Like TPCK, TAME also eliminates hormonal stimulation of adenylate cyclase without affecting hormone binding (Table 5).

We considered the possibility that the inhibition by TAME and TPCK might be unrelated to their well-documented effects as protease inhibitors. Both of these inhibitors possess a common structural feature (a p-toluene substituent) and toluene has been shown to

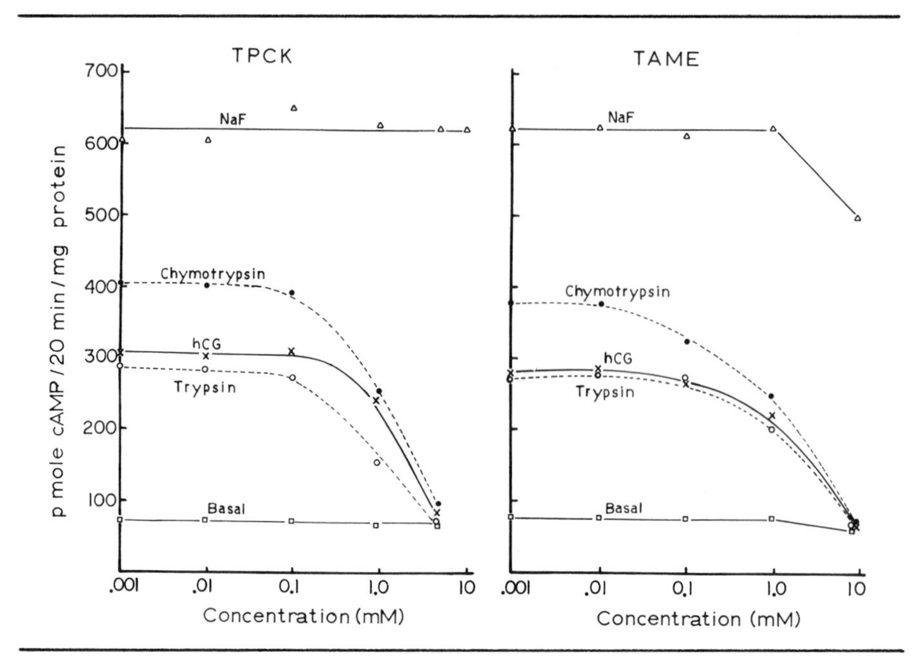

Fig. 9. The effects on increasing doses of the chymotrypsin inhibitor (TPCK) and the trypsin substrate (TAME) on basal, NaF, hCG, trypsin and chymotrypsin stimulation of rat ovarian adenylate cyclase. The assay was conducted for 20 min at 30°. NaF was 20 mM, hCG 2 μg/ml, trypsin and chymotrypsin 5 μg/ml. From Richert and Ryan, 1977c.

Table 5. Dose Response Effect of Protease Inhibitors on Basal and Stimulated Adenylate Cyclase

Inhibitors	Enzymes† Inhibited	Maximum Inhibitor Dose Tested (mM)	ID_{50} for Adenylate Cyclase Stimulation					[^{125}I]hCG Binding‡ (% of Control
			Basal	NaF	hCG	Trypsin	Chymo-trypsin	
Irreversible								
PMSF	1-5	5	3.0	3.5	0.8	1.5	0.7	56§
TLCK	1,4,5,6	5	0.8	1.0	0.4	0.3	0.3	111
TPCK	2,6	5	*	*	2.5	1.5	1.0	107
Competitive								
TAME	1	10	10.0	(40.0)	2.0	2.0	2.0	94
BAME	1	10	*	(40.0)	6.0	2.5	9.0	NT
Benzamidine	1	100	100.0	100.0	25.0	5.0	25.0	105
LLβN‡	7	2	1.0	2.0	0.2	0.2	0.2	10§

†Enzyme Code: trypsin(1); chymotrypsin(2); elastase(3); plasmin(4); thrombin(5); sulfhydryl proteases(6); leucine aminopeptidase(7)

‡LLβN L-leucyl-β-napthalamide

‡Inhibition of [^{125}I]hCG binding at maximal dose tested.

*No effect at highest dose tested.

§No inhibition of [^{125}I]hCG binding at dose effective for cyclase stimulation.

increase nucleotide permeability in mammalian (Hilderman and Deutscher, 1974), and bacterial cells (Harwood and Peterkovsky, 1975). When cyclase assays were performed with toluene and p-toluene sulfonic acid (TSA), however, there was no inhibition of hCG or protease stimulation (Table 6). Instead, the toluene dose-response curves showed a slight increase in hormone, protease and NaF stimulation of adenylate cyclase (data not shown).

Table 6. Effect of Protease Inhibitors on Basal and Stimulated Adenylate Cyclase. Assays were performed for 20 min at 37°. NaF, 20 mM; hCG 2 μg/ml; trypsin, 5 μg/ml; chymotrypsin, 5 μg/ml.

Inhibitors	Dose (mM)	Adenylate Cyclase Activity (cAMP pmoles/20 min/mg protein)				
		Basal	NaF	hCG	Trypsin	Chymotrypsin
Controls (± SE)		58 ± 3	557 ± 39	226 ± 8	227 ± 8	289 ± 16
Toluene	10	63	519	321	242	274
TSA	10	69	492	220	236	275
Irreversible						
PMSF	5	15	164	13	11	12
TLCK	1	28	268	44	14	81
TPCK	5	62	611	62	27	115
Competitive						
TAME	10	36	429	37	9	42
BAME	10	44	372	98	16	113
Benzamidine	100	35	339	35	33	63
LLβN	1	22	215	34	47	54
Natural	(mg/ml)					
Ovomucoid	1	63	534	190	73	NT
LBTI	1	30	396	119	30	29
PTI	1	39	474	158	41	37
α-1-antitrypsin	1	55	535	194	56	51
Microbial						
Leupeptin	0.1	78	676	284	92	290
Antipain	0.1	79	657	279	72	301
Chymostatin	0.1	62	506	229	244	62

The data in Tables 5 and 6 summarize the effects of a variety of protease inhibitors on ovarian adenylate cyclase. In general, low concentrations of the synthetic inhibitors block hCG and protease activation of adenylate cyclase. At higher doses, basal and NaF stimulation are affected. Even at the highest inhibitor concentrations, however, the fluoride-stimulated activity remains at least 7- to 12-fold higher than basal. A variety of natural protease inhibitors were ineffective in blocking hCG stimulation of adenylate cyclase.

The results in Table 7 demonstrate that the synthetic protease inhibitors affect cholera toxin and epinephrine stimulation of adenylate cyclase in nearly the same manner as hCG stimulation.

Table 7. Effect of Protease Inhibitors on Cholera Toxin and Epinephrine Stimulation of Adenylate Cyclase. Incubations were at 30° for 60 min. NaF, 20 mM; hCG, 2 μg/ml; epinephrine, 0.2 mM; cholera toxin, 10 μg/ml + 1 mM NAD+. From Richert and Ryan, 1977c.

Inhibitors	Dose (mM)	Adenylate Cyclase Activity (cAMP pmoles/min/mg protein) Basal	NaF	hCG	Epinephrine	Cholera Toxin
Control	—	4.7	49.7	18.5	21.2	15.8
PMSF	5	1.7	12.9	1.7	4.1	12.0
TLCK	1	0.7	21.3	0.9	0.8	0.9
TPCK	5	5.2	35.4	5.1	11.8	2.8
TAME	10	2.0	22.7	2.9	6.4	6.4
BAME	10	4.2	35.6	8.4	12.3	10.3
Benzamidine	100	1.8	18.2	2.4	6.9	4.0

VIII. Model for Hormonal Regulation of Adenylate Cyclase

The results demonstrate that adenylate cyclase in the rat ovary can be stimulated in vitro by exogenous serine proteases. The protease stimulation is comparable to, or greater than, the stimulation observed with hCG, cholera toxin, or epinephrine.

The phenomenon does not appear to be restricted to the rat ovary nor is it an artifactual consequence of using broken cell preparations to measure adenylate cyclase activity. Protease stimulation of adenylate cyclase has been recently reported in rat liver membranes (Lacombe et al., 1977), as well as in whole cells—e.g., rat embryo fibroblasts (Ryan, Short et al., 1975), and most recently in a KB cell line (Guiraud-Simplot and Colobert, 1977). Our studies also indicate that proteolytic enzyme stimulation of adenylate cyclase can ultimately lead to increased steroidogenesis in rat ovarian tissue.

The physiological significance of protease stimulation of adenylate cyclase is suggested by the observation that a variety of serine protease inhibitors block protease stimulation and hormonal stimulation of adenylate cyclase in the same manner. These results indicate that membrane proteases may be intimately involved in the regulation of adenylate cyclase activity.

Although the precise role of membrane proteases is not clear, the necessity for some form of "messenger" or "transducer" to serve as a linkage between hormone receptor and adenylate cyclase is becoming increasingly apparent. Thus, the original model of adenylate cyclase consisting of a hormone receptor subunit on the exterior cell surface physically coupled to a catalytic subunit on the inside of the cell membrane (Robison et al., 1967) has been modified by Birnbaumer et al. (1970) to include a transducer element between the two subunits. Other models have proposed that the hormone receptor portion is localized in the cell membrane at a different site than the catalytic subunit of the enzyme. According to the mobile receptor hypothesis of Cuatrecasas (1974), hormone binding at the cell surface would cause the receptor to diffuse through the lipid bilayer and bind to and activate the catalytic subunit.

The physical separation of receptor and adenylate cyclase was suggested by the findings of Catt and Dufau (1973) that there are "spare receptors" capable of hormone binding which are not required for maximum activation of adenylate cyclase or for maximum steroidogenesis. Recent studies from our own and other laboratories suggest that the hormone receptor and adenylate cyclase are localized on two different membrane fractions in the rat ovary (Bramley and Ryan, 1977) and the toad erythrocyte (Sahyoun et al., 1977). Schramm et al. (1977) have demonstrated that the adenylate cyclase enzyme of one cell type can be activated by a β-adrenergic receptor from a heterologous cell type after cell fusion by Sendai virus.

In view of these present concepts of adenylate cyclase activation, membrane proteases could fulfill a role in the activation process either as 1) diffusable "transducers" released at the hormone receptor site after hormone binding, or 2) as proteolytic agents capable of "mobilizing" the receptors to diffuse through the membrane. The concept of the mobile receptor has been discussed previously (Cuatrecasas, (1974) and will not be further considered here. One potential model for protease activation is presented in Fig. 10.

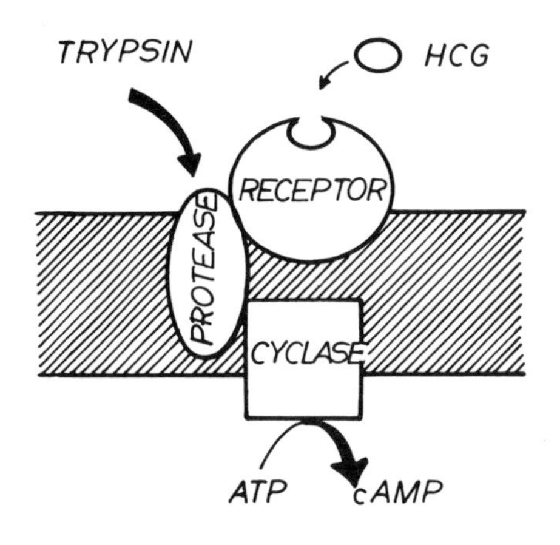

Fig. 10. An initial model depicting a membrane protease as an intermediate between hormone binding and activation of adenylate cyclase in rat ovarian membranes. From Richert and Ryan, 1977c.

The model assumes that the ovarian membrane protease would exist as a zymogen, or in a relatively inactive state. A low level of protease activity could be responsible for maintaining basal cyclase activity. Activation of the membrane protease could be accomplished by hormone binding to its receptor or by addition of exogenous serine proteases. By suppressing the activated membrane protease, TAME, TPCK, and the other protease inhibitors would effectively block hormonal and proteolytic enzyme stimulation of cyclase, and at high concentrations could affect even basal activity. NaF stimulation could result from a dephosphorylation of the catalytic subunit (Constantopoulos and Najjar, 1973), causing it to become more susceptible to the effects of active membrane proteases or other processes. Dephosphorylation has been shown to enhance the sensitivity of phosphorylate kinase to tryptic digestion (Cohen, Antoniw et al., 1976).

The model suggests that a protease, as a diffusible transducer, would serve as a positive modulator of adenylate cyclase activity. In the absence of proteolytic activity, hormone stimulated adenylate cyclase levels could not be elevated over basal. The mechanism for this transduction could be either a direct activation of the catalytic subunit or an indirect effect.

Guirand-Simplot and Colobert (1977) suggested that proteases directly activate the catalytic subunit of adenylate cyclase in cultured KB cells. This conclusion was based on the observation that proteases lowered the K_m of the cyclase enzyme for ATP and altered its Mg^{++} dependence. However, preliminary data from our laboratory, as well as the studies by Hanoune et al. (1977) have failed to confirm these observations. Furthermore, neither trypsin, chymotrypsin (see Table 4) nor several other commercial proteases, were capable of directly activating cyclase activity in Lubrol solubilized membranes. The demonstration of the direct activation of the catalytic subunit by proteases may require isolation of the membrane protease involved and the definition of suitable incubation conditions.

If membrane proteases, as mobile transducers, do not directly alter the catalytic subunit of the cyclase, then it is possible that they indirectly stimulate the cyclase by affecting nucleotide activation of the enzyme. The Rodbell model of adenylate cyclase regulation proposes that nucleotides (e.g., ATP or GTP) function as primary modulators of cyclase activity (Rendell et al., 1975). Nucleotide binding to the catalytic subunit raises the enzyme to a higher activity state. Hormones serve only in a secondary capacity to modify the rate at which the transition from a lower to a higher state of activity occurs. Recent studies by Levinson and Blume (1977) suggest that cholera toxin stimulation of adenylate cyclase is mediated by GTP and that the conversion of cyclase-bound GTP to GDP, and thus a lower activity state, is prevented by the toxin.

In this model of nucleotide regulation of adenylate cyclase, membrane proteases might serve to maintain the concentrations of nucleotides that are required to keep the cyclase enzyme in the high activity state. This could be accomplished by proteolytic degradation of nucleotide hydrolases (e.g., ATPase or GTPase) or by activating a nucleotide regenerating system.

A potential relationship between proteases and nucleotides can be envisioned even if one postulates a direct activation of the cyclase enzyme by the protease. Limited proteolysis of the catalytic subunit might lead to its activation and also its susceptibility to further proteolysis and inactivation. Nucleotide binding to the cyclase enzyme, either at the substrate site or some other site, might prevent further proteolysis and thus preserve the active state. Release of the nucleotide would then lead to further proteolysis and inactivation. An analogous protective mechanism for pyridoxal enzymes as a result of co-enzyme binding has been reported by Katunuma (1975) in studying mitochondrial membranes.

The basic feature of all of these models is that gonadotropin binding to the ovarian hormone receptor serves to activate a membrane protease. Evidence to support this concept comes from studies by Reichert (1962) which demonstrated that in vivo injection of pituitary gonadotropins stimulated alkaline (pH 7-8) protease activity in ovarian homogenates when hemoglobin was used as a substrate. More recent evidence by Strickland and Beers (1976) has demonstrated that in vitro exposure of granulosa cells to FSH or LH will elicit the production of a plasminogen activator. Although the plasminogen activator is thought to be responsible for the disruption of the follicle, the properties of this enzyme suggest that it may resemble the protease involved in adenylate cyclase activation. All plasminogen activators which have been isolated thus far are serine proteases (Christman et al., 1975). Furthermore, the plasminogen activator in transformed mouse embryo fibroblasts has a trypsin-like substrate specificity but is insensitive to a variety of natural trypsin inhibitors—e.g., lima bean and soybean trypsin inhibitors, and bovine pancreatic trypsin inhibitor (Danø and Reich, 1975). Further studies are needed to determine the sensitivity of the ovarian plasminogen activator to the protease inhibitors which affect adenylate cyclase activation.

It is also possible that membrane protease regulation of adenylate cyclase is not a function of a single enzyme but rather of multiple enzymes arranged in a cascade system. This could explain why a chymotrypsin inhibitor (e.g., TPCK) can block adenylate cyclase activation as effectively as trypsin inhibitors (TAME, TLCK). Alternatively, there may be a cellular membrane enzyme which possesses both trypsin and chymotrypsin-like specificities. A membrane enzyme having this unusual property has recently been isolated from rat liver by Jusic et al. (1976).

While further work is needed to elucidate the properties of the ovarian membrane protease, its mode of regulation, and its role in adenylate cyclase stimulation, the concept of protease involvement is interesting in another regard. Protease involvement in adenylate cyclase regulation could also explain the phenomenon of desensitization which has been described in the ovary and in other tissues (see Ryan, Birnbaumer et al., 1977, for review). If proteases serve to link the hormone receptor and the catalytic subunit of the cyclase, it is conceivable that hormonal activation of membrane proteases could ultimately lead to the alteration or destruction of the hormone receptor itself. Thus, "down-regulation" of hormone receptors would occur concomitantly with activation of adenylate cyclase.

With the increasing evidence that cellular membrane proteases are involved in a variety of functions—e.g., regulation of pyridoxal enzymes (Katunuma, 1975); hormone biosynthesis (Steiner et al., 1975); cell growth control and oncogenic transformation (Reich, 1975; Schnebli, 1975; Troll et al., 1975); as well as in lectin-induced blastogenesis (Saito et al., 1977), it would appear that we are only at the threshold of understanding the diverse roles of membrane proteases in controlling cellular function.

ACKNOWLEDGMENT

This work was supported by NIH Grants HD 01444 and HD -9140 and the Mayo Foundation. We are grateful to Evelyn Gardner for preparation of the manuscript and Kenneth Peters for excellent technical assistance.

REFERENCES

Birnbaumer, L., Pohl, S. L., Krans, M. L., and Rodbell, M. (1970). The actions of hormones on the adenyl cyclase system. Adv. in Biochem. Psychopharmacology 3, 185-208.

Boethling, R. S. (1975). Purification and properties of a serine protease from Pseudomonas maltophilia. J. Bacteriol. 121, 933-941.

Bramley, T. A., and Ryan, R. J. (1977). Interactions of gonadotropins with corpus luteum membranes. III. The identification of two distinct surface membrane fractions from superovulated rat ovaries. Submitted for publication.

Catt, K. J. and Dufau, M. L. (1973). Spare gonadotropin receptors in rat testis. Nature New Biology 244, 219-221.

Catt, K. J., Dufau, M. L., and Tsuruhara, T. (1973). Absence of intrinsic biologic activity in LH and hCG subunits. J. Clin. Endocrinol. Metab. 36, 73-80.

Christman, J. K., Acs, G., Silagi, S. and Silverstein, S. C. (1975). Plasminogen activator: biochemical characterization and correlation with tumorigenicity. In Proteases and Biological Control (Eds. E. Reich, D. B. Rifkin, and E. Shaw) pp 827-839. Cold Spring Harbor Laboratory, Cold Spring Harbor, New York.

Cohen, H. and Strampp, A. (1976). Bacterial synthesis of substance similar to human chorionic gonadotropin. Proc. Sco. Exp. Biol. Med. 152, 408-410.

Cohen, P., Antoniw, J. F., Nimmo, H. G., and Yeaman, S. J. (1976). Protein phosphorylation and hormone action. In Polypeptide Hormones: Molecular and Cellular Aspects. Ciba Foundation Symposium No. 41, pp 281-295, Elsevier/Excerpta Medica/North Holland, Amsterdam.

Constantopoulos, A. and Najjar, V. A. (1973). The activation of adenylate cyclase: II. The postulated presence of (A) adenylate cyclase in a phospho (inhibited) form (B) a dephospho (activated) form with a cyclic adenylate stimulated membrane protein kinase. Biochem. Biophys. Res. Commun. 53, 794-805.

Cuatrecasas, P. (1974). Membrane receptors. Ann. Rev. Biochem. 43, 169-214.

Danø, K., and Reich, E. (1975). Inhibitors of plasminogen activation. In Proteases and Biological Control (Eds. E. Reich, D. B. Rifkin, and E. Shaw). pp 357-366. Cold Spring Harbor Laboratory, Cold Spring Harbor, New York.

Good, A., Ramos-Uribe, M., Ryan, R. J. and Kempers, R. D. (1977). Molecular forms of hCG in serum, urine, and placental extracts. Fertil. & Steril. 28, 846-850.

Guiraud-Simplot, A., and Colobert, L. (1977). Adenylate cyclase activation by trypsin in KB cell cultures. Experentia 33, 899-901.

Hanoune, J., Stengel, D., Lacombe, M. L., Feldmann, G. and Coudrier, E. (1977). Proteolytic activation of rat liver adenylate cyclase by a contaminant of crude collagenase from Clostridium histolyticum. J. Biol. Chem. 252, 2039-2045.

Harwood, J. P. and Peterkofsky, A. (1975). Glucose-sensitive adenylate cyclase in toluene-treated cells of Escherichia coli B. J. Biol. Chem. 250, 4656-4662.

Hilderman, R. H. and Deutscher, M. P. (1974). Aminoacyl transfer ribonucleic acid synthesis in toluene-treated liver cells. J. Biol. Chem. 249, 5346-5348.

Jusic, M., Seifert, S., Weiss, E., Haas, R., and Heinrich, P. C. (1976). Isolation and characterization of a membrane-bound proteinase from rat liver. Arch. Biochem. Biophys. 177, 355-363.

Katunuma, N. (1975). Regulation of intracellular enzyme levels by limited proteolysis. Rev. Physiol. Biochem. Pharmacol. 72, 84-104.

Kurosky, A., Markel, D. E., Peterson, J. W., and Fitch, W. M. (1977). Primary structure of cholera toxin B-chain; a glycoprotein hormone analog? Science 195, 299-301.

Lacombe, M. L., Stengel, D., and Hanoune, J. (1977). Proteolytic activation of adenylate cyclase from rat liver plasma membranes. FEBS Letters 77, 159-163.

Laskowski, M. (1955). Trypsinogen and trypsin. In Methods in Enzymology. Vol. II. (Eds. S. P. Colowick and N. O. Kaplan). pp 26-36. Academic Press, Inc., New York.

Levinson, S. L. and Blume, A. J. (1977). Altered guanine nucleotide hydrolysis as basis for increased adenylate cyclase activity after cholera toxin treatment. J. Biol. Chem. 252, 3766-3774.

Markowitz, A. S. (1976). Protoplasmic and plasma membrane relationships. Trends in Biochem. Sci. 1, 161-163.

Reich, E. (1975). Plasminogen activator: secretion by neoplastic cells and macrophages. In Proteases and Biological Control (Eds. E. Reich, D. F. Rifkin, and E. Shaw) pp 333-341. Cold Spring Harbor Laboratory, Cold Spring Harbor, New York.

Reichert, L. E. (1962). Endocrine influences on rat ovarian proteinase activity. Endocrinology 70, 697-700.

Rendell, M., Salomon, Y., Lin, M. C., Rodbell, M., and Berman, M. (1975). The hepatic adenylate cyclase system. III. A mathematical model for the steady state kinetics of catalysis and nucleotide regulation. J. Biol. Chem. 250, 4253-4260.

Richert, N. D. and Ryan, R. J. (1977a). Specific gonadotropin binding to Pseudomonas maltophilia. Proc. Nat. Acad. Sci. 74, 878-882.

Richert, N. D. and Ryan, R. J. (1977b). Proteolytic enzyme activation of rat ovarian adenylate cyclase. Proc. Nat. Acad. Sci., in press.

Richert, N. D. and Ryan, R. J. (1977c). Protease inhibitors block hormonal activation of adenylate cyclase. Biochem. Biophys. Res. Comm., in press.

Richert, N. D. and Ryan, R. J. (1977d). Protease stimulation of steroidogenesis in the rat ovary. Unpublished data.

Robison, G. A., Butcher, R. W., and Sutherland, E. W. (1967). Adenyl cyclase as an adrenergic receptor. Ann. New York Acad. Sci. 130, 703-723.

Ryan, R. J., Birnbaumer, L., Lee, C. Y., and Hunzicker-Dunn, M. (1977). Gonadotropin interactions with the gonad as assessed by receptor binding and adenylyl cyclase activity. In Reproductive Physiology II. (Ed. R. O. Greep). Internat. Rev. Physiol. Vol. 13, pp 85-152, University Park Press, Baltimore.

Ryan, W. L., Short, N. A., and Curtis, G. L. (1975). Adenylate cyclase stimulation by trypsin. Proc. Soc. Exp. Biol. Med. 150, 699-702.

Sahyoun, N., Hollenberg, M.D., Bennett, V., and Cuatrecasas, P. (1977). Topographic separation of adenylate cyclase and hormone receptors in the plasma membrane of toad erythrocyte ghosts. Proc. Natl. Acad. Sci. 74, 2860-2864.

Saito, M., Aoyagi, T., Umezawa, H. and Nagai, Y. (1977). Bestatin, a new specific inhibitor of aminopeptidases enhances activation of small lymphocytes by concanavalin A. Biochem. Biophys. Res. Commun. 76, 526-533.

Schnebli, H. P. (1975). The effects of protease inhibitors on cells in vitro. In Proteases and Biological Control (Eds. E. Reich, D. F. Rifkin, and E. Shaw). pp 785-794, Cold Spring Harbor Laboratory, Cold Spring Harbor, New York.

Schoellmann, G. and Shaw, E. (1963). Direct evidence for the presence of histidine in the active center of chymotrypsin. Biochemistry 2, 252-255.

Schramm, M., Orly, J., Eimerl, S. and Korner, M. (1977). Coupling of hormone receptors to adenylate cyclase of different cells by cell fusion. Nature 268, 310-313.

Schwert, G. W., Neurath, H., Kaufman, S., and Snoke, J. E. (1948). The specific esterase activity of trypsin. J. Biol. Chem. 172, 221-239.

Steiner, D. F., Kemmler, W., Tager, H. S., Rubenstein, A. H., Lernmark, H., and Zuhlke, H. (1975). Proteolytic mechanisms in the biosynthesis of polypeptide hormones. In Proteases and Biological Control (Eds. E. Reich, D. F. Rifkin, and E. Shaw) pp 531-549. Cold Spring Harbor Laboratory, Cold Spring Harbor, New York.

Strickland, S. and Beers, W. H. (1976). Studies on the role of plasminogen activator in ovulation. J. Biol. Chem. 251, 5694-5702.

Troll, W., Rossman, T., Katz, J., Levitz, M. and Sugimura, T. (1975). Proteinases in tumor promotion and hormone action. In Proteases and Biological Control (Eds. E. Reich, D. F. Rifkin, and E. Shaw) pp 977-987. Cold Spring Harbor Laboratory, Cold Spring Harbor, New York.

Weintraub, B. D. and Rosen, S. W. (1971). Ectopic production of human somatomammotropin by nontrophoblastic cancers. J. Clin Endocrinol. Metab. 32, 94-101.

Wuerthele-Caspe Livingston, V. and Livingston, A. M. (1974). Some cultural, immunological and biochemical properties of Progenitor cryptocides. Trans. New York Acad. Sci. 36(2), 569-582.

DISCUSSION

Lobl: If you degrade part of your receptor through protease action couldn't that stimulate cyclase?

Ryan: Exogenous proteases do appear to destroy the hormone receptor, but we do not think that this is the mechanism for activation of the cyclase enzyme, because of the additive effects and the activation in the hCG desensitized ovary. It may also be possible that the postulated membrane protease destroys the receptor as well as activating the cyclase enzyme, but we do not know if it is necessary to link these functions. It may also be possible that the "receptor" is a multi-subunit substance with one portion of the molecule concerned with hormone binding while another portion is capable of exercising proteolytic activity when it is properly activated.

Denker: There is good evidence from a number of studies that several proteinases can stimulate cells to enter mitosis and to change synthetic activities. Interestingly, even the highly specific proteinase, thrombin acquired some attention very recently in this context. Do you think that general activation of cells could be involved in the phenomena which you have seen or do you think a specific response can be triggered by proteinases?

Ryan: You are quite correct that proteases have been implicated recently in general cellular functions such as blast transformation of lymphocytes and contact inhibition. These kinds of activities are inhibited by the microbial protease inhibitors, which are not effective in our system. We tend to believe that these general cellular activities are mediated through proteases on the external surface of the cells, that are accessible to these inhibitors, while the cyclase activation is mediated by an intramembranous protease that is not accessible to these inhibitors. Of course, it is also possible that our postulated membrane protease has a different inhibitor specificity than the microbial inhibitors that we have tested. Because of the intramembranous localization that we postulate, we are inclined to believe that the action is relatively specific rather than generalized, but we do not have direct data on this point.

Beers: Are the substrates such as TAME hydrolyzed in these experiments?

Ryan: This is very difficult to answer. The sensitivity of the TAME assay is not sufficient using a particulate membrane preparation. If you solubilize the membrane preparation with something like Triton, you have so much protease activity you can't really tell anything. Perhaps this is due to release of contaminating lysosomal enzymes. We need to find the right substrate, and the right assay conditions to be able to get at this question.

5

The Control of Proliferation of Ovarian Cells by the Epidermal and Fibroblast Growth Factors

DENIS GOSPODAROWICZ, ISRAEL VLODAVSKY, HUGH BIALECKI, and KENNETH D. BROWN

Ovarian cells maintained in tissue culture have been used extensively to study the agents and mechanisms involved in luteinization.[1] However, little attention has been given to agents controlling the proliferation of these cells. While gonadotropins (LH and FSH) are believed to influence ovarian cell proliferation *in vivo*, they have been shown to have little or no effect on the growth of these cells in vitro[1-3].

In the few instances in which ovarian cells have been observed to proliferate after the addition of gonadotropins to the culture medium[2,3] the response may have been due to a potent mitogenic agent, distinct from known pituitary hormones, that is a common contaminant of gonadotropin preparations[4].

Recently, several factors have been shown to stimulate the proliferation of a variety of cells in tissue culture. Two of these are epidermal growth factor (EGF) and fibroblast growth factor (FGF).

EGF is found at a high concentration in the submaxillary gland of the adult male mouse[5]. However, it is not produced solely in the submaxillary gland, since extirpation of the gland, while lowering the level of EGF in plasma, does not make it disappear completely[6]. First demonstrated as a mitogen for epidermal cells[7], EGF was later shown to stimulate fibroblast proliferation as well[8,9] and to have a primary structure similar to that of urogastrone[10].

FGF has been isolated from the brain and pituitary of mammals[11,12]. First shown to be a mitogenic agent for fibroblasts[13], it was later found to promote the proliferation of a wide variety of mesoderm-derived cells (Table 1). Since ovarian cells are also derived from the mesoderm and the mitogenic agents for these cells are unknown, we investigated the effect of FGF, as well as EGF, on these cells *in vitro*.

Table 1

Comparison of the Mitogenic Effects of FGF and EGF

Cell Type	Species	FGF	EGF	References
Balb/c 3T3	mouse	+++	++	13, 14, 15, 17
Swiss 3T3	mouse	++	++	17, 18
3T3 thermosensitive mutant	permissive temperature	-	NT	19
foreskin fibroblasts	human	++	± +	20, 21
glial cells	human	+	++	22, 23
kidney fibroblasts	bovine	+	NT	unpublished
amniotic cells (fibroblasts)	human, bovine	+++	+	24

NT = not tested; - = not active; ++ = active (ng/ml); +++ = very active (pg/fg/ml).

Table 1 (continued)

Comparison of the Mitogenic Effects of FGF and EGF

Cell Type	Species	FGF	EGF	References
chondrocytes	rabbit ear and articular chondrocytes	++	+++	24, 25
myoblasts	bovine fetus	++	-	25, 26, 27
vascular smooth muscle	primate bovine	++	+	24
vascular endothelial cells	bovine umbilical vein	++	-	12, 30
	aortic arch fetal and adult	++	-	29, 30 31, 32
	fetal heart	++	-	30
human endothelial cells		++	+	30, 33
cornea endothelial cells	bovine	+++	+++	34, 35
cornea epithelial cells	bovine	+++	-	35, 36
lens epithelial cells	bovine	+++	-	35, 36
cornea epithelium *in vivo* organ culture or with feeder layer	rabbit human	++	+++	37, 36, 38, 39
Y1 adrenal cortex cells	mouse	++	-	40
adrenal cortex cells	bovine	++	-	41
granulosa cells	bovine	++	+++	42
luteal cells	bovine	++	-	43

NT = not tested; - = not active; ++ = active (ng/ml); +++ = very active (pg/fg/ml)

Table 1 (continued)

Cell Type	Species	FGF	EGF	References
liver cells	rat (endoderm)	-	-	32
anterior pituitary cell	rat (endoderm)	-	NT	32
thyroid cells	cow (endoderm)	-	-	unpublished
epidermal cells	human, rabbit chick (ectoderm)	-	+++	44, 32, 45
pancreatic cells	rat (endoderm)	-	NT	32
blastemal cells	frog	++	NT	
blastemal cells	*Triturus viridescens*	++	-	26

NT = not tested; - = not active; ++ = active (ng/ml); +++ - very active (pg/fg/ml)

Three different aspects of the effects of these mitogens on luteal and granulosa cells have been investigated.

A) Their mitogenic effect on granulosa and luteal cell cultures has been compared. The mitogenic effect of EGF has been correlated with its binding properties to these cell types. Since granulosa and luteal cells are interrelated cell types (the luteal cells being derived by cytomorphosis from granulosa cells through the process of luteinization), and since EGF is a mitogen for granulosa cells but not for luteal cells, the comparison of its binding properties to these cell types could lead to some clues regarding what is relevant and irrelevant in the initial mitogenic stimulus.

B) The effect of these mitogens on the replicative lifespan of granulosa cell cultures and their effect on the terminal differentiation of these cells in culture have been examined.

C) The rapidity of the growth of the follicles and of the rise and decline of the corpora lutea, caused by the brevity of the reproductive cycle, imposes remarkable demands on the capacity of the ovarian blood vessels to proliferate and subsequently regress. We have investigated what control the ovarian tissues may exert on such a process and whether they, like tumors, can secrete an angiogenic factor responsible for the control of their own vascularization.

I. Control of the proliferation of granulosa and luteal cells by EGF and FGF

The ovarian cycle is a two-part process. The initial period is one of follicular growth and development. The mature follicles are composed of a central cavity surrounded by multiple layers of granulosa cells and two external cell layers, the richly vascularized thecae interna and externa. The granulosa cells are separated from the outer layers of the follicle by a basement membrane and and are themselves maintained in an avascular environment (Fig. 1).

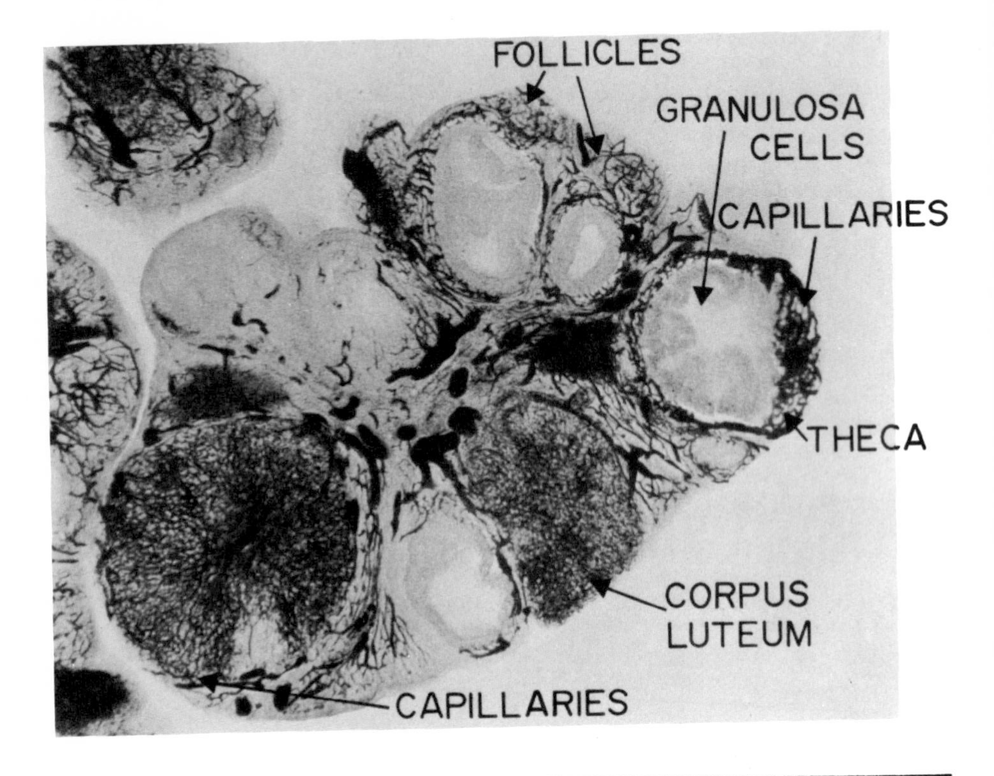

Fig. 1. A thick section of a rat ovary in which the blood vessels were perfused with carmine-gelatin. Notice the rich vasculature of the corpora lutea while the granulosa cell layers within the follicles are avascular.

The second part of the ovarian cycle is the formation, after ovulation, of the corpus luteum. The layers of granulosa cells become vascularized and the nutrient limitation upon granulosa cell proliferation is relieved. Cytomorphosis of granulosa cells into luteal cells occurs and leads to the establishment of the corpus luteum. Granulosa and luteal cells are thus intimately related cell types and, since both are easily maintained in tissue culture, they provide an excellent system with which to study the effects of cytomorphosis on the response to growth factors *in vitro*.

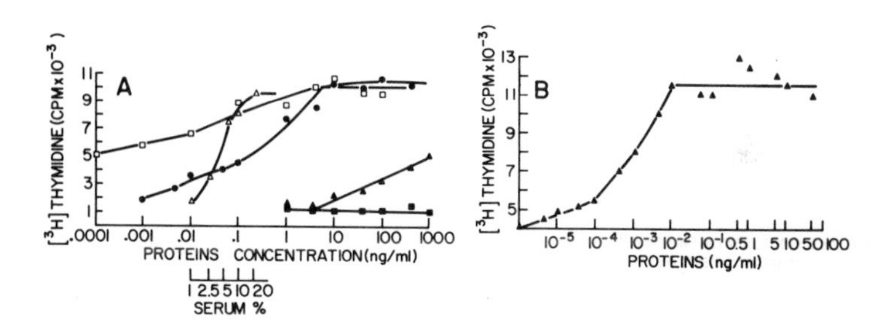

Fig. 2. Initiation of DNA synthesis in subcultures of granulosa cells in response to increasing concentrations of EGF, FGF, and serum. (A) Granulosa cells were plated as described[42]. FGF (• -- •), EGF (□ -- □), highly purified LH (■ -- ■), NIH-LH-B9 (▲ ▲), and calf serum (△-- △) were added to the plates in various concentrations. The control values were 1200 ± 130 cpm. NIH-FSH-B9 and insulin from 1 ng to 1 ug/ml gave 1200 ± 250 cpm. Every point was done in triplicate. (B) Same as (A), but the EGF (▲ -- ▲) was tested in a different experiment since, due to its potency, it was necessary to go to lower concentrations than with FGF. Control values were 3800 ± 280 cpm.

A) Effects of EGF and FGF on the proliferation of bovine granulosa cells. EGF and FGF proved to be the most potent mitogenic agents yet found, when their activity was analyzed using bovine granulosa cells as a model system. As little as 5 x 10^{-2} ng/ml of FGF was enough to induce initiation of DNA synthesis in resting populations of granulosa cells. Saturation was observed at 5 ng/ml[42]. EGF was even more potent; its minimal effective dose was 10^{-5} ng/ml, and saturation was reached at 10^{-1} ng/ml. Optimal concentration of serum (20%) stimulated DNA synthesis to the same extent as did 10^{-1} ng/ml of EGF or 5 ng/ml of FGF[42] (Fig. 2).

Since NIH-LH-B9 has been reported to contain FGF-like activity[13,16], we compared its effect to that of purified LH and to that of purified FGF and EGF. One microgram of NIH-LH-B9/ml stimulated DNA synthesis in granulosa cells as effectively as 10^{-1} ng/ml of FGF or 10^{-3} ng/ml of EGF. In contrast, highly purified LH did not stimulate DNA synthesis. This demonstrates that LH is not mitogenic for granulosa cells and indicates that a contaminant present in some LH preparations must be responsible for the mitogenic effects observed. Neither NIH-FSH-S9 (10 μg/ml) had any effect on DNA synthesis by granulosa cells[42].

Autoradiography of granulosa cell cultures maintained in the presence of optimal concentration of either EGF or FGF had a labelling index of 89%[42]. This rules out the possibility that EGF or FGF has an additive effect such that each stimulates a smaller proportion of the cells (Fig. 3).

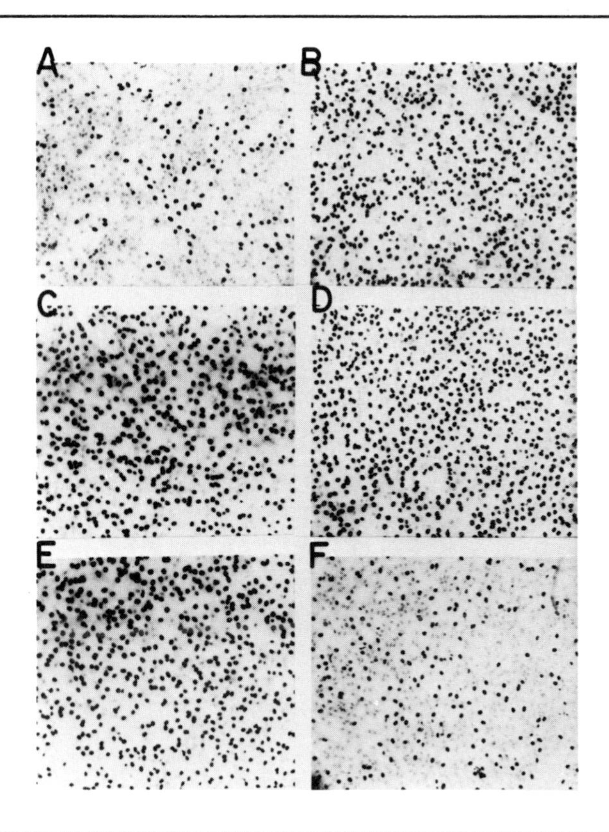

Fig. 3. Granulosa cell autoradiography. (A) control, (B) 10% calf serum, (C) 10 ng/ml FGF, (D) 10 ng/ml EGF, (E) 10 μg/ml NIH-LH-B9, (F) 10 μg/ml highly purified LH.

Since EGF and FGF stimulated the initiation of DNA synthesis at very low concentrations, we also determined the minimal concentration required for a maximal increase in cell number. EGF was active at 2×10^{-4} ng/ml (3×10^{-14} M) and saturation was observed at 0.2 ng/ml (3×10^{-11} M) (Fig. 4C). The cell density obtained at saturating concentrations of FGF was higher than that obtained with saturating concentrations of EGF. This may reflect the longer average cell cycle of cells maintained with EGF (Fig. 4A and B).

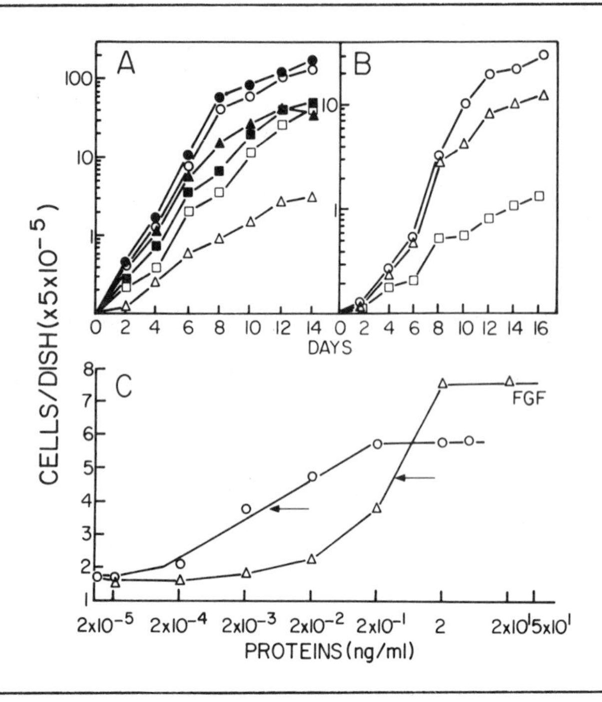

Fig. 4. Effects of FGF and EGF on the proliferation of bovine granulosa cells. (A) Granulosa cells were plated at 3000 cells per 6 cm dish in 5 ml F12 medium with 10% calf serum. Sixteen hours later they were changed to 1% calf serum without (△ -- △) or with 100 ng/ml of EGF (□ -- □) or FGF (■ -- ■). The proliferation effects of FGF and EGF were compared to that of 10% calf serum (▲-- ▲) to which 100 ng/ml of EGF (○ -- ○) or 100 ng/ml of FGF (--) were added. Every point was done in triplicate. Standard deviation did not exceed 10% of the mean. (B) Similarly to "A", the cells were maintained in 1% calf serum without (□ -- □) or with 100 ng/ml EGF (△ -- △) or FGF (○ -- ○). Every point was done in triplicate. Standard deviation did not exceed the mean. (C) Effect of increasing concentrations of EGF (○ -- ○) and FGF (△ -- △) on the proliferation of granulosa cells. The cells were plated as described in "A" and maintained in the presence of increasing concentrations of EGF and FGF. At day 7 the cells were trypsinized and counted. Control gave 30,000 ± 1800 cells. Arrows show the half-maximal response.

The mitogenic effect of FGF and EGF was dependent on the serum concentration in which the cells were maintained. Increasing concentrations of serum (from 0.1% to 10%) enhanced the growth of cells (Fig. 5) maintained in the presence of FGF or EGF.

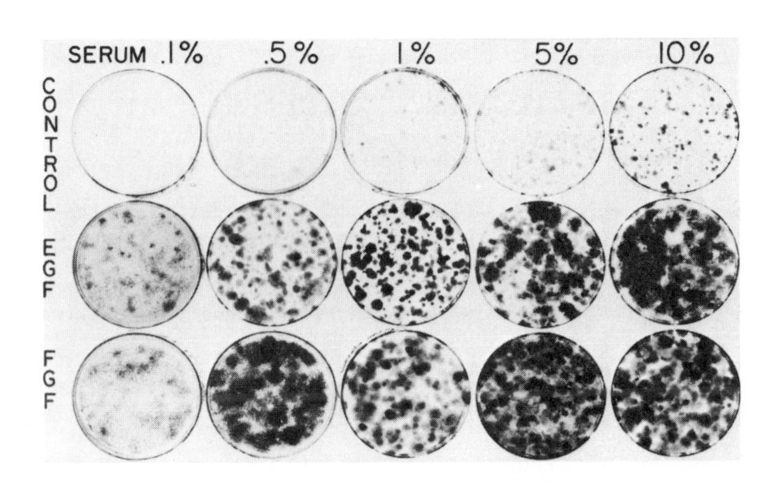

Fig. 5. Correlation between EGF, FGF, and the serum concentration on culture growth. Bovine granulosa cells were plated at 3000 cells per 6 cm dish in 5 ml of DME with 10% calf serum. Sixteen hours later the medium was changed and the cells were maintained in the presence of a constant concentration of calf serum (from 0.1 to 10%) with or without the addition of mitogens (100 ng/ml). The medium was renewed every other day. Plates maintained in the presence of 10% and 5% serum were fixed with 10% formalin and stained with 0.1% Giemsa on day 10, while plates maintained in the presence of 1, 0.5, and 0.1% serum were submitted to the same treatment on day 15.

In all cases there was a positive correlation between the amount of growth observed and the serum concentration in the medium. A mitogenic effect of EGF or FGF could be observed at serum concentrations as low as 0.1% (60 μg serum protein/ml). The main difference between the response of the cells maintained in the presence of 0.1% serum and those maintained in 10% serum with FGF or EGF was that in 0.1% serum the cells divided with a cycle of 3 days, while in 10% serum the doubling time was 18 hours.

The relationship between the rate of proliferation of the granulosa cell cultures and the concentration of EGF, FGF, and serum in the culture medium is shown in Fig. 4A and B. Cultures in 1% serum alone grew slowly, doubling in cell number every 2-3 days (Fig. 4A and B). When FGF or EGF (100 ng/ml) was added, the doubling time became 20 hours for EGF and 16 hours for FGF. Cultures grown with 1% serum plus EGF or FGF reached cell densities 20 times higher than those reached with 1% serum alone. These densities were similar to those obtained with 10% serum alone (Fig. 4A). It can therefore be said that 1% serum plus either EGF or FGF is as potent for promoting granulosa cell growth as is 10% serum. The cells maintained in 10% serum doubled in number every 18 hours. The addition of FGF or EGF (10 ng/ml) to 10% serum only slightly accelerated the average cell cycle which, at 18 hours, is probably limited by the speed of essential cellular anabolic processes. Cells maintained in 10% serum plus either FGF or EGF, however, reached a final cell density four times higher than that observed with 10% serum alone (Fig. 4A). The morphology of the cells maintained in 10% serum with and without mitogens is shown in Fig. 6.

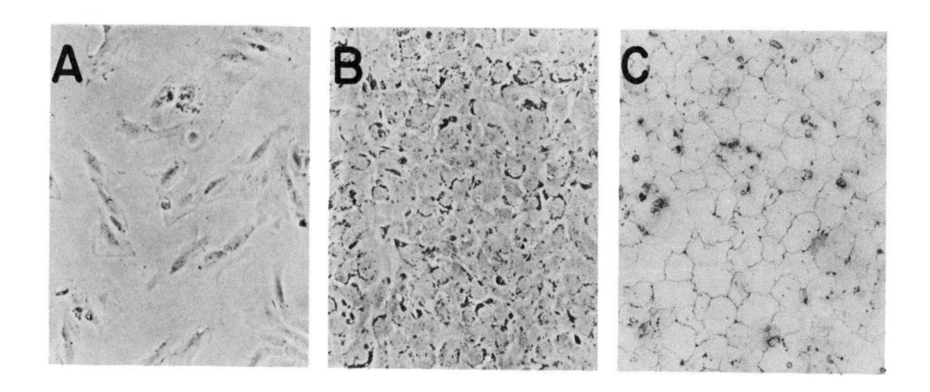

Fig. 6. Appearance of bovine granulosa cells in 1% serum. (A) Control. (B) With 100 ng/ml FGF. (Phase contrast optics 150X.) Similar results were obtained with 100 ng/ml of EGF. (C) Same as B but stained with silver nitrate to show cell borders. (200 X)

Similar results were obtained with pig granulosa cells (Figs. 7 and 8), thus demonstrating that the mitotic effects of EGF and FGF were not limited to bovine granulosa cells.

Our results for granulosa cells demonstrate, then:

1) That a given cell type can be sensitive to more than one growth factor.

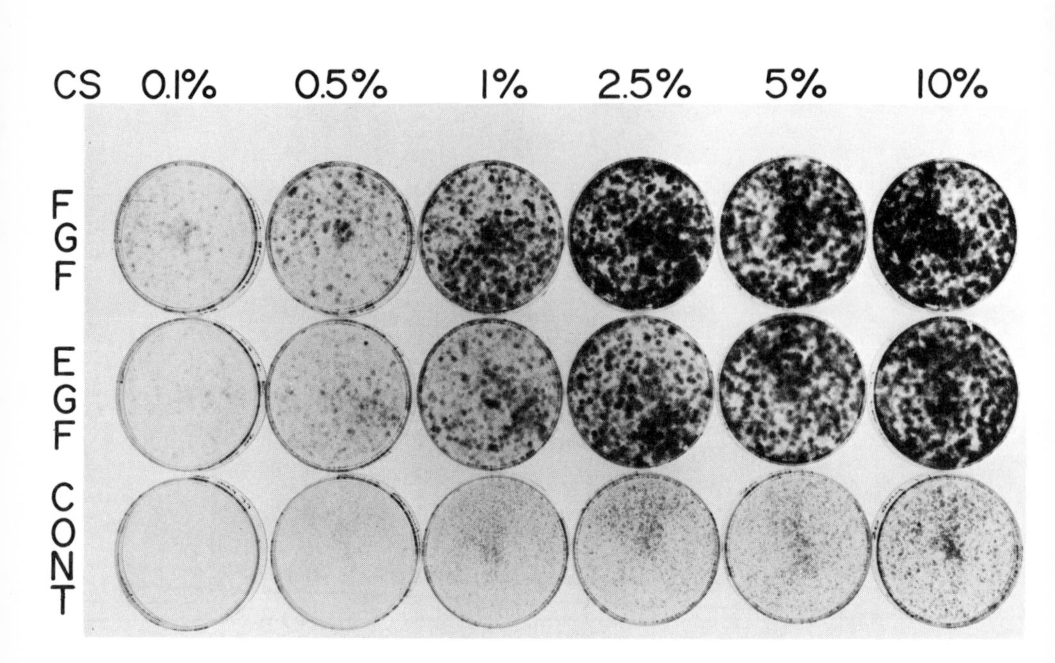

Fig. 7. Correlation between EGF, FGF, and the serum concentration on the growth of pig granulosa cell cultures. Pig granulosa cells were plated and maintained as described in Fig. 5.

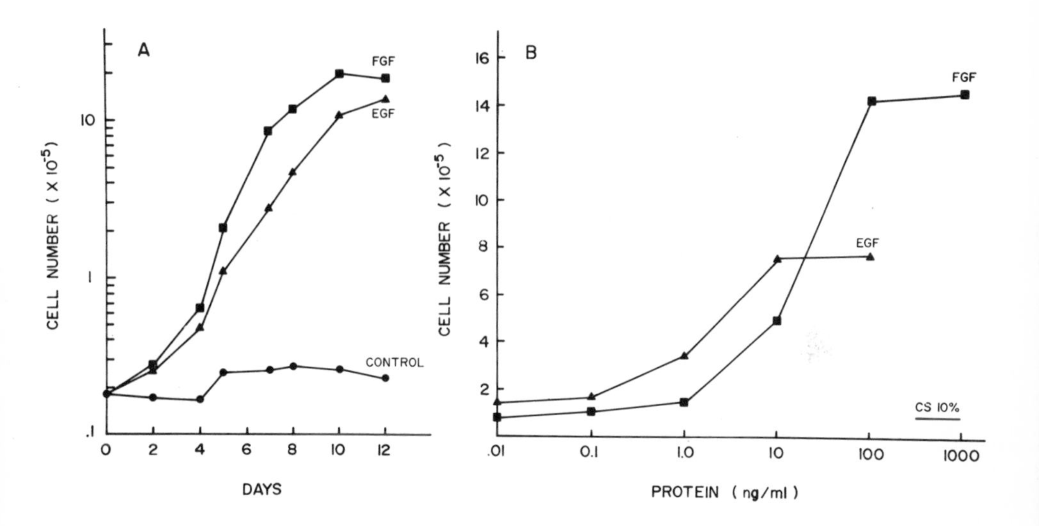

Fig. 8. Effect of FGF and EGF on the proliferation of pig granulosa cells. (A) Pig granulosa cells were plated at a density of 20,000 cells per 6 cm and maintained in the presence of 10% calf serum with or without EGF (10 ng/ml) or FGF (100 ng/ml) added every other day. (B) Porcine granulosa cell proliferation in response to increasing concentrations of EGF or FGF. Pig granulosa cells in contrast with bovine granulosa cells did not grow at all in 10% calf serum. Thus their dependency on FGF or EGF was more pronounced than for bovine granulosa cells (Fig. 4). This was reflected by a need for a higher concentration of either mitogen.

2) That mitogens can be active at very low concentrations (10^{-13} to 10^{-11} M).

3) That the mitogenic effect of the growth factor is dependent on the serum concentration in which the cells are maintained. This suggests that other factors in serum besides growth factors could be involved in the control of cell proliferation and perhaps regulate the cellular sensitivity to a given mitogen.

The observation that primary cultures of granulosa cells, in contrast to other cell types, survived quite well in low serum can be best explained by their origin. *In vivo* granulosa cells are maintained in an avascular environment, and this could account for their lower dependence on serum than cells derived from richly vascularized organs. Thus, since granulosa cells can be induced to divide by very small amounts of mitogens (pg to ng/ml), in a virtually serumless condition (60 μg of serum protein/ml), they should be an ideal cell type with which to study the mechanisms by which cells are induced to proliferate *in vitro*.

It remains to be seen whether or not FGF and EGF have similar effects on the proliferation of granulosa cells *in vivo*. Factors involved in follicular growth are poorly understood. It seems that during the early stage of ovarian development in rats the marked granulosa cell proliferation which takes place is independent of gonadotropin, since it is not suppressed by daily administration of rat gonadotropin antibodies (51). Studies by Pedersen (52) and others (53) have indicated that the DNA labelling index of granulosa cells could be affected by gonadotropins but that this depends on the age of the follicle. Gonadotropins are only active at a late stage of follicular development. Estrogens are also involved in the control of proliferation of granulosa cells and exert a direct effect on the growing follicles (54) since, in hypophysectomized animals, they increase the number of small and medium-sized follicles (54). Thus, the regulation of follicular development and, consequently, the control of granulosa cell proliferation seem to be extremely complex and may depend on the age of the individual as well as on the endocrine composition at a given time. However, since there is a general consensus that the factors regulating the transformation of primordial follicles into growing follicles do not involve the intervention of gonadotropin (55, 56), this phase of follicular development should be an ideal period in which to study the possible *in vivo* effect of EGF or FGF on proliferation of granulosa cells.

B) Effects of EGF and FGF on the proliferation of luteal cells.

Since luteal cells are derived from granulosa cells by cytomorphosis, it was of interest to see whether these cells respond to the same set of proliferative signals as do granulosa cells. Our results show that luteal cells are no longer sensitive to EGF but retain their sensitivity to FGF. It is thereby demonstrated that during luteinization granulosa cells shift their sensitivity not only to gonadotropins but also to mitogens.

Luteal cells maintained in the presence of serum grew poorly. A high concentration of serum (2.5 to 10%) was required to produce macroscopically visible clones (Fig. 9). If EGF (100 ng/ml) was added to the culture, an increase in cell number could be observed from the increased staining of the dish, but it only corresponded to a roughly two-fold increase in cell density. FGF, unlike EGF, has a marked effect on the increase in cell number. Addition of 100 ng/ml of FGF to cells maintained in serum concentration as low as 0.1% brings an increase in staining (Fig. 9).

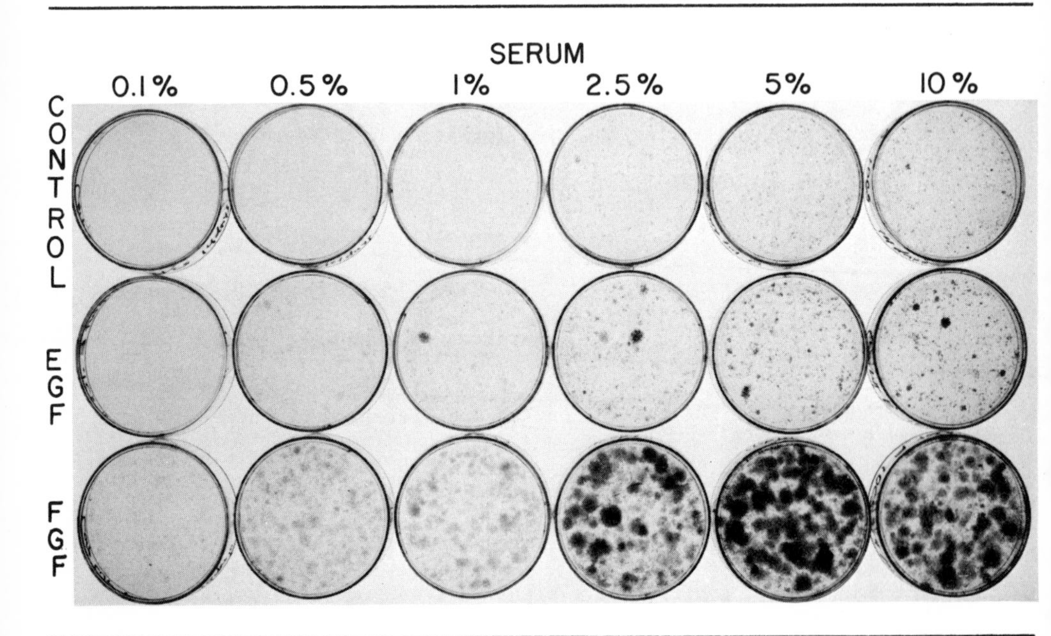

Fig. 9. Correlation between EGF, FGF, and the serum concentration on culture growth. Luteal cells were plated at 3000 cells per cm dish and maintained in the presence of a constant concentration of EGF (100 ng/ml) or FGF (100 ng/ml) and with an increasing concentration of calf serum (from 0.1-10%). The medium was renewed every other day. Plates were fixed with 10% formalin and stained with 0.1% Giemsa on day 14.[43]

This effect was obtained with cells maintained in 0.1-10% serum and was proportional to the serum concentration in which the cells were maintained.

The relationship between the rate of cell proliferation and the concentrations of EGF, FGF, and serum was investigated using luteal cells of corpora lutea from pregnant sources. Cultures maintained in 1% serum grew slowly, doubling in cell number in 7 days (Fig. 10), while in 10% serum they doubled in 3 days (Fig. 11). Addition of EGF (100 ng/ml) to cells maintained in either 1% or 10% serum allowed for another doubling, after which the cells stopped dividing. However, addition of FGF (100 ng/ml) to cells maintained in 1% serum made them divide with a cycle of 48 hours, and the final density observed was higher than that seen with 10% serum (Fig. 11). When FGF was added to cells

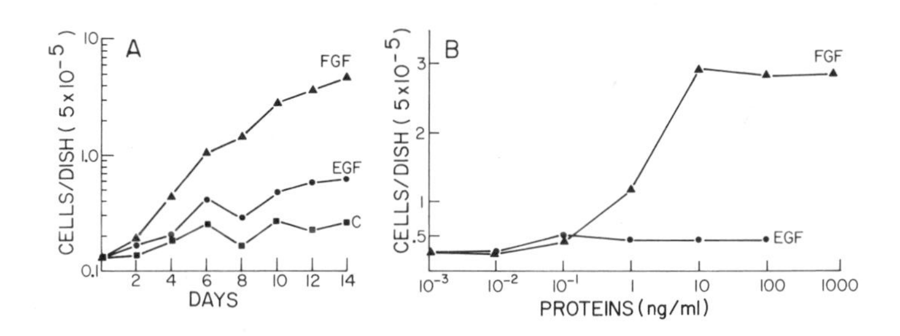

Fig. 10. Effect of FGF and EGF on the proliferation of luteal cells maintained in low serum. (A) Luteal cells were plated at 3000 cells per 6 cm dish. They were then maintained in the presence of 1% calf serum without (■ -- ■) or with 100 ng/ml of EGF or FGF (● -- ▲). (B) Effect of increasing concentrations of EGF (● -- ●) and FGF (▲ -- ▲) on the proliferation of luteal cells. The EGF cells were maintained in the presence of 1% serum and increasing concentrations of FGF and EGF. At day 14 the cells were trypsinized and counted. Control had 5100 ± 480 cells.

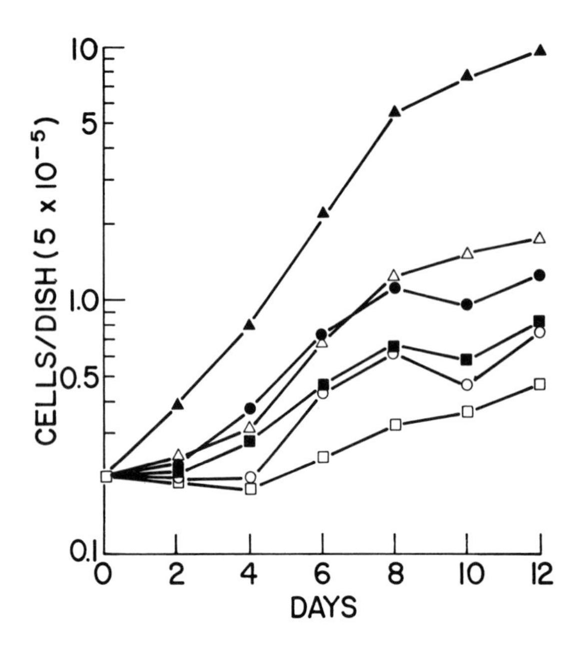

Fig. 11. Effect of FGF and EGF on the proliferation of luteal cells maintained in low and high serum concentrations. Luteal cells were plated at 3000 cells as described. They were then maintained in the presence of 1% calf serum without (□ -- □) or with 100 ng/ml EGF (○ -- ○) or FGF (△ -- △). The proliferation effect of FGF and EGF was compared to that of 10% calf serum (■ -- ■) to which 100 ng/ml EGF (● -- ●) or 100 ng/ml FGF (▲ -- ▲) were added.[43]

maintained in 10% serum, the doubling time of the luteal cells during exponential growth (days 4-8) dropped at 24 hours (Fig. 11). The doubling time of the cells in response to FGF thus depends on the serum concentration in which the cells are maintained.

These results demonstrate:

1) That FGF is mitogenic for luteal cells, while EGF has little effect upon them.

2) That, as already observed with granulosa cells, the effect of either EGF or FGF on luteal cells depends on the serum concentration in which the cells rest. While the effect is observed at serum concentrations as low as 0.1%, it is most accentuated at high serum concentrations. As with granulosa cells, the main difference in their response to FGF between cells maintained in low serum and those maintained in high serum was that in low serum the doubling time of the cells was extremely long (up to 4 days), while in high serum the doubling time could drop to 24 hours.

Our results demonstrate, then, that although luteal cells and granulosa cells are interrelated cell types, the control of their proliferation *in vitro* is quite different; the sensitivity to EGF is lost during luteinization, while the sensitivity to FGF is retained. FGF is the first known mitogen for luteal cells *in vitro*, although it may not necessarily be the only one. Although the mechanisms by which luteal cells lose their sensitivity to EGF are not known, one can speculate that this loss could be due to a loss of receptor sites or to a lower affinity of the receptor sites for EGF, since high concentrations of EGF induce only a doubling.

While the physiological meaning of the loss of sensitivity to EGF of luteal cells is unclear and will require more research, our results demonstrate that mitogens, distinct from known gonadotropins, can be shown to be active for ovarian cells *in vitro*. Furthermore, the susceptibility of the ovarian cells to mitogens varies and depends on the stage of maturation of the tissue. While our observations are restricted only to the *in vitro* condition, it is possible that similar sensitivities of ovarian cells to mitogens occur *in vivo*.

II. A comparison of the binding of epidermal growth factor to granulosa cells and luteal cells in tissue culture

In previous studies[42] it has been shown that cultured bovine granulosa cells derived from medium-sized ovarian follicles, are highly responsive to the mitogenic stimnulus provided by both epidermal and fibroblast growth factor. In contrast, bovine luteal cells, which are derived from granulosa cells through the process of cytomorphosis, are no longer sensitive to EGF, although they do respond quite well to FGF[43].

In the present study we have looked at the quantitative and kinetic aspects of EGF binding to granulosa and luteal cells. Our results demonstrate that both granulosa and luteal cells are capable of binding EGF in a highly specific manner. Thus, the lack of response of luteal cells to EGF cannot be explained on the basis of a loss of EGF receptor sites. Since the cytomorphosis of granulosa cells to luteal cells is associated with a loss of sensitivity but not of EGF receptor sites, the in vitro proliferative response of these cell types to EGF and FGF provides a direct way to study the relationship between mitogenicity, binding capacity, and inactivation ("down regulation") of receptor sites for EGF.

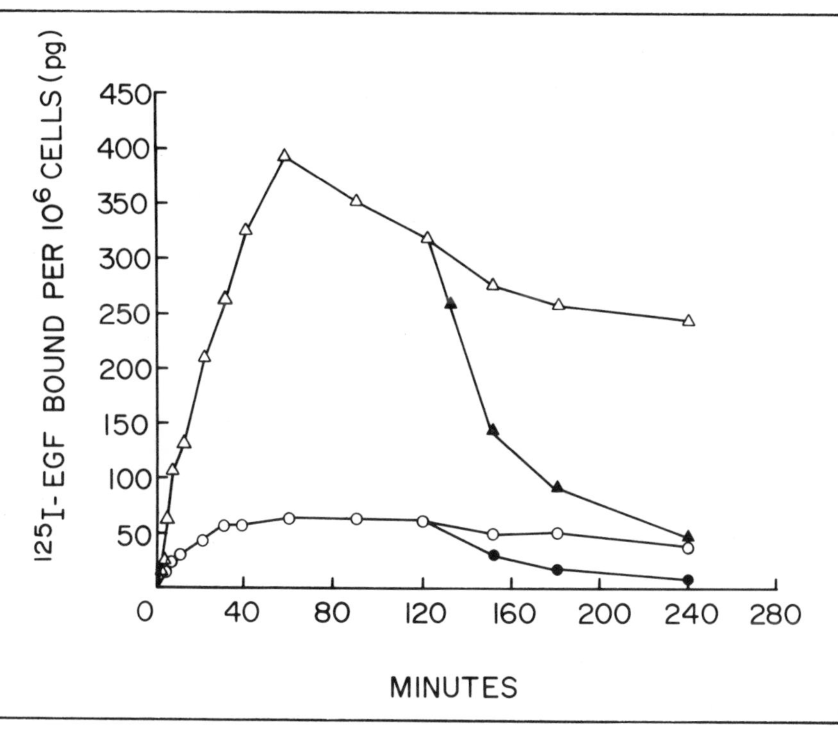

Fig. 12. Time-course of ^{125}I-EGF binding to granulosa and luteal cells. The granulosa (○) and luteal (△) cell densities were 31,000 cells/cm^2. Two ml of binding medium, containing 2.4 ng/ml of ^{125}I-EGF (45,000 cpm/ng), were added to each dish and the dishes were incubated at 37°C. At the indicated time interval the specific cell-bound radioactivity (○,△) was determined. Non-specific binding was measured by adding 250 ng/ml of unlabelled EGF to the incubation medium at 0 min. Displacement of bound EGF from the cells was followed after the addition of 250 ng/ml of unlabelled EGF to the incubation medium at 120 min. (●,▲).

Binding properties of ^{125}I-EGF to granulosa and luteal cells

The time-course of binding of ^{125}I-EGF to bovine granulosa and luteal cells is shown in Fig. 12. Maximal binding was reached after 30 min. at 37°C for the granulosa cells and at 1 hour for the luteal cells. The maximal amount of hormone specifically bound by the luteal cells was eight-fold higher than the amount bound at saturation by the granulosa cells. Upon continued incubation of luteal cells with the labelled hormone the amount of cell-bound radioactivity decreased in 4 hours to 62% of the initial maximal amount bound at one hour. The same observation was made with granulosa cells; by 4 hours the amount of cell-bound radioactivity was 55% of that observed at 30 minutes.

The effect of increasing concentrations of EGF on the binding of ^{125}I-EGF to granulosa and luteal cells is shown in Fig. 13. The binding reaction is a saturable process with a maximal binding at 2 ng/ml (3×10^{-10}M) and half-maximal binding at 0.75 ng/ml (1.1×10^{-10}M) for the granulosa cells. With luteal cells maximal binding was observed at 12 ng/ml (1.8×10^{-9}M) (Fig. 13). A Scatchard plot of these data is shown in Fig. 14A. While for the granulosa cells the dissociation constant (K_D) was 2.4×10^{-10}M and the maximal binding was 234 pg of EGF per 10^6 cells, for the luteal cells the K_D was 7.0×10^{-10}M, and the binding at saturation was 1036 pg of EGF per 10^6 cells. In a separate experiment done with different cultures of granulosa and luteal cells (Fig. 14B) the granulosa cells had a K_D of 3.96×10^{-10}M and 10^6 cells bound 229 pg of EGF, whereas luteal cells had a K_D of 6.54×10^{-10}M and 1117 pg were bound per 10^6 cells.

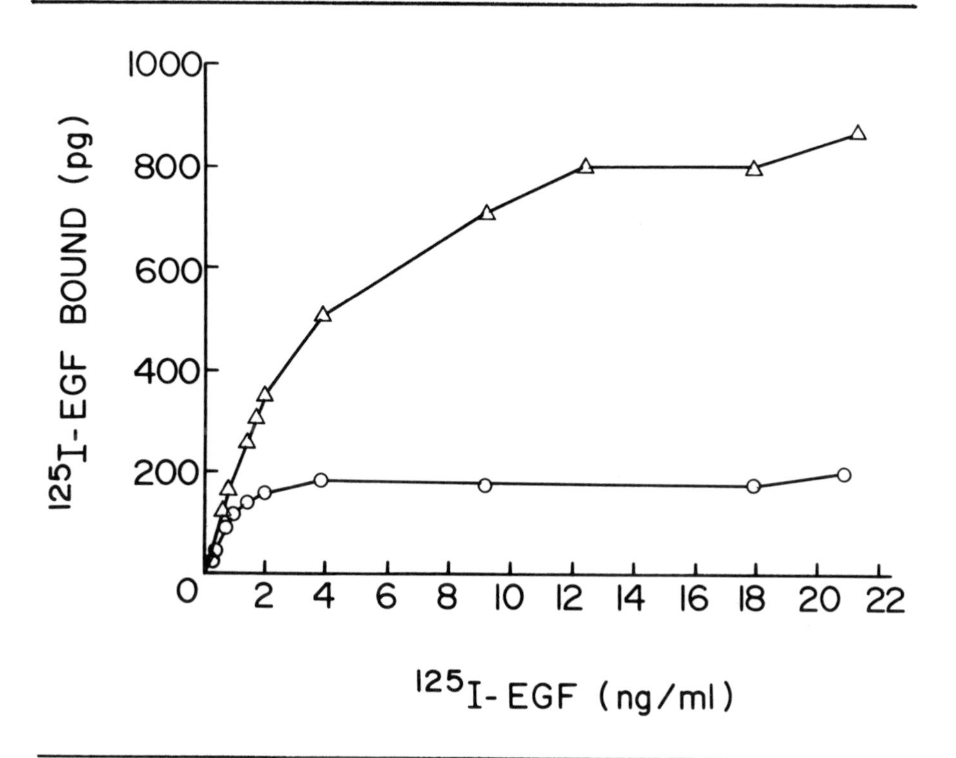

Fig. 13. Effect of ^{125}I-EGF concentration on binding to granulosa and luteal cell cultures. Increasing amounts of ^{125}I-EGF (126,000 cpm/ng) were added to granulosa (O--O) or luteal cell cultures (△ -- △). Non-specific binding was measured by adding 500 ng/ml of unlabelled EGF to the incubation medium at 0 minutes. The specific binding (total binding minus non-specific binding) was determined after a 40 min. incubation at 37°C.

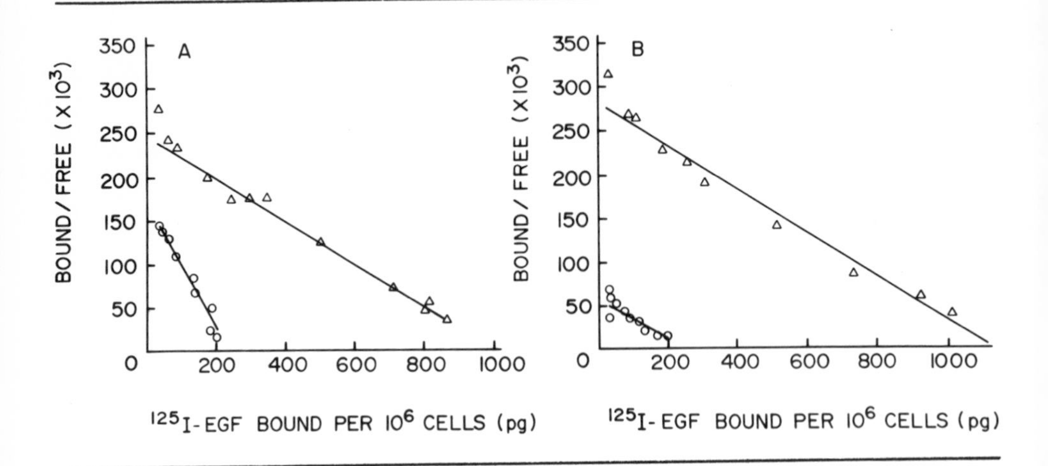

Fig. 14. Scatchard plots showing the effect of ^{125}I-EGF concentration on binding to granulosa and luteal cells. Various concentrations of ^{125}I-EGF (0.02-30 ng/ml, 40,000 cpm/ng) in 2 ml of binding medium were added to dishes, and the specific cell-bound radioactivity was determined after a 40 min. incubation at 37°C. "A" and "B" represent two separate experiments done with different cultures of granulosa (○--○) and luteal cells (△ -- △).

Relationship between EGF-binding capacity and mitogenicity

When the binding of EGF to granulosa and luteal cell cultures was compared to its mitogenic effect on these cultures, it was observed, as previously reported, that although EGF did bind to luteal cells it had no mitogenic effect, as indicated by no significant increase in cell-number. By contrast, EGF was a strong mitogen for granulosa cell cultures and gave a maximal (up to 20 fold) increase in cell-number at concentrations as low as 0.2 ng/ml (3 x 10^{-11}M) and a half-maximal increase at 15 pg/ml (2.5 x 10^{-12}M)[42].

These results indicate that the presence of EGF receptor sites is by itself not sufficient for the induction of cell proliferation by EGF.

Interaction of ^{125}I-EGF with granulosa cells maintained in culture for short- and long-term

Granulosa cells harvested from large preovulatory follicles undergo cytomorphosis and luteinize spontaneously in culture devoid of pituitary hormones. In contrast, granulosa cells obtained from small follicles fail to luteinize spontaneously in culture.

Using granulosa cells maintained in tissue culture, one can study the dependency or loss of dependency on growth factors which could be developed during the process of cytomorphosis, i.e., when granulosa cells luteinize to give a population of luteal cells.

The following experiments were undertaken to determine whether a long-term, luteinizing culture of granulosa cells becomes unresponsive to EGF and whether change in its responsiveness is associated with changes in EGF-binding capacity, internalization, and release and/or in the EGF-induced loss and recovery of surface receptors for EGF.

A) Mitogenic activity and binding capacity

Freshly isolated granulosa cells and late passages of cultured granulosa cells were tested for sensitivity to EGF and for EGF-binding capacity. The results indicate that cells from a late passage undergo luteinization *in vitro* and like luteal cells, show no response to EGF. This change was associated with an

EGF-binding capacity 6 fold higher than the binding to cells freshly isolated (Fig. 15).

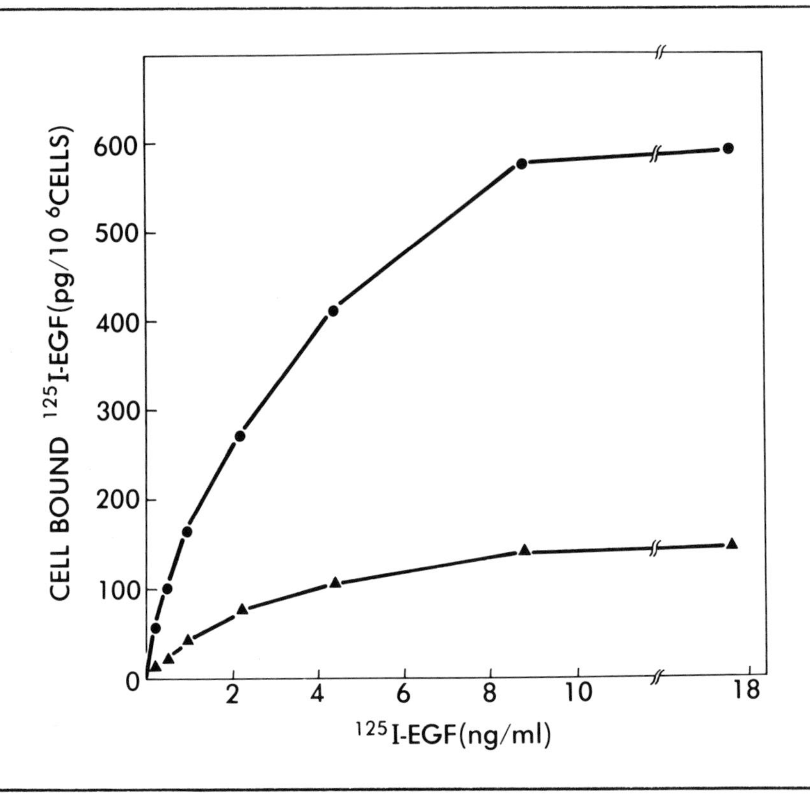

Fig. 15. Effect of ^{125}I-EGF concentration on binding to freshly isolated and late passage granulosa cells. Cells derived from 10 mm follicles were tested for EGF-binding capacity immediately after being isolated from the follicles (▲, EGF-responsive cells) and after seven cell passages (●, non-responsive cells). The specific binding of labelled EGF (0.2-18 ng/ml, 67,000 cpm/ng) was determined after a 45 min. incubation at 37°C.

Freshly isolated granulosa cells which are EGF-responsive show a saturation of EGF binding at 4 ng/ml, whereas cultures from late passages which no longer respond to EGF and show signs of terminal differentiation saturate at higher concentrations (10 ng/ml) (Fig. 15).

In the following experiments EGF-responsive cells were obtained from small follicles (3-5 mm) and cultured in the presence of FGF for 1-3 weeks (2-4 cell passages). The non-responsive cells were derived from big follicles (10-15 mm), cultured (2-3 weeks) in the presence of FGF and then in its absence for at least 4 weeks before use in the experiments. These cells were considerably enlarged and show the morphology of luteal cells.

B) Release of cell-bound ^{125}I-EGF

To measure the release of cell-bound EGF, monolayers were incubated at 4°C with labelled EGF, washed extensively to remove the unbound hormone, and the amount of cell-bound radioactivity determined at various times after transferring to 37°C. The data (Fig. 16) indicate that with both EGF-responsive and non-responsive granulosa cells the bound radioactivity decreased rapidly ($t1/2$ = 60 min) during the subsequent incubation at 37°C. As with other cell types, 85-95% of the initial cell-bound radioactivity was not associated with the cells after two hours incubation in these conditions. A 60-80% inhibition of this release was obtained in the presence of chloroquine (0.1 mM) or 2,4-dinitrophenol (0.2 mM). This indicates that the loss of cell-bound radioactivity is dependent on the prior internalization and degradation of the EGF molecules which require both the generation of a metabolic energy and the proteolytic activity of lysosomal enzymes.

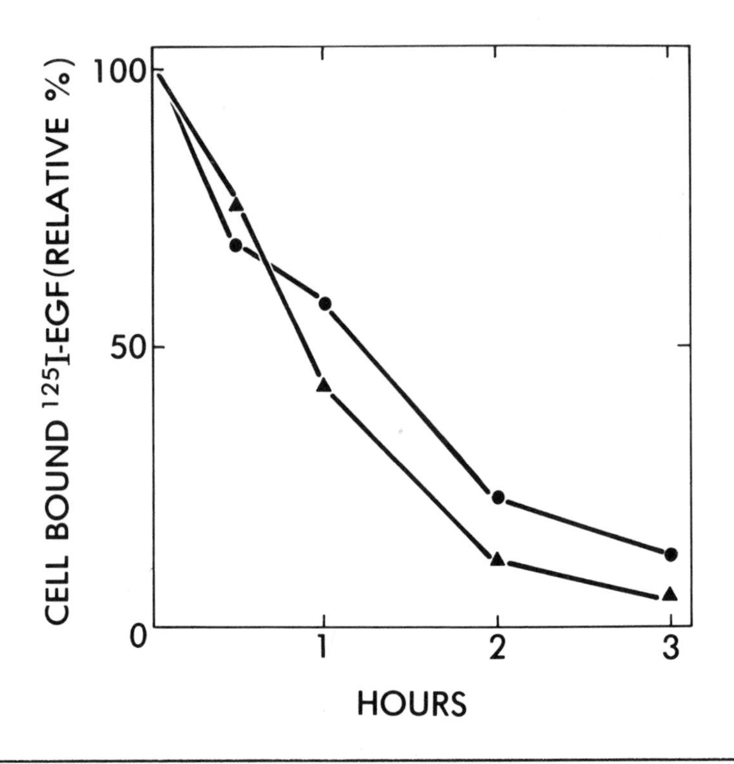

Fig. 16. Release of cell-bound ^{125}I-EGF from responsive and non-responsive cultures of granulosa cells. ^{125}I-EGF (5 ng/ml, 370,000 cpm/ng) was allowed to bind (75 min, 4°C) to confluent cultures of granulosa cells that respond (▲), or lost response (●) to EGF. The cells were then washed 6 times at 4°C, 2 ml of the growth medium added, and the culture dishes transferred to 37°C. The cell-bound radioactivity was determined at the indicated intervals. The results are expressed as the relative percentage of cell-bound radioactivity remaining at the indicated times, taking as 100% the amount of radioactivity present after the initial binding period at 4°C.

C) Loss of EGF receptor sites due to preincubation with EGF

The following experiments were undertaken to determine whether exposure of ovarian cells to EGF is, as in other cell systems, associated with a decreased capacity of the cells to rebind EGF and whether EGF-responsive and non-responsive granulosa cells differ in the EGF-induced regulation of the concentration and/or availability of its own surface receptor sites. For this purpose cells were preincubated for 10 hours with saturating and non-saturating concentrations of unlabelled EGF, washed, incubated (37°C) for three hours to allow the membrane-bound EGF to degrade, and tested for their ability to rebind fresh ^{125}I-EGF. As shown in Fig. 17A preincubation with physiological concentrations of EGF will induce in both EGF-responsive and non-responsive cells up to an 80% reduction in EGF binding capacity. A significant decrease in binding was obtained in cells that were pre-incubated with EGF concentrations as low as 0.25 ng/ml and a 40-50% reduction in binding was obtained by preincubation with EGF at 1 ng/ml. With cells that respond to EGF these concentrations are enough to obtain a maximal mitogenic response. Saturating concentrations of EGF are therefore not required for the induction of both cell proliferation and the loss of surface receptor sites for EGF.

Any molecular explanation for the mitogenic effect of EGF must take into account the observation that EGF, like other mitogens, must be present in the medium for an extended period of time to stimulate DNA synthesis and subsequent cell division. We have therefore studied the time-course of the loss of EGF receptor sites in EGF-responsive and non-responsive granulosa cells that were preincubated with EGF. Cells were first incubated for different time periods with unlabelled EGF (5 ng/ml), unbound hormone was

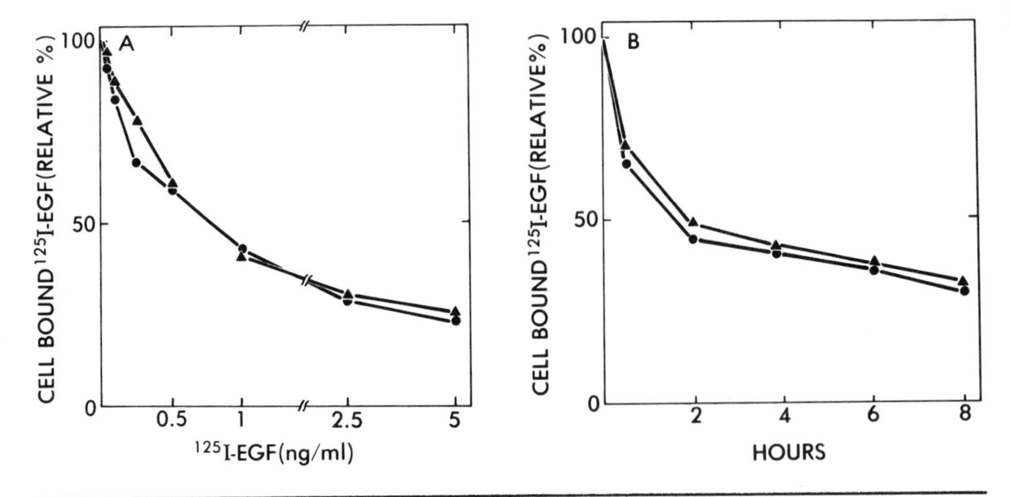

Fig. 17. Loss of EGF-binding capacity induced by preincubation with unlabelled EGF. (A) Ten hours preincubation with saturating and non-saturating concentrations of unlabelled EGF. (B) Preincubation for different time periods with unlabelled EGF (5 ng/ml). Cultures of granulosa cells that respond (▲), or lost response (●) to EGF were incubated at 37°C with unlabelled EGF in F-12 medium containing 1% calf serum. The cells were then washed free of unbound EGF and incubated in the binding medium (3 hrs., 37°C) to permit degradation of the bound hormone. Cells were washed four times, transferred to 4°C, and their capacity to rebind ^{125}I-EGF (5 ng/ml, 370,000 cpm/ng) was determined under standard conditions (4°C, 75 min.). The results are expressed as the relative percentage of cell-bound radioactivity, taking as 100% the amount of bound radioactivity in cells that were not exposed to unlabelled EGF.

removed, and the cell-bound EGF was allowed to degrade. The results (Fig. 17B) indicate that with both EGF-responsive and cells that are no longer responsive to EGF a two hour preincubation with EGF was sufficient to induce a 50% decrease in the number of surface receptor sites available for EGF binding. This is a much shorter time than the period in which EGF must be present in the medium to stimulate DNA synthesis. With both types of cells 15-25% of the receptor sites for EGF still remain available for binding even after an extended exposure (up to 24 hours) to excess quantities of EGF (up to 100 ng/ml). It is of interest that in the case of granulosa cells which do not luteinize less than 10% of the receptor sites for EGF have to be occupied in order to obtain a maximum mitogenic response[42].

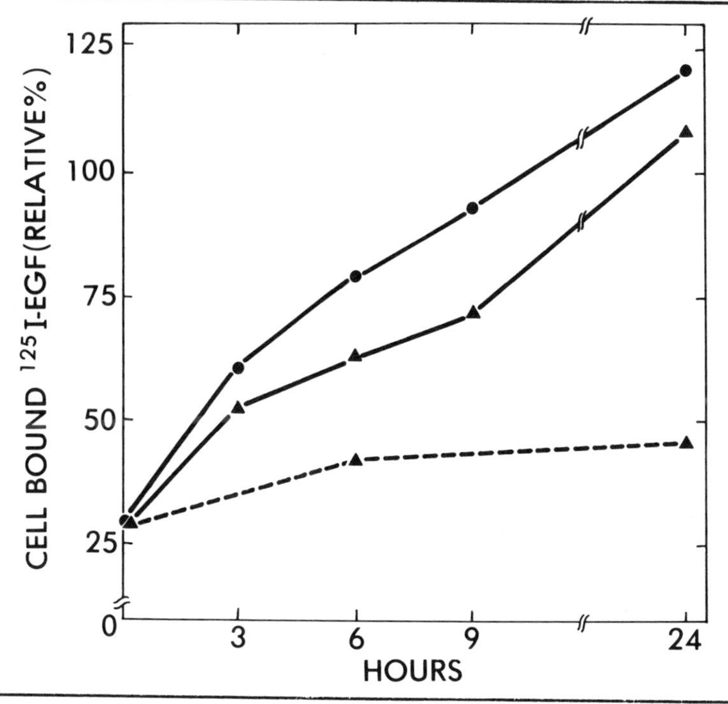

Fig. 18. Recovery of EGF-binding capacity in EGF-responsive and non-responsive cells that were first exposed to EGF. EGF responsive (▲ -- ▲) and non-responsive (● -- ●) granulosa cultures were pre-incubated with unlabelled EGF (5 ng/ml, 10 hrs.) in F-12 medium containing 1% calf serum. The cells were then washed free of unbound EGF and incubated in the binding medium for 3 hrs. at 37°C to permit degradation of the bound hormone. The cells were washed extensively, and the medium replaced with F-12 medium containing 10% calf serum (——) or 0.1% bovine serum albumin (---). At the indicated times cells were transferred to 4°C and assayed for ^{125}I-EGF (5 ng/ml, 270,000 cpm/ng) binding under standard conditions. The extent of EGF-binding to cells that were similarly incubated but not exposed to unlabelled EGF was taken as 100%, and the results are expressed as the relative percentage of cell-bound radioactivity.

D) Recovery of EGF-binding capacity

Cells that were first incubated with EGF to induce a 70-80% decrease in EGF-binding capacity were incubated for different time periods in the absence of EGF, and their ability to rebind fresh EGF was tested. The results (Fig. 18) indicate that both luteinizing and non-luteinizing granulosa cells are capable of gaining the initial EGF-binding capacity within 12-24 hours of incubation in the presence, but not in the absence, of serum.

Our present results on the binding of EGF to granulosa and luteal cells as well as to early and late luteinizing passages of granulosa cells indicate a lack of correlation between EGF-binding capacity and mitogenic activity. Granulosa cells which proliferate in response to EGF bind ten times more EGF than is needed to induce a maximal proliferative response, whereas luteal cells which do not respond at all can even bind 4-6 times more EGF than granulosa cells. These observations lead us to conclude that 1) binding studies should be correlated with biological activity as expressed by the increase in cell number, since it is often implied that binding of a given agent will induce a biological effect. The example of EGF-binding to granulosa cells versus that to luteal cells demonstrates that this is not always the case; 2) the injection of a ^{125}I-labelled mitogen *in vivo* and its subsequent distribution in the body can lead to false conclusions as to where the mitogen acts. For example, with ^{125}I-EGF the *in vivo* binding to granulosa cells will be barely detectable because of the high sensitivity of the cells, while luteal cells which do not respond to EGF will show, in comparison to granulosa cells, a detectable binding. This could lead to the wrong conclusion that luteal, and not granulosa, cells are the target cells for EGF.

We have shown that during the *in vitro* luteinization of granulosa cells the number of surface receptors for EGF is increased and, as with EGF-responsive cells, can be down regulated by the extracellular concentration of EGF. Our experiments on the binding of EGF clearly indicate that the loss of a mitogenic response to EGF in cells that undergo luteinization is not due to a defect in the internalization and release of cell-bound EGF and/or in the EGF-induced loss and recovery of surface receptor sites for EGF.

The induction of cell division is accompanied by various changes in cellular physiology and properties of the cell surface. It is, however, not clear which of the induced alterations are correlative and which are both necessary and sufficient for the G_0-G_1 transition. Our results indicate that the internalization and release of cell-bound EGF as well as the induced loss and recovery of EGF receptor sites are by themselves not sufficient to commit the cells to undergo cell division. These factors might, however, have a role in transmitting the mitogenic signal. In view of our results it is more likely that the inactivation of both the cell-bound EGF and its appropriate surface receptors simply serves as a control mechanism by which the cell can degrade and release the cell-bound EGF and regulate the concentration of its receptor sites, so that an excess of hormone can specifically decrease the number of its own functional receptors.

III. The effects of the epidermal and fibroblast growth factors on the replicative lifespan of cultured bovine granulosa cells

Most mammalian cells in tissue culture have a limited replicative lifespan. The cells may lose the ability to divide because the culture environment is inadequate to sustain growth or because an inherent, intracellular program dictates their senescence. They may also lose the ability to divide as a consequence of terminal differentiation.

An example of terminal differentiation is provided by keratinocytes, which stop proliferating in culture when they become trapped in extracellular keratin that they themselves synthesize. It has recently been shown that EGF can enable cultured keratinocytes to undergo an increased number of population doublings between the time they are established in culture and the time of their terminal differentiation, thus extending the life of the culture both in terms of generations and of absolute time[45]. This observation is not only of theoretical interest but may also be of practical use by permitting the cultivation of very large numbers of progeny cells from small numbers of parent cells.

Since EGF and FGF are both potent mitogens for granulosa cells, and since granulosa cells, when they luteinize, also undergo terminal differentiation, we have investigated the abilities of EGF and FGF to retard luteinization and thus prolong the replicative lifespan of these cultures. We have compared the effects of the two factors and have also correlated the culture lifespan with the developmental stage of the follicles from which the cultures were derived.

Effects of EGF and FGF on the lifespan in culture of granulosa cells

Granulosa cell strains originating from follicles ranging from 4 mm to 15 mm in diameter were serially transferred by inoculating 2×10^4 cells per 6 cm dish and subculturing 7 to 13 days later. When the lifespans of cultures originating from follicles of various sizes were compared, it was observed that the lifespans of cultures originating from small follicles were longer than those from large follicles. Granulosa cell cultures from 4 and 7 mm follicles had a lifespan of 11 to 12 generations, while the lifespan of cultures originating from 13 to 15 mm follicles did not exceed 7 generations (Fig. 19). Granulosa cell cultures derived from small or large follicles had the same macroscopic appearance in their first generation. The cells were small and tightly packed, and there were discrete granules within the cytoplasm. The appearance of the cells changed rapidly during the first week of culture -- the cells became considerably enlarged, and under phase contrast a granular perinuclear region surrounded by a broad expanse of cytoplasm with prominent longitudinal ridges was visible. Discrete lipid inclusions, reflecting early luteinization, became apparent after staining with oil red O.

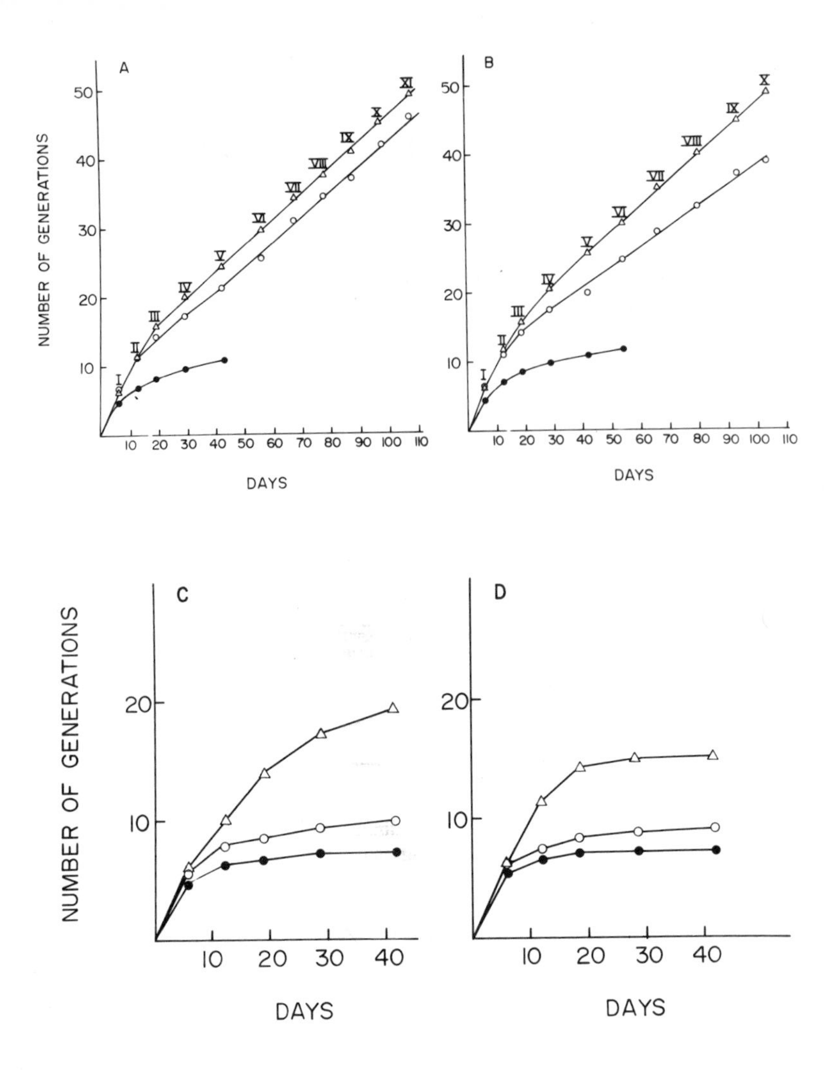

Fig. 19. Effects of EGF and FGF on the culture-lifetime of granulosa cells. Granulosa cell cultures were begun from different-sized follicles (A - 4 mm; B - 7 mm; C - 13 mm; D - 15 mm). The number of cell generations was calculated from the number of colonies formed and the number of cells harvested at each transfer. Each point represents a single transfer. The average growth rate is given by the slope. Cultures were grown in the presence of EGF (○ -- ○, 10 ng/ml), FGF (△--△, 100 ng/ml) or in their absence (●--●). Roman numerals indicate the passage-number.

The addition of EGF or FGF to the medium of cultures originating from small follicles greatly increased the lifetime of the cultures (Fig. 19 A, B). They have, at this writing, reached 50 generations. This is a 5-fold increase in longevity over control cultures. There has been no remarkable change in the appearance of the cultures after 50 generations when compared to the morphology of earlier generations. The cells form a monolayer of small, polygonal cells with a perinuclear granular region. A few cells can be seen to be enlarged. These represent only a small percentage of the cell population undergoing terminal differentiation.

The mean doubling-time of EGF- and FGF-maintained cultures was similar for the first 12 generations, after which the doubling-time of EGF-maintained cultures became longer than that of cultures maintained in the presence of FGF. This is reflected by changes in the slopes of the curves correlating generation-number with age (Fig. 19 A, B). Since our method of calculating culture lifetime does not allow discrimination of a small number of cells that terminally differentiate (luteinize) in growing colonies and are therefore removed from the proliferating pool, it is conceivable that the apparently longer doubling-time of EGF-maintained cultures could, in fact, be due to a larger proportion of granulosa cells terminally differentiating in cultures maintained in the presence of EGF than those maintained in the presence of FGF.

The lifespan of granulosa cell cultures depends upon the presence of FGF or EGF in the medium, as shown in Fig. 20. Cultures maintained in the presence of FGF for 24 generations were transferred

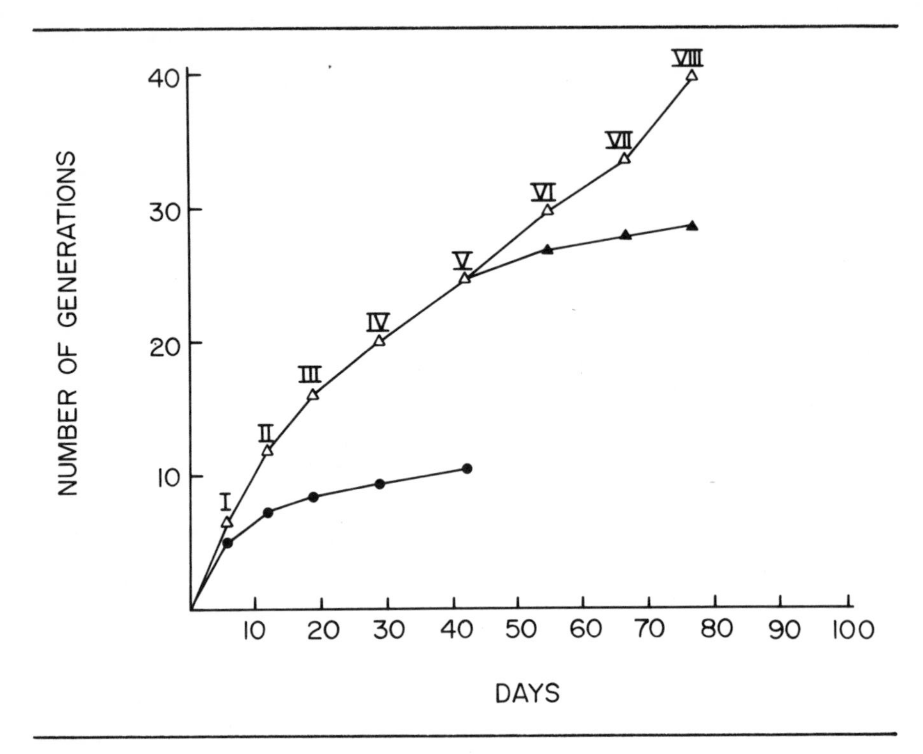

Fig. 20. The effect of deleting FGF from the culture medium on the proliferation of long-term cultures of granulosa cells originating from small-sized follicles (4 mm). Granulosa cell cultures were maintained in the absence (● -- ●) or in the presence of FGF (100 ng/ml) (△ -- △) for 14.5 generations. At 24.5 generations the cultures maintained in the presence of FGF were split. Half of the cultures were then maintained in the absence of FGF (▲ -- ▲), while the other half had FGF added to it. Roman numerals give transfer-numbers.

to a culture medium lacking FGF. Within the first passage the doubling time of the cultures increased considerably, and at the second passage the cells stopped proliferating and terminally differentiated. Cultures continually maintained with FGF, however, proliferated actively. This demonstrates that the continuous presence of the mitogen is required to prevent the terminal differentiation of the cells.

The difference in apparent doubling-time between cultures maintained with EGF and those maintained with FGF was even more pronounced with cultures started from large follicles (13 mm). The lifetime of control cultures did not exceed 6-7 generations (Fig. 19), after which the cells terminally differentiated. Addition of EGF to the medium did not greatly increase the lifetime of the cells, which stopped proliferating after 9 generations. The cells were then all terminally differentiated and exhibited a morphology similar to that of the control cultures after 6 generations. The addition of FGF, in contrast to EGF, prolonged the lifetime of the culture, but its effect was not as pronounced as in the case of cultures originating from small follicles, since, after 20 generations, terminal differentiation was observed. The terminal differentiation was even more pronounced with cultures originating from 15 mm follicles, for which neither EGF nor FGF could prolong the lifetime of the culture beyond 10 generations (Fig. 19).

The effect of EGF and FGF on plating efficiency

Cessation of growth in tissue culture could result either from an increase in doubling time, sometimes extending for weeks, or from a decrease in viability reflected in a reduced plating efficiency. Therefore, we investigated the effects of EGF and FGF on the plating efficiency of granulosa cell cultures originating from small, medium, and large follicles.

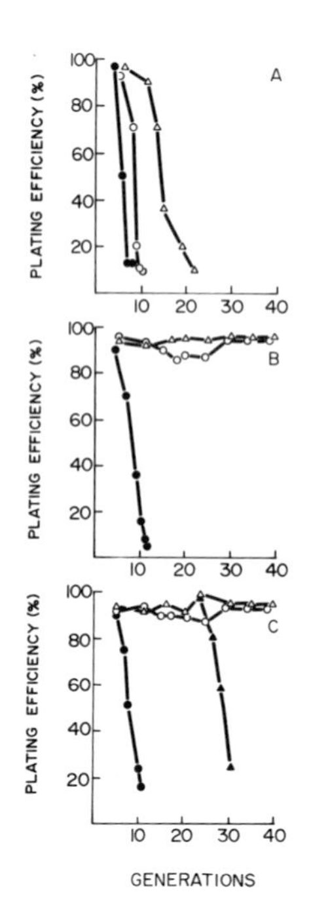

Fig. 21. The effects of FGF and of EGF on the plating efficiency of granulosa cells during the culture lifetime of granulosa cells derived from different-sized follicles. A) 15 mm; B) 7 mm; C) 4 mm follicles. Cultures grown in the presence of EGF (○--○); cultures grown in the presence of FGF (△ -- △); in the absence of mitogens (●--●). After 23 generations granulosa cells from 4 mm follicles maintained in the presence of FGF were maintained in the absence of mitogen (▲--▲). After 3 passages the cloning efficiency of the cultures dropped to 20%.

As shown in Fig. 21, control cultures originating from small follicles (4 mm) had a plating efficiency that dropped rapidly as the cultures aged (90% plating efficiency at generation 6 versus 8% at generation 11). This could reflect the terminal differentiation of the cultures which, by generation 11, affected most of the cells in the population. Cultures maintained with either EGF or FGF, on the other hand, still had a plating efficiency of 90% for FGF and 80% for EGF at the 50th generation. When cultures of granulosa cells maintained for 24 generations in the presence of FGF (Fig. 21C) were transferred to a medium lacking FGF, the plating efficiency dropped rapidly to as low as 20% by the 30th generation. This reflected the rapid terminal differentiation of the cells that takes place as soon as the mitogen is removed from the medium. Similar results were obtained with medium-sized follicles (7 mm), but, unlike small follicles, granulosa cell cultures maintained with EGF showed a constant decline in plating efficiency as the cultures grew older. FGF-maintained cultures kept a constant colony-forming efficiency of 90%. With granulosa cell cultures originating from large, preovulatory follicles (15 mm) neither EGF nor FGF slowed down the differentiation reflected by the rapid drop in the plating efficiency of the culture. In control cultures the plating efficiency dropped to 10% by the 7th generation. With EGF it dropped to 10% by the 10th generation. With FGF it dropped to 10% by the 22nd generation. This indicates that granulosa cells originating from large follicles are somehow committed to terminal differentiation and that neither EGF nor FGF can prevent it.

The effects of EGF and FGF on clonal growth of granulosa cells

The proliferation of mammalian cells in tissue culture depends on cell density. At very low cell density (clonal cell density) mammalian cells in early passage cultures no longer divide. However, if they are maintained on an irradiated layer of feeder cells that "condition" the medium, the cells can proliferate. This "conditioning" phenomenon probably reflects the inadequacy of most culture media to supply all the requirements for proliferation of any given cell type. The cells "condition" the medium by synthesizing the missing metabolites in the culture medium. If the cell density is high enough, adequate concentrations of those metabolites will be reached so that the cells can start proliferating. At low cell density, however, adequate concentrations cannot be achieved before the cells succumb to starvation in the inadequate medium. The feeder layer thus has the advantage of producing the required metabolites in high enough concentrations so that even very sparse cell populations can readily proliferate.

In view of the potent mitogenic effects of FGF and EGF on granulosa cell cultures originating from small follicles, their effect on cell proliferation of cultures seeded at low density was also investigated. As shown in Fig. 22, both EGF and FGF were potent mitogenic agents with cultures seeded at cell densities of 50, 100, or 500 cells per 6 cm dish.

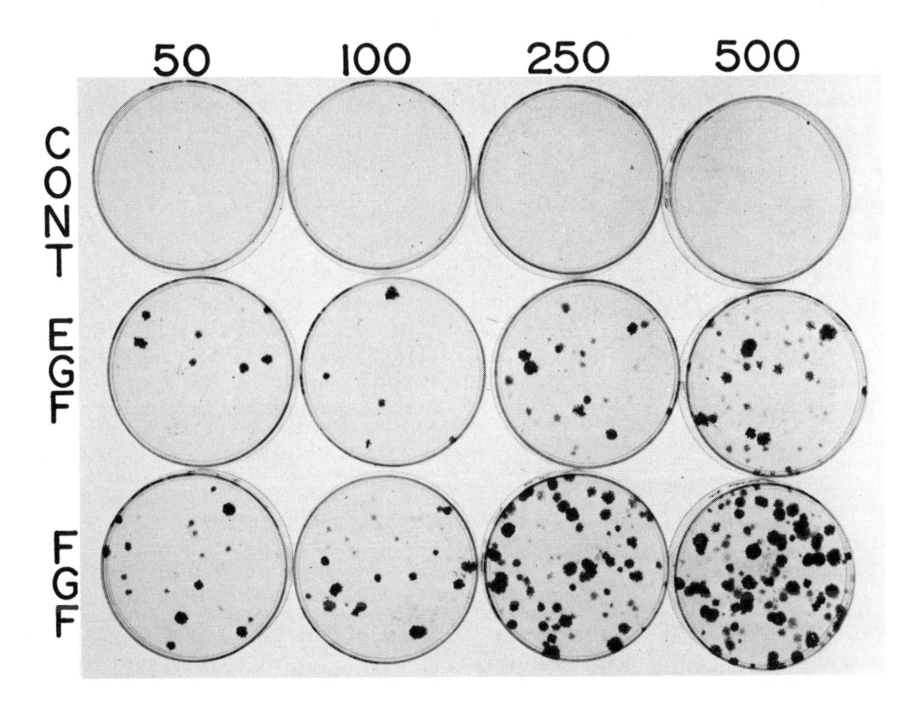

Fig. 22. Porcine granulosa cells plated at 50, 100, 250, and 500 cells per plate were maintained in the presence of DME 10% calf serum with or without EGF (10 ng/ml) or FGF (100 ng/ml) added every other day.

Since under these conditions all the cells in the control culture died, both FGF and EGF can be considered to be survival agents as well as mitogenic agents for low density granulosa cell cultures.

In this regard it should be kept in mind that survival and mitogenic agents are not always easy to distinguish. For example, a survival agent can stimulate mitosis in cultures maintained in media lacking only the survival agent. At the same time, a mitogenic agent may promote survival, if survival is dependent upon cell density.

Our results indicate, then, that both FGF and EGF can delay terminal differentiation of cultured granulosa cells originating from small and medium-sized follicles. Since granulosa cells from large follicles are already committed to terminal differentiation, the effects of EGF and FGF were not as pronounced as with cultures originating from small and medium-sized follicles, and FGF had a much greater effect than did EGF.

The finding that human diploid fibroblasts have a limited and predictable replicative lifespan in culture suggested an analogy to the process of aging and has led to many studies of the phenomenon as well as to theories of its basis. Although a reproducible lifetime can be obtained for fibroblasts under standard conditions, the lifetime can be increased by improving the culture conditions. This would not be expected if the lifetime were limited by a fixed rate of mutation. The dependence of the lifespan of diploid cells on culture conditions has been shown more clearly with keratinocytes, for which the addition of EGF to the culture medium results in a trebling of the lifetime of the culture. The lifespan of granulosa cell cultures is similarly extended by the medium. The practical implications of this finding can be easily appreciated, if one considers that a single granulosa cell from a small or medium-sized follicle in the absence of EGF or FGF does not proliferate beyond 11 generations and results ideally in 2×10^3 progeny cells. Beyond that point most, if not all, of the cells are committed to terminal differentiation. In contrast, if EGF or FGF is included in the medium, that single cell proliferates beyond 50 generations and yields a progeny of 1×10^{15} cells (a 5×10^{11}-fold increase), and the cell population, being far away from terminal differentiation, will still actively divide. Even this figure is a conservative estimate, since our experimental cultures continue to proliferate beyond the 50th generation.

How does the replicative lifespan of cells in our cultures compare with the lifetime of granulosa cell populations *in vivo*? The most reliable data are those of Pedersen[57], who has studied in detail the proliferation of granulosa cells in the ovarian follicles of mice. He found that primary follicles start with a follicular cell layer composed of 2 or 3 granulosa cells (type 2). During maturation these primary follicles develop into preovulatory follicles (type 8) with 50,000 granulosa cells. This represents 15 generations. Although these data clearly cannot be extrapolated to bovine follicles, which are far larger than mouse follicles, if one assumes that bovine follicles (primary follicle type 2), which have bigger oocytes than mouse ovaries, start with a follicular cell layer of 10 to 20 cells, preovulatory bovine follicles (type 8) would have a quantity of granulosa cells corresponding to 16 generations.

Since granulosa cultures obtained from small or medium-sized bovine follicles (type 5b and 6, according to the classification of Pedersen) can exceed 50 generations in tissue culture when maintained in media containing either EGF or FGF, it is clearly demonstrated that these mitogens can significantly delay terminal differentiation. Without mitogens these cells would undergo terminal differentiation within 10 generations.

A specific use for EGF and FGF in the study of granulosa cell physiology can now be suggested. Until now the main metabolic activity of granulosa cells that has been investigated is their ability to produce steroids. This is clearly an activity that occurs during the terminal differentiation of the cells and reflects the last part of the active life of the cells before luteinization or cell death or both. (This occurs in mice during the last two days of a lifespan of 20 days.) One may then wonder what functions the granulosa cells perform during the first 18 days of their existence. One obvious function is to provide nutrients for the oocytes, since during the initial 12 days of the follicular development the oocytes quintuple in volume (type 1 follicle to type 5a follicle). Another function during the later stage of follicular development is to form the liquor folliculi (from day 12 to day 18 in the mouse, follicle type 5 to type 6). Such a function requires the synthesis of great amounts of protein and proteoglycan and is reflected in the morphology of cells when they are scanned by electron microscopy, whether *in vivo* or *in vitro*. During this period the granulosa cell layers, instead of being full of lipids, as in the later stage (luteinization =

terminal differentiation), have a cytoplasm rich in rough, endoplasmic reticula with dilated cisternae as big as .1 u in diameter in some cases (Fig. 23).

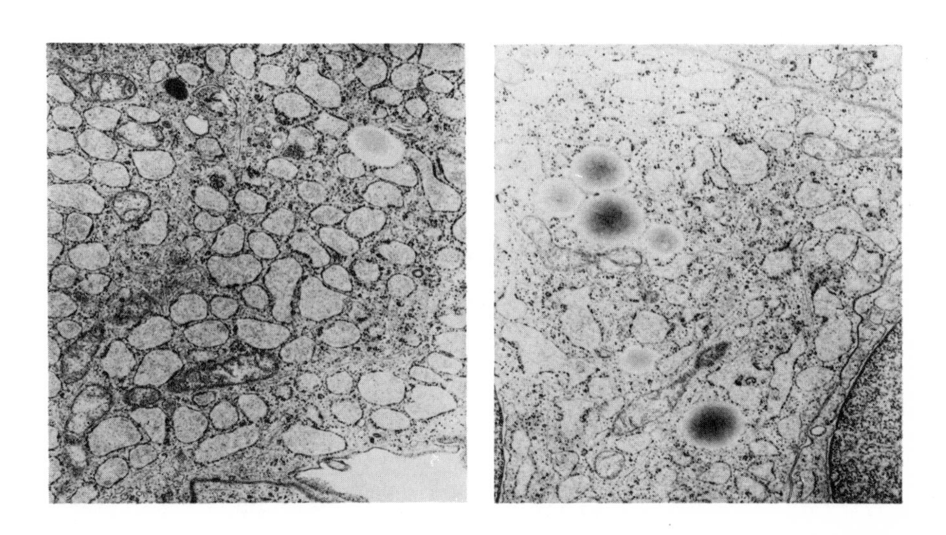

Fig. 23. Electron micrographs of granulosa cells obtained from 4 mm follicles. In A, the cytoplasm is filled with granular endoplasmic reticulum consisting of very dilated cisternae. In B, lipids in clusters can be seen. (25,000 x)

Both the nature and the function of the proteins and proteoglycan synthesized by granulosa cells during the earlier stage of follicular development are difficult to investigate *in vivo* due to the few cells available. This function could more easily be investigated *in vitro* by using mitogens such as EGF or FGF which, by preventing terminal differentiation, allow considerable amounts of cellular material to be obtained.

IV. The production of a corpus luteum angiogenic factor responsible for the proliferation of capillaries and the neovascularization of the corpus luteum

The factors controlling the changes in the vascular pattern of the ovary that occur during the normal reproductive cycle and during pregnancy have not been investigated. Nonetheless, the ovaries provide an excellent model for the study of the physiological adaptations of blood vessels in tissues that undergo rapid and repeated growth and involution. The rapidity of follicular growth and of the rise and decline of the corpora lutea caused by the brevity of the reproductive cycle imposes remarkable demands on the ability of blood vessels to proliferate and, subsequently, to regress. The cyclic changes also involve concomitant physiological adjustments in blood flow, capillary permeability, and other functional behavior.

Particularly impressive are the extremely rapid and radical vascular changes that take place following ovulation. Preovulatory follicles are composed of a central cavity surrounded by multiple layers of granulosa cells and 2 external cell layers-- the richly vascularized thecae interna and externa. The granulosa cell layers are separated from the thecal layers of the follicles by a basement membrane and are themselves avascular (Fig. 1).

Coincident with ovulation and collapse of the follicle the

most impressive changes occur in the capillary wreath surrounding the follicle. From the endothelium of these vessels capillary sprouts begin to grow into the granulosa cell layer. Rapidly growing capillaries develop in 48 hours into a totally new, complex network of sinusoidal vessels which invade the previously avascular granulosa cell layers, induce luteinization, and later nourish the parenchyma of the corpus luteum (Fig. 1).

The rapid proliferation of capillaries that takes place during the early phase of the development of the corpus luteum can be compared to the vascularization of solid tumors. Solid tumor growth has been associated with the proliferation of capillaries in surrounding host tissues. Algire and Chalkley[58] were the first to appreciate that growing tumors elicit capillary ingrowth from the host. They suggested that this might be an underlying factor responsible for the autonomous growth of these tumors. Greenblatt and Shuzik[59] later demonstrated that the new capillary growth is elicited by some diffusible material, which later was partially purified by Folkman and his colleagues[60] and named "tumor angiogenesis factor" (TAF).

Since the early vascular changes taking place during the development of the corpus luteum are strikingly similar to the capillary proliferation induced by tumors, we have investigated the possibility that the early phase corpus luteum produces a diffusible substance similar to TAF.

This report describes the host's angiogenic response to implants of follicles and corpus luteum in the cornea. As already pointed out by Gimbrone *et al.*[61], the cornea provides a transparent, avascular substratum in which the relationship between survival, differentiation, and growth of a given tissue and neovascularization can be continuously observed *in vivo*. Implantation at a distance from the circumferential vessels of the limbus produces an anatomic separation of the graft from responding host vessels. This arrangement allows independent observation of the behavior of both elements.

A) Comparison of the neovascularization induced by follicular and corpus luteum implants

Of 12 follicular implants in the rabbit cornea, 8 failed to stimulate neovascularization when transplanted into the cornea of untreated females or of females treated with PMSG. Of 4 follicles transplanted into pseudopregnant rabbits (i.e., female rabbits injected with PMSG followed by an LH injection), all 4 induced neovascularization. Histological examination showed that these follicles had luteinized in the rabbit cornea (Table 2). This is in agreement with the results of others who have shown that preovulatory follicles implanted under the kidney capsules of pseudopregnant animals luteinize spontaneously[62]. Of 15 corpus luteum implants, 13 induced neovascularization (Table 3). In two ovarian stroma implants containing small follicles, but no corpus luteum, no neovascularization of the cornea was observed (unpublished data). This demonstrates that early phase corpora lutea, unlike follicles, have the capacity to elicit sprouting and proliferation of capillaries from the host. This property was also shared by follicular implants which spontaneously luteinized when transplanted into the rabbit cornea of pseudo-pregnant rabbits.

TABLE 2

Effect of Ovarian Follicular Implants on the Growth of Capillaries in the Rabbit Cornea

Follicular Implants	Inflam*	Lutein**	Vessel Length (mm)	Length of Experiment (days)
autologous +	-	-	-	16
autologous multiple follicles +	-	-	one vessel (3,5 mm)	16
homologous +	-	-	-	16
autologous multiple follicles +	-	-	-	16
autologous +	-	-	-	13
autologous +	-	-	-	13
homologous multiple follicles +	-	-	-	13
homologous multiple follicles +	-	-	-	13
homologous §	-	+	7	19
homologous §	-	+	7	19
homologous §	-	+	3,5	19
homologous §	-	+	7,5	19

* inflammation
** luteinization
\+ implants in normal female rabbit
§ implanted in pseudopregnant female rabbit

TABLE 3

Effect of Corpus Luteum Implant on the Growth of Capillaries in the Rabbit Cornea

Corpus Luteum Implant	Inflam	Luteiniz	Vessel Length (mm)	Length of Experiment (days)
homologous	-	+	8	20
homologous	-	+	7	20
homologous	-	+	8	20
homologous	-	+	8	20
autologous	-	+	-	20
autologous	-	+	-	20
homologous	-	+	2	15
homologous	-	+	3,5	15
homologous	-	+	3,5	15
homologous	-	+	2,5	15
autologous	-	+	5,5	12
autologous	-	+	3,5	12
homologous	-	+	2,5	12
homologous	-	+	3,5	12
homologous	-	+	0,5	12

B) Morphology of the host's neovascular response

1) Follicular implant

When follicles were implanted in the rabbit eye, whether far from (7-8 mm) or near (2-4 mm) the limbus, no vascular response on the part of the host was observed, provided that the follicles did not luteinize. In a few cases a few delicate capillary loops originated from the limbus during the first two days, but they never grew, nor did they make contact with the follicular implant (Fig. 24A). By day 14 the follicles had collapsed and the cells exhibited pyknosis. No edema or inflammatory cells were observed (Fig. 24B).

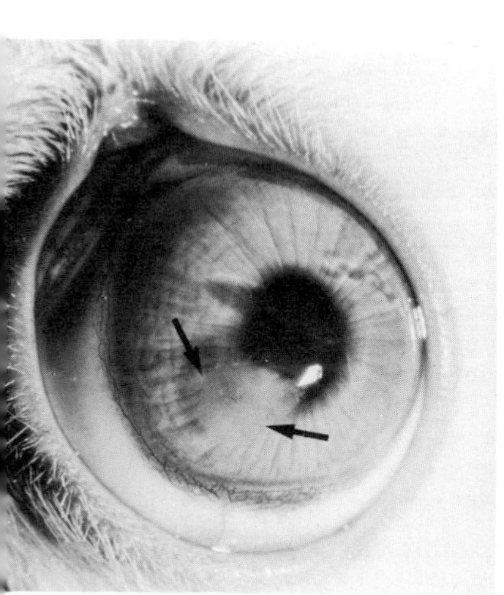

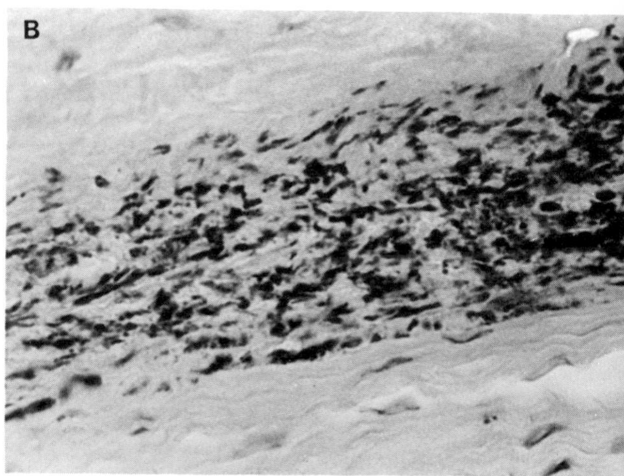

Fig. 24. Lack of proliferation of capillaries induced by implant of rabbit follicles in the rabbit cornea. (A) The follicular tissues, consisting of 3 follicles, have been implanted in the center of the eye (arrow). No capillary proliferation was observed 14 days later. (B) Histological section of the cornea stained with hematoxylin-eosin and containing the follicular implant. The cellular structures have disappeared. The nuclei exhibit nuclear condensation (pyknosis), and, although the tissues are clearly resorbing, no inflammatory cells or capillaries were observed near the implant.

2) Corpus luteum

When corpora lutea were implanted near the limbus (3 mm) (Fig. 25B), after 24 hours newly-formed vessels penetrated the cornea centripetally from the adjacent limbal area and grew toward the implant. Delicate capillary loops originated perpendicularly from small venules in the limbal plexus and then grew as elongated hairpins. Linear growth of these hairpin vessels initially proceeded at approximately 0.50 mm/day.

As soon as the capillaries reached the implant (by 4 to 5 days), hemorrhage was observed around it (Fig. 25B). This indicates that the pre-existing vessels present in the corpus luteum did not disintegrate and reattached to the host vessels by anastomosis. The revascularization of corpora lutea implanted near the limbus results, then, from the capillary proliferation in the host as well as from the re-attachment of the capillaries by anastomosis to pre-existing corpus luteum capillaries. Following the neovascularization of the implant, capillaries which did not invade or contact the implant regressed, leaving ghost vessels in the cornea.

When the corpus luteum was implanted far away from the limbus (7-8 mm) (Fig. 25A, C), newly-formed capillaries, similar to those observed in previous experiments with corpora lutea implanted near the limbus, appeared within 48 to 72 hours and were fully developed within 4 to 5 days. Then secondary and tertiary branches rapidly developed, converting the initial sprouts into dense, vascular brushes. A red pannus with very little space between each blood vessel resulted (Fig. 25A). Vascular proliferation was greatest in the anterior layers of the corneal stroma but also occurred in deeper regions.

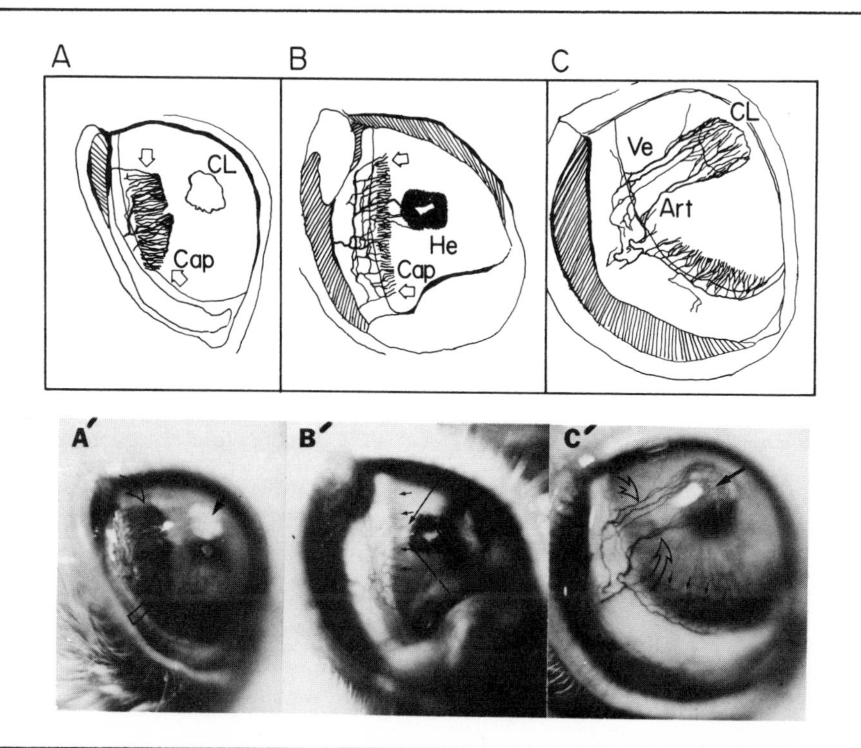

Fig. 25. Proliferation of capillaries induced by implant of rabbit corpus luteum in the rabbit cornea. (A) The corpus luteum (CL) has been implanted near the center of the eye. Six days following the implantation capillaries (cap) proliferate as a brush border from the limbus toward the implant. (B) Corpus luteum implanted near the limbus. By day 3 capillaries (Cap) derived from the limbus have connected with the implant and anastomose with luteal capillaries leading to a hemorrhage (He) around the implant. (C) By day 14 the capillaries have invaded the corpus luteum and reorganization of the vascularization has taken place. The capillaries connecting with the corpus luteum have differentiated into arterioles (Art) and venules (Ve) forming 3 main vascular trunks. The capillaries not connecting are regressing, thereby leaving ghost vessels. ("A," "B," and "C" are graphic illustrations of replicate photographs A', B', C'.)

When the capillaries reached the corpus luteum (9-11 days), they invaded the tissue (Fig. 25C). Since by that time all of the pre-existing capillaries of the corpus luteum had degenerated, its vascularization could only be accounted for by neovascularization dependent upon the host. After neovascularization of the corpus luteum a spectacular rearrangement of the dense vascular network that proliferated in the cornea took place. All capillaries which did not connect to the corpus luteum regressed rapidly (Fig. 25C). Capillaries connecting with the implant started to differentiate into arterioles, while others simultaneously became venules, leaving two or three main vascular trunks reaching from the limbus to the corpus luteum and forming afferent and efferent vessels (Fig. 25C). Since it can be expected that the angiogenesis factor released by the corpus luteum was now carried off by the bloodstream and was no longer available for diffusion through the cornea, the gradient of angiogenesis factor that supported the capillary proliferation was probably no longer present. This could explain why all capillaries which do not infiltrate the corpus luteum regress.

C) Histological study of the corpus luteum implants

A histological section of a 15-day old central implant of corpus luteum is shown in Fig. 26. Extensive luteinization could be observed through the corpus luteum (Fig. 26A). Large cells with light cytoplasm, when stained with hematoxylin and eosin, were clearly seen (Fig. 26B). Histological sections of growing capillaries approaching the corpus luteum implant showed that the surrounding stroma was essentially free of inflammatory cells (Fig. 26C); the vessels proper showed no margination or exudation of leukocytes. Numerous small capillaries were visible through

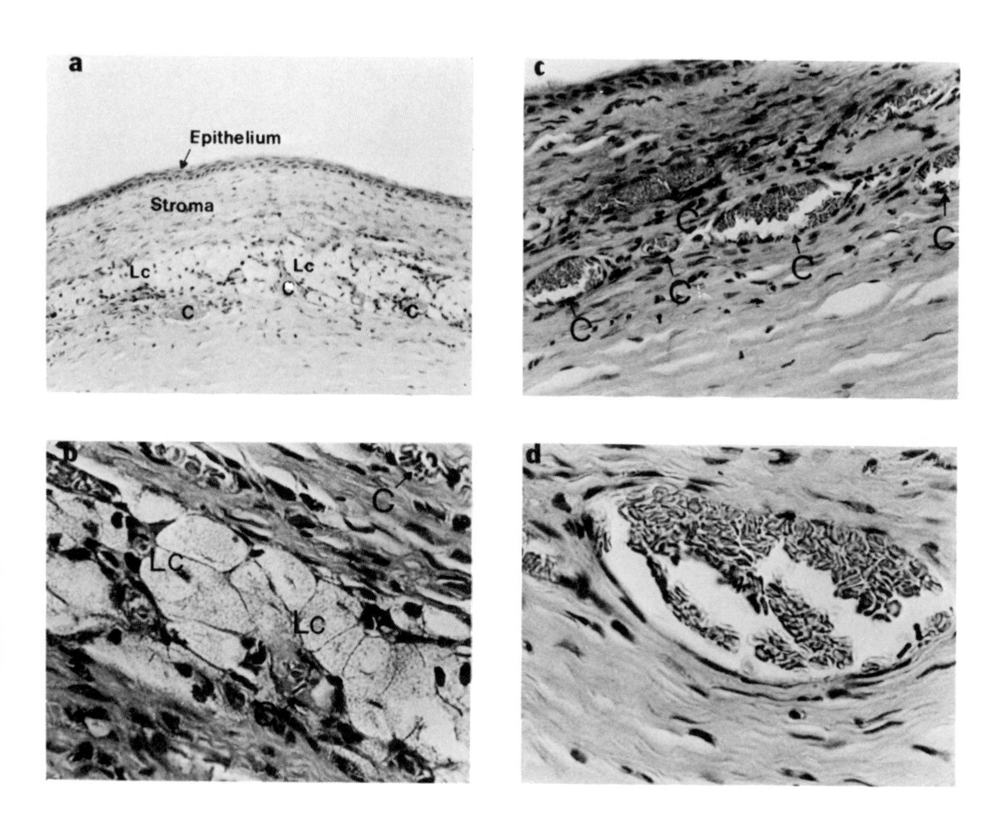

Fig. 26. Corpus luteum 15 days after implantation into the rabbit cornea. (A) The luteal cells can be seen to be localized deep in the stroma. Luteal cells (Lc), capillaries (c) (hematoxylin-eosin staining, 100 X). (B) Capillaries growing deep in the stroma and invading the corpus luteum implant. Luteal cells (Lc), capillaries (c) (400 X). (C) Capillaries (arrows) growing deep in the stroma and full of red blood cells (100 X). (D) Fully differentiated blood vessels filled with red blood cells (400 X).

the implant (Fig. 26). Numerous differentiated blood vessels could be seen deep in the stroma (Fig. 26C). There was no sign of edema, no degeneration or necrosis of luteal cells was observed, and there was no accumulation of polymorphonuclear leukocytes (Fig. 26C).

D) The correlation between the angiogenic activities of the corpus luteum, follicles, ovarian stroma, and their mitotic activity is reflected by the stimulation of the initiation of DNA synthesis in resting Balb/c 3T3 cell cultures.

Since implanted corpora lutea were capable of stimulating the proliferation of capillaries in the reabbit cornea, while follicular implants were inactive, we have compared their relative mitotic activities _in vitro_. We used the Balb/c 3T3 cell, which has been shown to be sensitive to the same mitogens _in vitro_ as vascular endothelial cells, as a model[26,29,30-33]. As shown in Fig. 5, corpus luteum crude extracts were potent stimulators of the initiation of DNA synthesis; the minimal effective dose was found to be 200 ng/ml in two separate determinations and a plateau was observed at 1 ug/ml. Follicular extracts, on the other hand, were inactive when compared to the corpus luteum crude extracts. As shown in Fig. 27, corpus luteum crude extracts were potent stimulators of the initiation of DNA synthesis; the minimal effective dose was found to be 200 ng/ml in two separate determinations and a plateau was observed at 1 μg/ml. Follicular extracts, on the other hand, were inactive when compared to the corpus luteum crude extracts. As shown in Fig. 27 A and B, a marginal stimulation was observed at

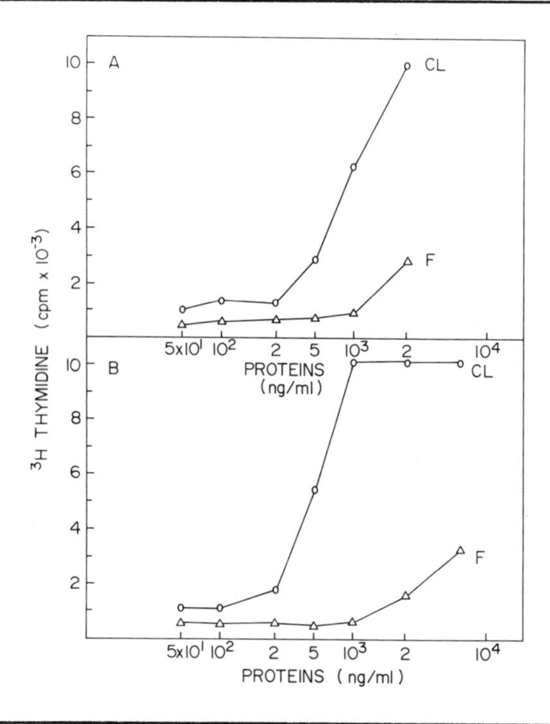

Fig. 27. Effect of crude extract of ovarian follicles and corpus luteum on the initiation of DNA in resting populations of Balb/c 3T3 cells. Increasing concentrations of ovarian follicles crude extract (F) or corpus luteum crude extract (CL) were added to resting populations of Balb/c 3T3 cells. Twelve hours later the cells were pulsed with thymidine, and at 24 hours the cultures were washed, lysed, and determination of ^{3}H thymidine incorporated into DNA. When increasing concentrations of ovarian stroma crude extract were used, the results were similar to those obtained with follicular crude extract.

2 ug/ml, while at 10 μg/ml the degree of stimulation was equivalent to that observed with 300 ng/ml of corpus luteum crude extract. Similar results were obtained whether rabbit ovarian tissues or bovine ovarian tissues were used.

Our results demonstrate that ovarian corpus luteum, unlike follicles or ovarian stroma, is capable of inducing neovascularization. Using the rabbit cornea, which is a naturally transparent avascular structure, one can easily study the rates of new vessel formation and the parts played by the host and the ovarian implant, respectively, in its vascularization.

Although it has been reported by others[63] that adult tissue does not stimulate capillary proliferation in the host, we find, using the corpus luteum, that a normal adult tissue does have this capacity. Although this does not necessarily mean that all adult tissue has the same capacity, it is probable that all adult tissues that are actively invaded by capillaries have this capacity at some stage of their development.

Corpora lutea implanted near the limbus behave similarly to an embryonic graft, since pre-existing vessels do not distintegrate but are reattached by anastomosis to the host vessels. In contrast to the embryonic graft, however, in which there is either minimal or no vascularization on the part of the host vessels[64], corpora lutea are capable of inducing neovascularization, as is demonstrated by the distance that the capillaries grow through the cornea before reaching the implant. The extent of this neovascularization was even more marked when the implants were far away from the limbus. Since, by the time the corneal capillaries reached the implant, the pre-existing vessels within the implant had disintegrated, the result was a <u>total</u> neovascularization on the part of the host.

In contrast to corpora lutea implants, follicles did not stimulate neovascularization. This is in agreement with the *in vivo* situation in which active capillary proliferation is *only* observed with corpus luteum, whereas follicles are an avascular structure. Similarly, cornea vessels were not formed in response to a sham procedure. Since recent studies have shown that corneal tissue becomes vascularized as a consequence of leukocytic infiltration[65], it was of paramount importance to demonstrate that leukocytes did not invade the corpus luteum implant and that inflammation, which is known to be a potent inducer of corneal vascularization, did not develop. In our histological study we could not find any evidence of leukocytic infiltration in either early or late corpus luteum implants. Further proof that leukocytic infiltration is not a factor in inducing the vascularization observed with implants of corpus luteum is the observation that follicular implants did not induce neovascularization unless they luteinized. Since one would expect that infiltration of the follicle by leukocytes would stimulate the proliferation of blood vessels, the lack of angiogenesis following follicular implants demonstrates that such infiltration did not take place.

The observations that capillary budding always originated from a limited sector of the limbal plexus adjacent to the implant and developed into elongated hairpin loops directed toward the corpus luteum and that peripheral implants (1-2 mm from the limbus) elicited vessel formation as early as the first day, whereas centrally located implants elicited vessel formation after 3 to 4 days, demonstrate that the vascularization induced by the corpus luteum is mediated by diffusible substances.

Further proof that a stimulatory factor is being released from the corpus luteum is provided by the spectacular reorganization of vascularization once the capillaries have invaded the corpus luteum. What was before a very dense, reactive, cherry pannus with very little space between each blood vessel developed into 2 or 3 main arterioles and venules. All capillaries not contacting the corpus luteum regress, thus suggesting that the stimulating factor is now released into the bloodstream provided by the neovascularization and no longer diffuses through the cornea.

It may seem disturbing that, besides synthesizing steroids, the corpus luteum should also be able to produce factors that stimulate vascular endothelial cell proliferation. However, the late corpus luteum of pigs, guinea pigs, rabbits, mice, rats, and humans can synthesize and release the polypeptide relaxin[66]. This indicates that the endocrine function of the corpus luteum is not strictly limited to the production of steroids.

If one speculates that corpora lutea, like tumor cells, contain a substance capable of stimulating the proliferation of capillaries, it may be possible to purify this substance by using the growth of cultured vascular endothelial cells of Balb/c 3T3 cells as an assay system. Using such a model, we have compared the mitotic activities of crude extracts of corpora lutea, follicles, and ovaries from which the corpora lutea were removed. When we compared the mitotic activities for vascular endothelial cells of these 3 extracts, only the corpus luteum extract was capable of stimulating the proliferation of vascular endothelial cells maintained *in vitro*. Further support was thus lent to the hypothesis that corpora lutea can secrete an angiogenesis factor which could be named "corpus luteum angiogenic factor" (CLAF).

Although the corpus luteum factor involved in the formation of new capillaries may, in all probability, be similar to factors made by tumor cells, one might wonder what the new situation is that causes the luteal cells to synthesize and release an angiogenic factor. As pointed out by Bassett[62], who made the only known study of the changes of the vascular pattern of the rat ovary during the estrous cycle, one of the main factors dictating the degree of vascularization of the various components of the ovaries is their metabolic requirement. This requirement is not peculiar to the ovaries, however, since capillary proliferation within a tissue seems, in most cases, to be correlated with an increased oxygen demand by that tissue. There is, for example, direct evidence suggesting that the initial stimulus attracting blood vessels in the fetal retina is an increased demand for O_2 arising from the inner layer of the retina which, in turn, directly or indirectly stimulates capillary proliferation[68]. A dramatic example of the modulation of capillary proliferation by O_2 is offered by the pathology of retrolental fibroplasia. This form of blindness is caused by the formation of fibrous tissue behind the lens and appears when premature children are kept in incubators under high O_2 tension. Under these conditions, hyperoxia produces vascular spasm, and a total regression of retinal capillaries occurs. When the infant is later exposed to normal O_2 tension, however, there is a wild regrowth of the vessels, which, in some cases, leads to extensive hemorrhage and formation of fibrous scar tissue[69].

Considering the ovary in light of the retinal model, one notes that luteal cells secrete huge amounts of steroids (primarily progesterone) which are synthesized from acetate, a process requiring great amounts of energy and, therefore, of O_2. This high requirement for energy is reflected by the high mitochondrial content observed in luteal cells. Granulosa cells, in contrast, are relatively dormant metabolically, and this dormancy is reflected by their low mitochondrial content. It is then reasonable that capillary proliferation should accompany the granulosa-luteal conversion as a response to the increased demand for oxygen by the luteal cells. An increased demand for O_2 could consequently stimulate the production of other factors such as TAF or CLAF which could, in turn, directly stimulate the proliferation of capillaries.

In conclusion, we would like to propose that the corpus luteum actively participates in its own neovascularization, which develops from capillaries sprouting from the vascular wreath present in the theca interna. While the initial stimulus is, in all likelihood, the increased demand for O_2 of the luteal cells caused by their high metabolic activity, the factor involved in inducing the proliferation of capillaries is a mitogenic factor (CLAF) produced by the corpus luteum.

Acknowledgements

This work was supported by Grants HL 20-197 and HD 11082 from the National Institutes of Health, VC-194 from the American Cancer Society, and by the Rockefeller Foundation. I. Vlodavsky is a recipient of the Chaim Weizmann Research Training Fellowship.

REFERENCES

1. Channing, C.P., Recent Prog Horm Res 26: 589, 1970.
2. Channing, C.P., Endocrinology 87: 49, 1970.
3. Channing, C.P., In McKerns, K.W. (ed.), The Gonads, Appleton-Century-Crofts, New York. 1966, p. 245.
4. Gospodarowicz, D., Nature 249: 123, 1974.
5. Cohen, S., J Biol Chem 237: 123, 1974
6. Bynny, R.L., D.N. Orth, S. Cohen, and E.S. Doyne, Endocrinology 95: 776, 1974.
7. Cohen, S., and C.R. Savage, Recent Prog Horm Res 30: 551, 1974.
8. Cohen, S., and G. Carpenter, Proc Natl Acad Sci USA 72: 1317, 1975.
9. Cohen, S., G. Carpenter, and K. Lembach, Adv Metab Disord 8: 265, 1975.
10. Gregory, H., Nature 257: 325, 1975.
11. Gospodarowicz, D., J Biol Chem 250: 2515, 1975.
12. Gospodarowicz, D., J. Moran, and H. Bialecki, In Growth Hormone and Related Peptides, Excerpta Medica International Congress 381, Elsevier, New York, 1976, p. 141.
13. Gospodarowicz, D., and J. Moran, Proc Natl Acad Sci USA 71: 4648, 1974.
14. Armelin, H.A., Proc Natl Acad Sci USA 70: 2702, 1973.
15. Gospodarowicz, D., J Biol Chem 250: 2515, 1975.

16. Gospodarowicz, D. and Moran, J.S., Proc Natl Acad Sci USA 71: 4584, 1974.

17. Rudland, P.S., Seifert, W., and Gospodarowicz, D., Proc Natl Acad Sci USA 71: 2600, 1974.

18. Rose, S., Pruss, R., and Herschman, H., J Cell Physiol 86: 593.

19. Rudland, P.S., Eckhart, W., Gospodarowicz, D., and Seifert, W., Nature 250: 337, 1974.

20. Gospodarowicz, D. and Moran, J., J Cell Biol 66: 451, 1975.

21. Lembach, K.J., Proc Natl Acad Sci USA 73: 183, 1976.

22. Westermark, B. and Wasteson, A., Adv Metab Disord 8: 85, 1975.

23. Westermark, B., Biochem Biophys Res Com 69: 304, 1976.

24. Gospodarowicz, D., Moran, J., and Owashi, N., J Clin Endocr and Metab 44: 651, 1977.

25. Gospodarowicz, D. and Mescher, A.L., J Cell Physiol, in press, 1977.

26. Gospodarowicz, D., Ill, C., Mescher, A.L., and Moran, J., In Molecular Control of Proliferation and Cytodifferentiation, 35th Symposium for the Society for Developmental Biology, Academic Press, New York, in press, 1977.

27. Gospodarowicz, D., Weseman, J., Moran, J.S., and Lindstrom, J., J Cell Biol 70: 395, 1976.

28. Ross, R. and Glomset, J., J Cell Biol, in press, 1976.

29. Gospodarowicz, D., Moran, J., and Braun, D., J Cell Physiol, 91: 377, 1976.

30. Gospodarowicz, D., Brown, K.D., Birdwell, C.R., and Zetter, B.R., J Cell Biol, in press, 1977.

31. Gospodarowicz, D., Moran, J., Braun, D., and Birdwell, C.R., Proc Natl Acad Sci USA 73: 4120, 1976.

32. Gospodarowicz, D., In Marchesi, V.T. (ed.), Prog in Clin. and Biol. Res. Membranes and Neoplasia: New Approaches and Strategies, A.R. Liss, Inc., New York. 1976, p. 1.

33. Zetter, B.R., Gospodarowicz, D., In Lundblad, R.L., (ed.), Chemistry and Biology of Thrombin, Ann Arbor Press. 1977.

34. Gospodarowicz, D., Zetter, B., World Health Org. Symp. Geneva, in press, 1977.

35. Gospodarowicz, D., Mescher, A.C., Brown, K.D., and Birdwell, C.R., Exp Eye Res, in press, 1977.

36. Gospodarowicz, D., Mescher, A.L., and Birdwell, C.R., Exp Eye Res, in press, 1977.

37. Savage, C.R., and Cohen, S., Exp Eye Res 15: 361, 1973.

38. Frati, I., Danieli, S., Delogu, A., and Coveli, I., Exp Eye Res 14: 135, 1972.

39. Green, H., Nat Cancer Inst Monogr, in press, 1977.

40. Gospodarowicz, D., and Handley, H.H., Endocrinology 97: 102, 1975.

41. Gospodarowicz, D., Ill, C.R., Hornsby, P., and Gill, G.N., Endocrinology 100: 1080, 1976.

42. Gospodarowicz, D., Ill, C.R., and Birdwell, C.R., Endocrinology 100: 1108, 1977.

43. Gospodarowicz, D., Ill, C.R., and Birdwell, C.R., Endocrinology 100:1121, 1977.

44. Cohen, S., Develop Biol 12: 394, 1965.

45. Rheinwald, J.C. and Green, H. Nature 265: 421, 1977.

46. Gospodarowicz, D., Rudland, P., Lindstron, J., and Benischke, K., Adv in Metab Disord 8: 301, 1975.

47. Hollenberg, M.D. and Cuatrecasas, P., J Biol Chem 250: 3845, 1975.

48. Cohen, S. and Carpenter, G., Proc Natl Acad Sci USA 72: 1317, 1975.

49. Lembach, K.J., Proc Natl Acad Sci USA 73: 183, 1976.

50. Hollenberg, M.D., Arch Biochem Biophys 171: 371, 1975.

51. Eshkol, L., and Lenenfeld, B., In Saxena, B., Belring, J., and Gandy, H.M. (eds.), Gonadotropins, Wiley, New York. 1972, p. 335.

52. Pedersen, T., Acta Endocrinol (Kbh) 62: 117, 1969.

53. Ryle, M., J Reprod Fertil 19: 87, 1969.

54. Croes-Buth, S., Paesi, J.A., and deJongh, S.E., Acta Endocrinol (Kbh) 32: 399, 1959.

55. Greenwald, G.S., In Geiger, S.R. (ed.), Handbook of Physiology, vol. IV, section 7, part 2, American Physiological Society, Washington, D.C. 1974, p. 293.

56. Blandau, R.J., In Greep, R.O., and Weiss, L., Histology, ed. 3, Academic Press, New York. 1973, p. 768.

57. Pedersen, T., Acta Endocrinol 64: 306, 1970.

58. Algire, G.H. and Chalkley, H.W., J Nat Cancer Inst 6: 73, 1945.

59. Greenblatt, M. and Shubik, P., J Nat Cancer Inst 41: 111, 1968.

60. Folkman, J., Merler, E., Abernathy, C., et al. J Exp Med 133: 275, 1971.

61. Gimbrone, M.A., Cotran, R.S., Leapman, S.B., and Folkman, J., J Nat Cancer Inst 52: 413, 1974.

62. Ellsworth, L.R. and Armstrong, D.T., Endocrinology 94: 892, 1974.

63. Ausprunk, D.H., Knighton, D.R., and Folkman, J. Am J Pathol 79: 597, 1975.

64. Fromer, C.H. and Klintworth, G.K., Am J Pathol 79: 437, 1975.

65. Fevold, H.L., Hisaw, F.L., and Meyer, R.K., Am Chem Soc 52: 3340, 1930.

66. Bassett, C.L., Am J Anat 73: 251, 1943.

67. Ashton, N., Ward, B., and Serpetll, G., British Journal of Opthal 38: 397, 1954.

68. Silverman, W.A., Scientific American 236: 100, 1977.

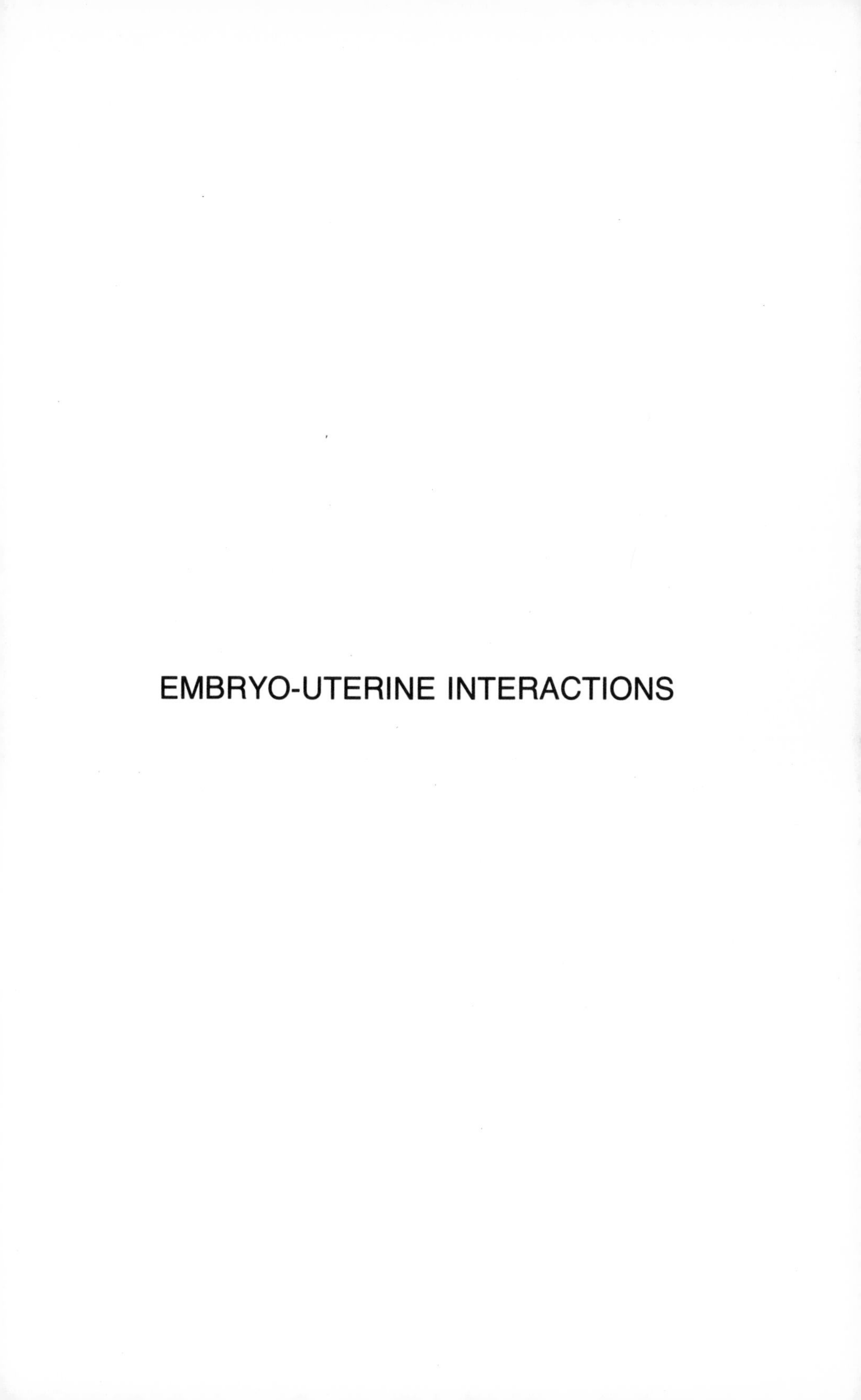

EMBRYO-UTERINE INTERACTIONS

6

The Role of Trophoblastic Factors in Implantation

HANS W. DENKER

Introduction

The molecular-biological mechanisms of implantation initiation are largely unknown so far. Most investigators interested in pre- and periimplantation physiology have centered their attention on the uterine milieu, which is mediating the hormonal regulation of implantation from the maternal side. However, experimental data available give strong evidence that the effects of uterine factors must be considered as mostly implantation-inhibiting rather than implantation-promoting: implantation gets easily started in extrauterine (ectopic) sites and, under certain circumstances, in non-pregnant (virgin) uteri (while here it does not occur in inappropriate stages of pregnancy or pseudopregnancy) or in uteri of actinomycin D-treated animals (Kirby, 1970; McLaren, 1973; Psychoyos, 1973; Finn, 1974). It might therefore be assumed that, in regular pregnancy, a blastocyst can start implantation as soon as either the inhibitory activity of the uterine milieu is reduced (hormonally controlled) or when it is overcome by an increased invasive activity of the ripened trophoblast. It seems to be very desirable, therefore, to accumulate data on the role of trophoblastic factors in this process.

Biochemical data on trophoblastic factors directly involved in initiation of attachment are very scarse, however. During the last 12 years, systematic studies on initiation of implantation have been performed in the rabbit (Denker, 1969 a, 1970 a and b, 1971 a-c,

1972, 1973, 1974 b, 1976 c). The rabbit is very suitable as an experimental animal for this purpose because its blastocysts are large enough to allow great series of combined histochemical and morphological studies, and they also provide sufficient material for biochemical investigations (although micro methods are still required). In spite of the fact that the rabbit represents the central type of implantation, its trophoblast is very invasive (forming a hemochorial placenta) so that it can be expected to be suitable as a model for certain molecular mechanisms likewise involved in implantation initiation of the human blastocyst no matter that the latter implants interstitially.

Morphological observations on the role of the trophoblast in implantation initiation in the rabbit

At the time when implantation starts, the surface of the rabbit blastocyst is still covered by thick extracellular coverings (blastocyst coverings) composed of glycoproteins which are derived from tubal (mucoprotein layer), uterine ("gloiolemma" after Böving, 1963) and, as we have recently found out, also trophoblastic secretion material (Denker and Gerdes, in preparation). By an unknown mechanism probably involving biochemical activities of the abembryonic trophoblast, the rabbit blastocyst orients itself in the uterus in such a way that normally the abembryonic pole (opposite to the embryonic disc) comes to lie against the antimesometrial endometrium. The mechanism must involve rotation of the whole blastocyst inside the uterus; rotation of the embryonic disc inside the trophoblast vesicle as proposed for the mouse by Kirby et al. (1967) is impossible in the rabbit because here at this stage the embryonic disc is already widely differentiated and intercalated in the trophoblastic layer. Since the orientation takes place with the blastocyst coverings still intact, we must postulate that the blastocyst alterates the physicochemical properties of its coverings at either the embryonic or the abembryonic pole. In fact, the coverings become stickier at this time at the abembryonic pole. This effect has been attributed to a local rise in pH (Böving, 1963).

In the following process of dissolution of the blastocyst coverings, attachment and invasion, the active elements seem to be, as judged from the morphological point of view, the trophoblastic knobs, syncytial elements of the abembryonic and lateral parts of the trophoblast. Their light and electron microscopical morphology has been investigated and disputed in detail (Böving and Larsen, 1973; Enders and Schlafke, 1971; Steer, 1970, 1971; Denker, 1970 a and b, 1976 c). The first sign of beginning implantation is that the various layers of the blastocyst coverings become indistinct, at first always where a trophoblastic knob is located, but this reaction spreads quickly to involve also the inter-knob areas. Subsequently, the coverings are being dissolved here, and the trophoblastic knobs attach to the uterine epithelium, fuse with it and penetrate it to reach the subepithelial capillaries always using the shortest possible route (see Böving, 1962; Enders and Schlafke, 1971; Denker, 1976 c). The abembryonic-antimesometrial yolk sac placenta which is formed by this way is followed by the establishment of a chorioallantoic placenta, both of which representing a hemochorial type of contact between embryonic and maternal tissues (for review of details see Denker, 1976 c).

The chemical composition of rabbit blastocyst coverings and of the surface coat of the uterine epithelium and the occurrence of enzymes possibly involved in their breakdown

Since no biochemical analysis of rabbit blastocyst coverings has been performed so far, knowledge of their chemical composition must be based exclusively on extensive histochemical investigations published (Denker, 1970 a and b). According to these investigations, the blastocyst coverings fulfill the criteria of epithelial mucins or glycoproteins with a protein backbone and attached carbohydrate side chains, strongly negatively charged due to the presence of terminal sialic acid and sulfate ester groups. Principally the same holds true for the composition of a cell surface glycoprotein coat which is especially well developed at the surface of the uterine epithelium, but which shows alternatingly more sialic acid-rich and more sulfate ester-rich regions.

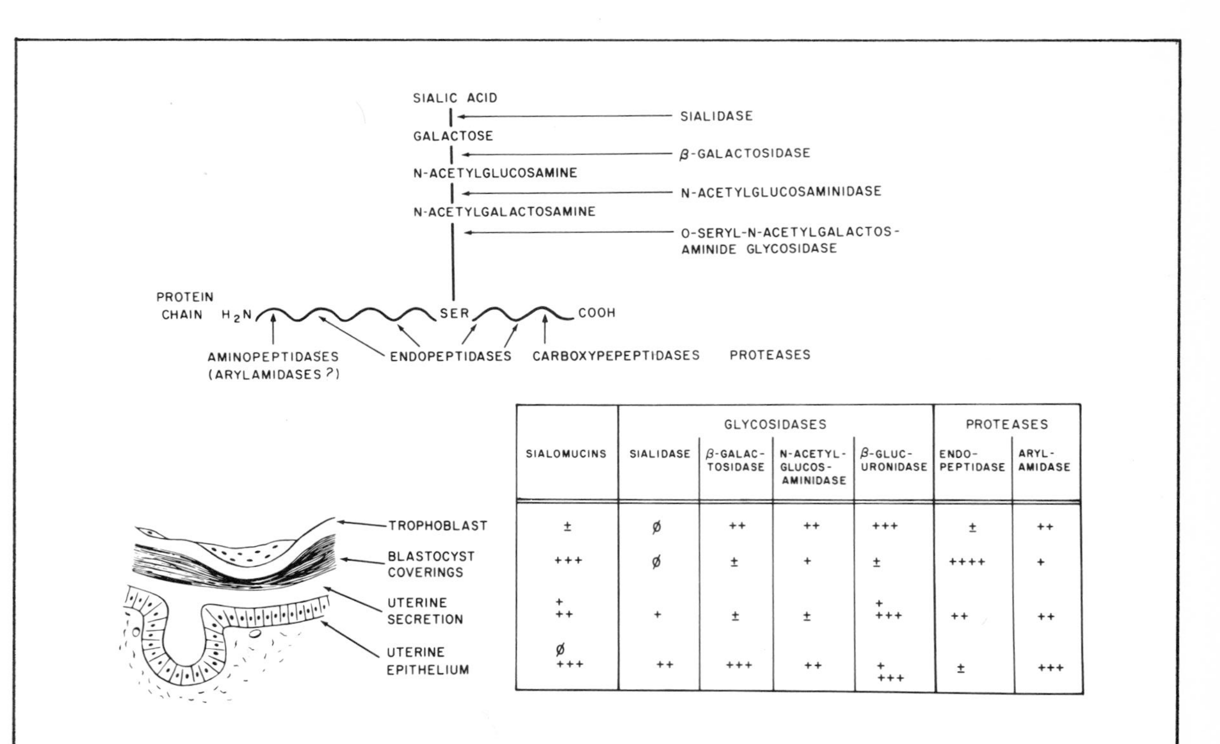

	SIALOMUCINS	GLYCOSIDASES				PROTEASES	
		SIALIDASE	β-GALACTOSIDASE	N-ACETYLGLUCOSAMINIDASE	β-GLUCURONIDASE	ENDOPEPTIDASE	ARYLAMIDASE
TROPHOBLAST	±	∅	++	++	+++	±	++
BLASTOCYST COVERINGS	+++	∅	±	+	±	++++	+
UTERINE SECRETION	+ ++	+	±	±	+ +++	++	++
UTERINE EPITHELIUM	∅ +++	++	+++	++	+ +++	±	+++

Fig. 1. The general principle of composition of an epithelial glycoprotein, and sites of attack of degrading enzymes, are shown here as a model for the composition and the dissolution of rabbit blastocyst coverings and the uterine epithelial surface coat. In the lower part of the diagram, histochemical findings on the distribution of sialic acid-rich glycoproteins and of corresponding hydrolytic enzymes as found in rabbit implantation sites are summarized.

Basing ourselves on known facts about the principal structure of epithelial mucins and cell surface glycoproteins (shown, in an arbitrarily chosen example, in Fig. 1), as well as on knowledge of general ways of degradation of such substances, we have postulated that both various glycosidases and proteases (endopeptidases and exopeptidases) are involved in the penetration of the trophoblast through the blastocyst coverings and the uterine epithelial surface coat, and we have at first accumulated information on the presence and distribution of such enzymes in these sites (Denker, 1969 a, 1970 a and b, 1973, 1976 c). Some of the results are outlined in Fig. 1. From these studies it became evident that implantation stage trophoblast and uterine epithelium show considerable activities of various glycosidases as well as exo- and endopeptidases which could possibly be involved in the penetration process.

Influences of the blastocysts on enzyme activities in the uterine secretion

The exopeptidase class of enzymes mentioned above is represented,in the rabbit blastocyst and uterus, by at least 3 individual aminopeptidases differing in properties such as substrate specificity (Denker and Stangl, 1976). One of these so-called amino acid arylamidases, arylamidase I, deserves special interest because it appears in the uterine secretion in particularly high concentration, showing a sharp and high peak of activity at 5 days post coitum (d p.c.) (Fig. 2; Denker, 1971 c; Denker, 1976 c; van Hoorn, unpublished). Interestingly, the presence of a blastocyst seems to stimulate the discharge of the enzyme from the uterine epithelium into the uterine lumen. As a comparison of pregnant and pseudopregnant uteri (Fig. 2) and of blastocyst-bearing segments and interblastocyst segments of the same uteri revealed, the extrusion appears to be locally accelerated and enhanced in the presence of blastocysts. At the same time, de novo synthesis seems to have stopped. As a result, the blastocyst site uterine epithelium appears "exhausted" after 6 2/3 d p.c. (Denker and van Hoorn, 1974; van Hoorn and Denker,1975; Denker, 1976 b).

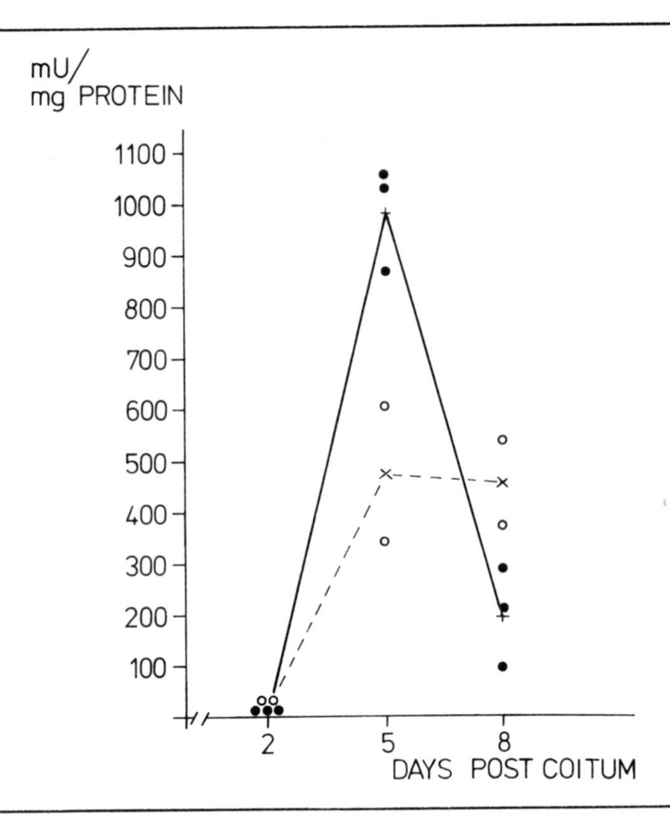

Fig. 2 Amino acid arylamidase activity in rabbit uterine flushings in the preimplantation and implantation phase of pregnancy (solid line and dots), and in pseudopregnancy (broken line and open circles). In the presence of blastocysts (regular pregnancy), there is a tremendous increase in enzyme activity in the uterine secretion showing a sharp peak at 5 d p.c. and a steep decrease thereafter, while in the absence of blastocysts (pseudopregnancy) the activity does not increase as much and the profile reminds of a plateau at a lower level. (Dots and circles represent individual values from different animals, lines interconnect mean values).

Preliminary observations seem to suggest that there is also a blastocyst-dependent increase in the activity of certain endopeptidases in the uterine secretion of the rabbit; however, these observations require confirmation by larger series of experiments (Denker and Petzoldt, 1977).

Trophoblast-dependent endopeptidase activity and its role in implantation initiation in the rabbit

Histochemical investigations using a gelatin substrate film test specifically designed for this purpose (Denker, 1969 b, 1974 a) revealed a peculiar endopeptidase activity of rabbit blastocysts which was found only at those stages at which the blastocyst coverings are being dissolved and the abembryonic trophoblast attaches onto the uterine epithelium (Denker, 1969 a, 1971 a, 1972, 1976 c). From the beginning, it was assumed that the enzyme might be involved in implantation initiation, particularly because peak activity is observed (at 7 - 7 1/2 d p.c.) in the abembryonic-antimesometrial region where the lysis of the blastocyst coverings is just under way and where attachment begins, and it disappears after completion of this process (Fig. 3). This assumption was substantiated by a number of physiological experiments (see below), so that it appeared justified to make attempts to characterize this enzyme (enzyme system) and its inhibitors in spite of the fact that the very low quantities of enzyme which can be obtained from early embryos meant that we would have to deal with particularly great difficulties.

Biochemical properties of trophoblast-dependent proteinase

Until recently, all attempts to study the biochemical properties of rabbit blastocyst proteinase were hampered by lack of a substrate suitable for quantitative photometric tests. The gelatin membranes used in the histochemical test system provide high sensitivity but allow only a rough estimation of the level of activity (Denker, 1977 a). So far most information was derived from a large series of experiments on the interaction with highly specific proteinase inhibitors (Denker, 1976 a and c) (Tab. 1). It became evident that we are dealing with a serine proteinase, the active center of which is closely related to that of trypsin. However, the enzyme differs from trypsin in several important properties including substrate specificity: its substrate specificity is much more restricted, so that commonly used trypsin substrates (synthetic substrates like benzoyl-arginine-p-nitroanilide, benzoyl-arginine-ß-naphthylamide (BANA) or benzoyl-arginine-ethylester; protein substrates like casein) are not hydrolyzed in any measurable rates (Denker, 1976 a and c). The same holds true for comparable chymotrypsin or elastase substrates. The enzyme is certainly not related to collagenase which does not belong to the trypsin family of enzymes but is a metalloenzyme inhibited by EDTA (while this chelating agent does not affect the trophoblast-dependent blastocyst proteinase).

Table 1.

Proteinase inhibitors which show strong inhibition of rabbit trophoblast-dependent blastocyst proteinase in vitro

(Histochemical substrate film test; stage tested: 7 d p.c.)

Inhibitor [1)]	Specificity [2)]
1. Soybean trypsin inhibitor (Kunitz)	Try, Chy
2. Aprotinin (Trasylol ^{R})	Try, Chy
3. Antipain	Try
4. Leupeptin	Try
5. Boar seminal plasma trypsin-acrosin inhibitor	Try
6. Bovine pancreatic secretory trypsin inhibitor	Try
7. α_1-Antitrypsin	Try, Chy
8. NPGB	Try

[1)] Listed in the order of efficacy.

[2)] Listed only in regard to trypsin or chymotrypsin inhibition

For further details and literature see Denker (1976 a and c)

As experiments with the fibrin plate method have revealed, fibrin is a poor substrate for the rabbit trophoblast-dependent proteinase, and the enzyme does also not act as a plasminogen activator (Denker, 1976 c and unpublished data). It therefore does not seem to be related to the plasminogen activator - plasmin system currently under intensive investigation in several laboratories (Rifkin and Pollack, 1976; for further details and discussion see Denker, 1976 c).

By electrophoretic experiments involving agar gel and micro disc gel electrophoresis it seems to be possible to separate blastocyst and uterine secretion proteinases (Kirchner, 1972; Denker and Petzoldt, 1977; Denker, 1976 c). In pregnant rabbit uterine secretion, a proteinase is found which differs from the blastocyst enzyme by possessing considerable BANA-splitting activity (thereby resembling trypsin more closely). Its mobility in micro disc electrophoresis is only slightly different (Denker and Petzoldt, 1977). For both the blastocyst proteinase and the trypsin-like enzyme from uterine secretion, biochemical work on isolation, further characterization and investigation of substrate specificity as well as interaction with inhibitors present at these sites is under way.

The dependence of implantation-associated proteinase activity and of dissolution of blastocyst coverings on the trophoblast or on uterine secretion

The respective role of trophoblast and endometrium in initiation of implantation has been much disputed. By some authors, proteinase activity has been regarded as solely of uterine origin (Mintz, 1971; Kirchner, 1972; Kirchner et al., 1971; Pinsker et al., 1974). Available evidence from investigations in the rabbit, however, points to a central role of the trophoblast in development of implantation-associated proteinase activity, in the dissolution of the blastocyst coverings and in initiation of implantation in this species. This does not negate that the endometrium probably provides important factors, too. Certainly, both tissues act in a concerted manner in this process.

In order to disclose the functional relationship of both partners, we have attempted to eliminate the trophoblast from the system. Models for blastocyst coverings without trophoblast were placed into the uterine cavity of pregnant or pseudopregnant rabbits (Denker and Hafez, 1975; Denker, 1976 c). These models did not develop proteinase activity that was in any way comparable to that seen in implanting blastocysts, and these coverings did not show signs of lysis.

Additional evidence for a dependence on the trophoblast of both the proteinase activity and the dissolution of the blastocyst coverings was derived from observations on blastocysts which, incidentally, started implanting in an abnormal orientation, namely with the abembryonic trophoblast facing the mesometrial instead of the anti-mesometrial endometrium. Since the dissolution of the rabbit blasto-cyst coverings does not start simultaneously at the whole circum-ference of the blastocyst but normally in the abembryonic-anti-mesometrial region where also the main proteinase activity is found, the upside-down blastocysts should enable us to reveal whether this process depends mainly on the trophoblast or on the antimesometrial endometrium. In fact, the main proteinase activity as well as the onset of dissolution of the blastocyst coverings were observed, in these inversely orientated blastocysts, always at the abembryonic pole no matter that it faced the mesometrial instead of the antimesometrial endometrium, by this way giving evidence for a dependence on the abembryonic trophoblast. In addition to the 3 cases reported initial-ly (Denker, 1974 b), we found 11 more of the dystopically implanting blastocysts which all showed consistently the same features (Denker, 1976 c). However, proteinase activity was sometimes remarkably high also in the cleft between the antimesometrial endometrium and the surface of the blastocyst (in this case the embryonic disc region). This might indicate that the antimesometrial endometrium does contribute essential factors (e.g. activators) in a higher concen-tration as the mesometrial endometrium does.

The possibility should be kept in mind that the known difference in pH values which exists between both parts of the endometrial surface (Zimmermann, 1961; Petzoldt, 1971) might play a role here. It seems attractive to imagine that, in normal implantation, factors or conditions provided by the antimesometrial endometrium (like more alkaline pH) might increase the activity of factors of the abembryonic trophoblast (like proteinase activity) (or v.v.) so that, when during rotation of the blastocyst the abembryonic pole comes incidentally to lie against the antimesometrial endometrium, dissolution of the blastocyst coverings starts here, the coverings become stickier at this pole, and the blastocyst becomes trapped in the correct orientation.

Inhibition of implantation through application of proteinase inhibitors *in vivo*

In order to prove the validity of the hypothesis that the described blastocyst proteinase plays a key role in dissolution of the blastocyst coverings and in initiation of implantation, we undertook a large series of experiments on the effects of an *in vivo* application of proteinase inhibitors on implantation. Basing ourselves on our list of inhibitors which inhibit the trophoblast-dependent blastocyst proteinase effectively *in vitro* (Denker, 1976 a and c; see also Tab.1), we selected for the majority of experiments the basic trypsin-kallikrein inhibitor from bovine organs (Aprotinin, Trasylol ^{R}) and antipain, both showing only very little toxic side effects. A more limited number of experiments were performed with p-nitrophenyl-p'-guanidinobenzoate (NPGB) and boar seminal plasma trypsin-acrosin inhibitor which both likewise inhibit the enzyme strongly *in vitro*. For comparison, ε-aminocaproic acid (EACA), an inhibitor of plasminogen activation which does not inhibit the rabbit blastocyst proteinase *in vitro*, was also applied. All these inhibitors were administered into the uterine lumen of pregnant rabbits at 6 d 12 hours p.c.; one uterine horn always received only an injection of vehicle fluid to serve as a control. Light and electron microscopical morphology and proteinase histochemistry were studied in the following stages of pregnancy until 11 1/2 d p.c.; for details see Denker (1976 c).

These experiments have impressively demonstrated that intrauterine application of proteinase inhibitors interferes strongly with

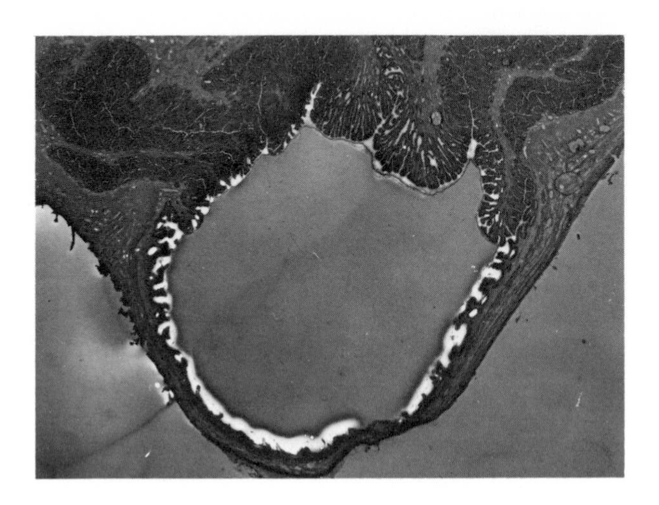

Fig. 3. Histochemical gelatin substrate film test showing trophoblast-dependent proteinase activity at an implantation site of the rabbit, 7 1/2 d p.c. Longitudinal section. Maximal activity (bright lysis zones) is found in the abembryonic-antimesometrial region. This picture represents the findings which are typical for regular implantation sites, although EACA had been injected into this uterus. x 8.6.

implantation in the rabbit. Of the inhibitors used, the most effective one proved to be Trasylol [R]. After intraluminal application of a dose of Trasylol [R] of 4 - 12 mg at 6 1/2 d p.c., there was no proteinase activity detectable by the substrate film test at 7 1/2 d p.c. (Fig. 3, 4), and only occasionally traces could be seen at or after 8 1/2 d p.c. Dissolution of the blastocyst coverings did not get started, so that at 7 1/2 d p.c. the blastocysts were still free in the uterine lumen and most of them were still completely surrounded by the blastocyst coverings which showed no signs of erosion but their several layers were still clearly visible as typical for preimplantation stages (Fig. 5, 6). Expansion of the blastocysts continued, although at a lower rate than in the control horns, and obviously as a result of this expansion in the absence of lysis the coverings were disrupted at several places in an increasing proportion of the blastocysts. By this way parts of the trophoblast

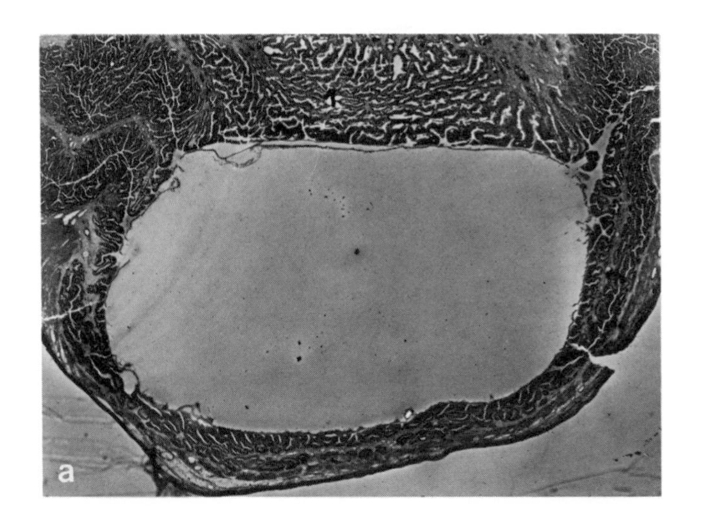

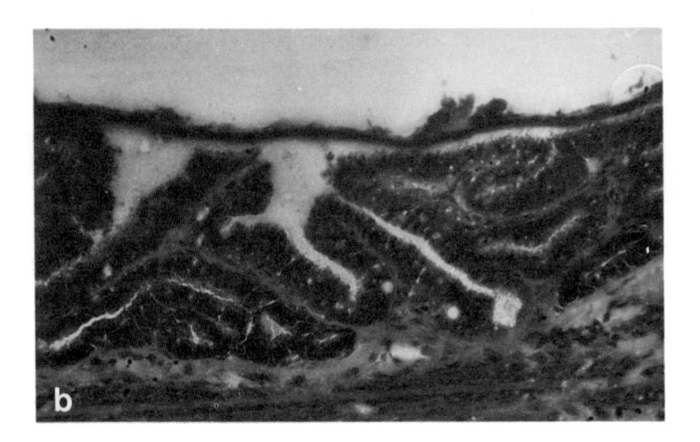

Fig. 4. Blockage of proteinase activity and of implantation in the rabbit through application of 0.6 mg of Trasylol[R] at 6 1/2 d p.c. Gelatin substrate film test as in Fig. 3; 7 1/2 d p.c.

(a) Proteinase activity is largely inhibited except that, at this very low dose, few trophoblastic knobs can show some activity. Compare with Fig. 3. x 9.3. (b) Higher magnification of the abembryonic-antimesometrial region. Non-dissolved blastocyst coverings appear as a dark band. No proteinase activity. x 170.

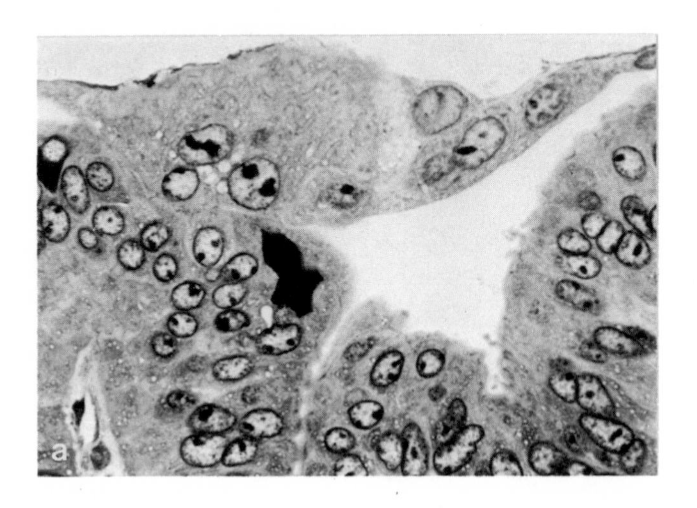

Fig. 5. (a) Light microscopical morphology of rabbit trophoblastic knob attachment to the uterine epithelium as seen in the controls at 7 1/2 d p.c.; as compared to (b) a knob which, after intrauterine application of 6 mg of Trasylol R, was not able to attach due to blockage of dissolution of the blastocyst coverings (dark band, below). Semithin sections, 7 1/2 d p.c., x 670.

could come into contact with the uterine epithelium while the remainders of the coverings were still interposed between both partners at other places (Fig. 7). From 8 1/2 d p.c. on, many blastocysts showed, therefore, a focal attachment of a few of their trophoblastic knobs at the endometrial surface (Fig. 8). Fusion did occur in these limited areas, but the trophoblast usually did not reach the

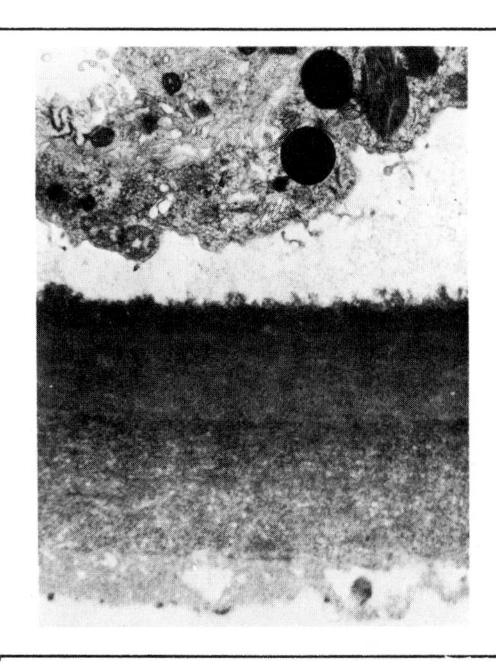

Fig. 6. Electron micrograph showing the undissolved blastocyst coverings and a part of the trophoblast of a rabbit blastocyst after intrauterine application of 6 mg of Trasylol R. The different layers of the blastocyst coverings can still easily be recognized. 7 1/2 d p.c., x 4300.

subepithelial capillary after this experimentally induced retardation of implantation (for a detailed discussion see Denker, 1976 c). There is a strong indication that the limited contact achieved remained functionally insufficient: the influx of large quantities of maternal plasma proteins into the blastocyst cavity (future yolk sac) which is typical for regular implantation and early post-implantation stages did not occur, but rather the blastocyst cavity showed only traces of protein after proteinase inhibitor treatment, contrasting sharply with a very high protein content of the surrounding uterine lumen which was often enlarged by copious secretion.

The fine structural analysis gave no indication for a toxic damage of either the trophoblast or the surrounding maternal tissues (Fig.13). However, the embryonic disc showed early degeneration (Fig. 9). The remaining trophoblastic vesicles (with an entoderm layer inside),

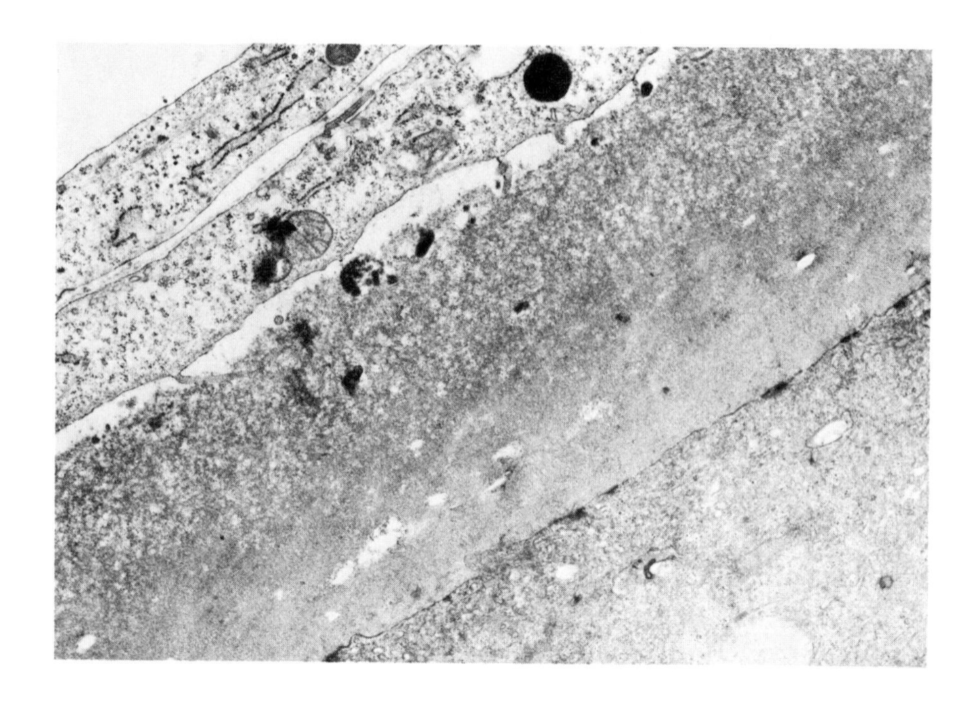

Fig. 7. Electron micrograph showing inhibition of rabbit trophoblast attachment after Trasylol R - application as in Fig. 6, but investigated one day later. Stretched and thinned-out remnants of the blastocyst coverings are still interposed between the trophoblast (above) and the symplasmatically transformed uterine cavum epithelium (lower right corner). Interestingly, multiple hemidesmosome-like junctions(dark structures) have formed between the apical surface of the uterine epithelial symplasm and the blastocyst coverings. Entoderm is seen in the upper left corner. 8 1/2 d p.c., x 14300.

although slowly continuing expansion for a few days, finally also degenerated, and most of them were resorbed until 11 1/2 d p.c.

Inhibition of implantation was less efficient in the case of antipain application. The dissolution of the blastocyst coverings was not completely inhibited but only slowed down and was confined to the neighborhood of trophoblastic knobs (Fig. 10). Still, the attachment

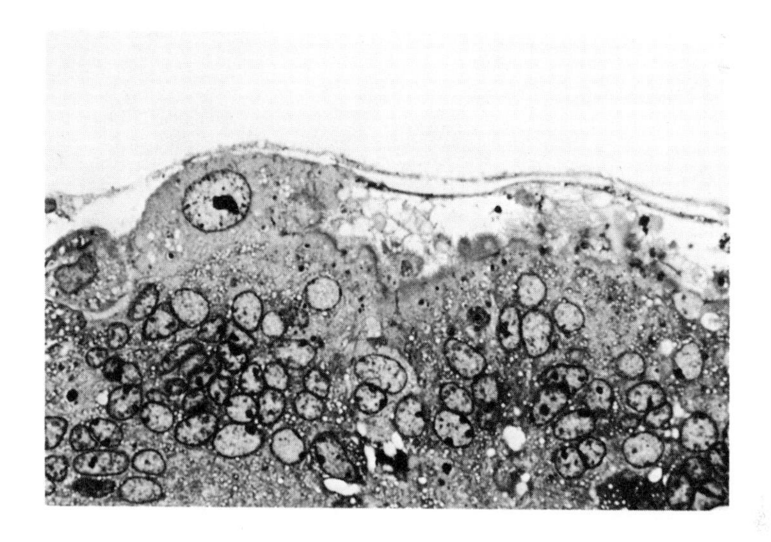

Fig. 8. When, after Trasylol R - application as in Figs. 5-7, the blastocyst coverings have ruptured due to the continuing expansion of the blastocyst, exposed parts of the trophoblast can manage a focal attachment. A trophoblastic giant cell is shown which has superficially attached to and fused with the uterine epithelial symplasm. The very thin line on top is the entoderm. Semithin section, 8 1/2 d p.c., x 660.

was delayed and locally restricted, transport of maternal plasma proteins into the blastocyst cavity was decreased, the embryonic discs degenerated and most conceptuses were resorbed until 11 1/2 d p.c. (Fig. 12).

It should be noted that in our experiments only a single dose of the respective inhibitor was injected. Determination of the inhibitor concentration in the uterine lumen at varying intervals after application showed a rapid decline within 24 hours, the decline being faster in case of antipain than in case of Trasylol R (for details see Denker, 1976 c). This might in part account for the differences in efficacy seen between both inhibitors. On the other hand, Trasylol R which is strongly basic might have accumulated in the blastocyst coverings (which are composed of acid mucosubstances, see above) the proteolytic dissolution of which could then be expected to be particularly heavily affected.

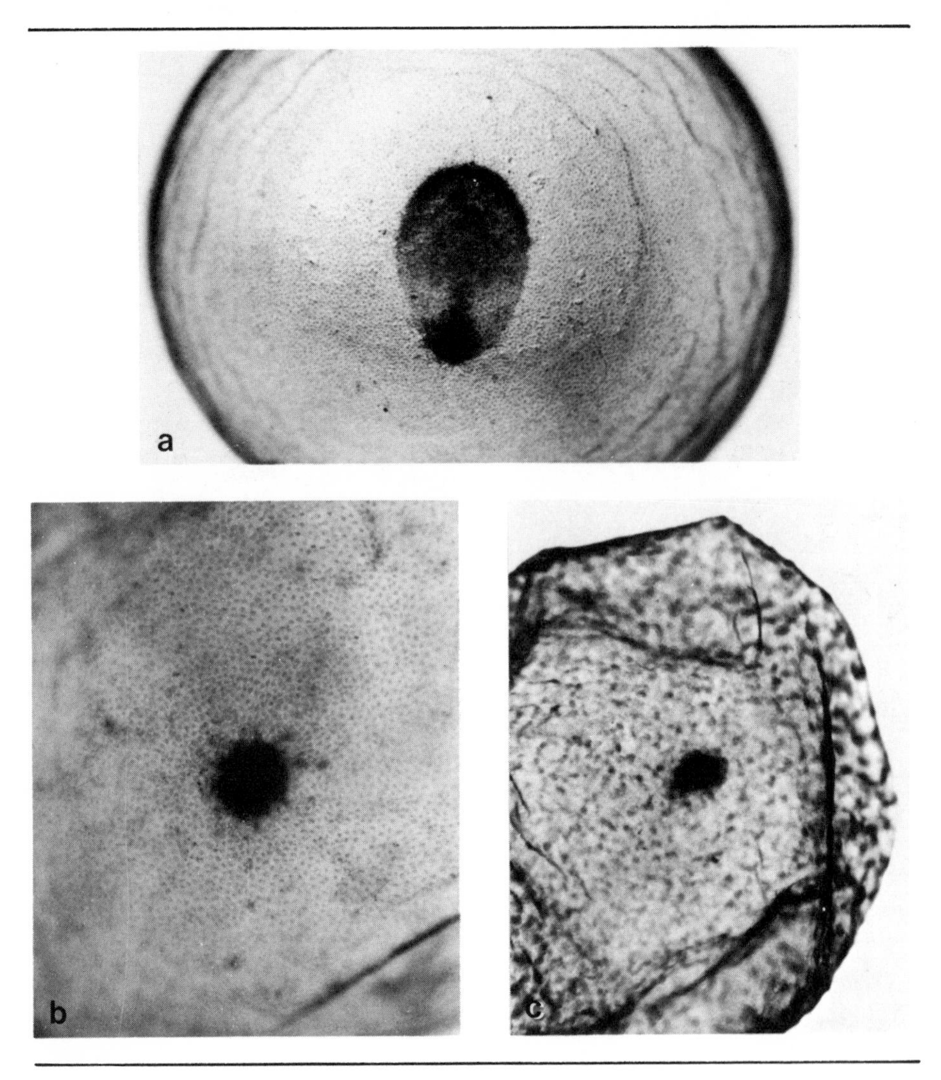

Fig. 9. The embryonic disc shows early degeneration after intra-uterine application of proteinase inhibitors. The photographs are taken from whole blastocysts flushed out from the uteri and stained with toluidine blue.
(a) Control uterus, 6 1/2 d p.c. The embryonic disc of this blastocyst, which is particularly advanced, shows a clear partition into an anterior field (above) and a posterior field with the primitive streak (below). x 20. (b) Degenerated embryonic disc of a blastocyst flushed out from a uterus at 7 1/2 d p.c., after application of 12 mg of Trasylol^{R}. In regular pregnancy, blastocysts are attached and cannot be flushed out at this stage. x 34. (c) Partially collapsed blastocyst flushed out from a uterus at 8 1/2 d p.c., after application of 12 mg of antipain. The embryonic disc is degenerated. The surrounding trophoblast shows numerous dark structures which appear to be non-attached trophoblastic knobs. x 17.

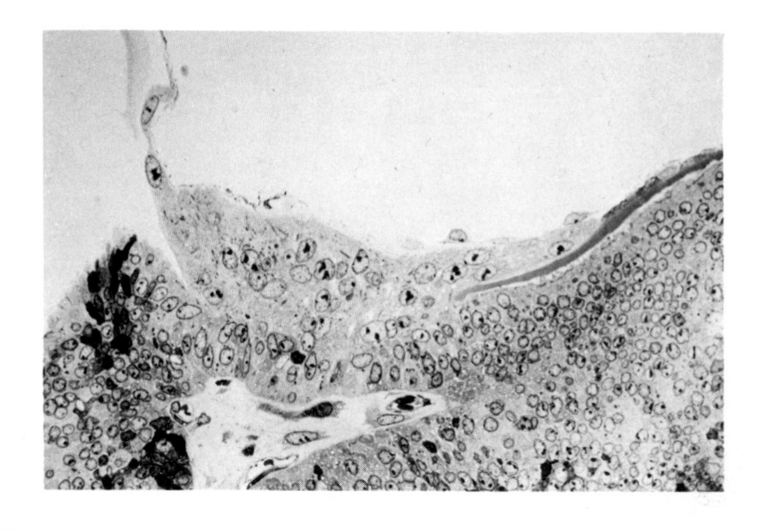

Fig. 10. After application of 6 mg of antipain, the trophoblastic knob shown here has still managed to induce local lysis of the blastocyst coverings and to attach to the uterine epithelium. Laterally of it, the coverings are not dissolved but remain interposed between trophoblast and endometrium (dark band). Semithin section, 7 1/2 d p.c., x 260.

One interesting detail seen after antipain treatment was that the trophoblastic knobs which were not attached enlarged and developed high proteinase activity. This activity unequivocally originated from the cytoplasm of these knobs (Fig. 11), whereas in regular pregnancy most activity is found in the disintegrating blastocyst coverings while the trophoblastic knobs show only traces, an observation which had previously caused some controversy about the origin of the enzyme from the trophoblast or the endometrium (Denker, 1969 a, 1971 c, 1974 b, 1976 c; Denker and Hafez, 1975; Kirchner, 1972; Kirchner et al., 1971). The possibility that we are dealing with a compensatory over-production of proteinase by the trophoblastic knobs after antipain treatment must be investigated further. The phenomenon is not antipain-specific: an activity of trophoblastic knobs is also seen after application of very low doses (0.6 mg/uterus) of Trasylol [R].

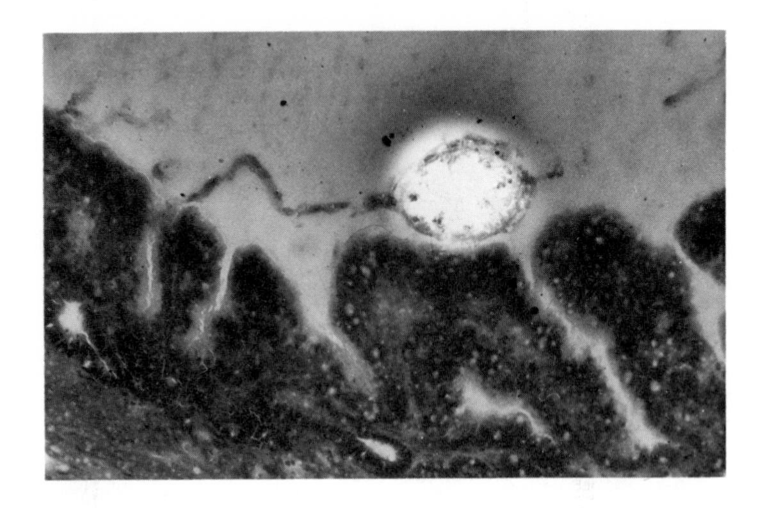

Fig. 11. High proteinase activity of a non-attached trophoblastic knob seen after intrauterine injection of 6 mg of antipain. Uterine secretion and endometrium show no reaction. Gelatin substrate film test, 8 1/2 d p.c., x 100.

Application of EACA which does not inhibit the trophoblast-dependent blastocyst proteinase was found not to interfere with implantation.

We conclude from these experiments that trypsin-like proteinase activity of the described type plays in fact a key role in initiation of implantation in the rabbit. A similar conclusion was drawn from experiments on the intrauterine application of proteinase inhibitors in mice (Dabich and Andary, 1974), although in that study it was not determined whether it was the dissolution of the zona pellucida and trophoblast attachment which were affected, but only implantation sites were counted at later stages.

On the basis of these experiments it is not possible, however, to judge about the respective role of enzymes of the trophoblast and those of endometrial origin, because both tissues possess trypsin-like proteinases which are sensitive to the used inhibitors (see above). In the mouse,a key role in initiation of implantation has been attributed to a uterine secretion proteinase (so-called "implantation initiating factor", Mintz, 1971; Pinsker et al., 1974). In the rabbit,

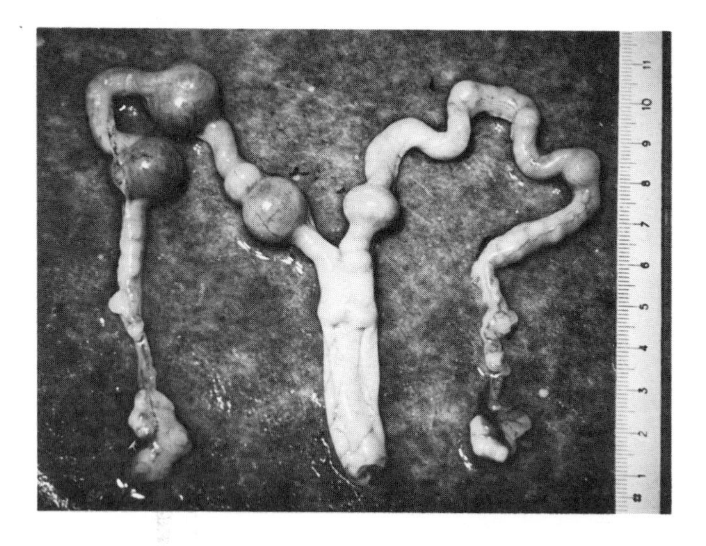

Fig. 12. Genital tract of a female rabbit at 11 1/2 d p.c., after injection of 12 mg of antipain into the lumen of the right uterus at 6 1/2 d p.c. The left uterus received an injection of saline solution to serve as a control. At the inhibitor side, 5 resorption sites were counted; only the 3 larger ones which still contained remnants of blastocysts are visible on the photograph. The control side shows 3 normal implantation sites and one resorption site.

however, there is strong evidence that the described proteinase activity of implanting blastocysts is trophoblast-dependent (see above). High activity of the enzyme can be identified in trophoblast homogenates after electrophoresis (Denker and Petzoldt, 1977; Denker, 1976 c). On the basis of the experiments on application of proteinase inhibitors in vivo we conclude that this proteinase has an important function in dissolution of the blastocyst coverings. It is not clear yet whether it is also involved in attachment of the trophoblast onto the uterine epithelium and in invasion. It should be investigated whether the delayed attachment of some trophoblastic knobs observed in the Trasylol [R] experiments after rupture of the blastocyst coverings could be also prevented when a sufficient concentration of inhibitor would be kept in the uterine lumen over a longer time.

We are also considering the possibility that both trophoblast-dependent and uterine secretion proteinases act in a concerted manner

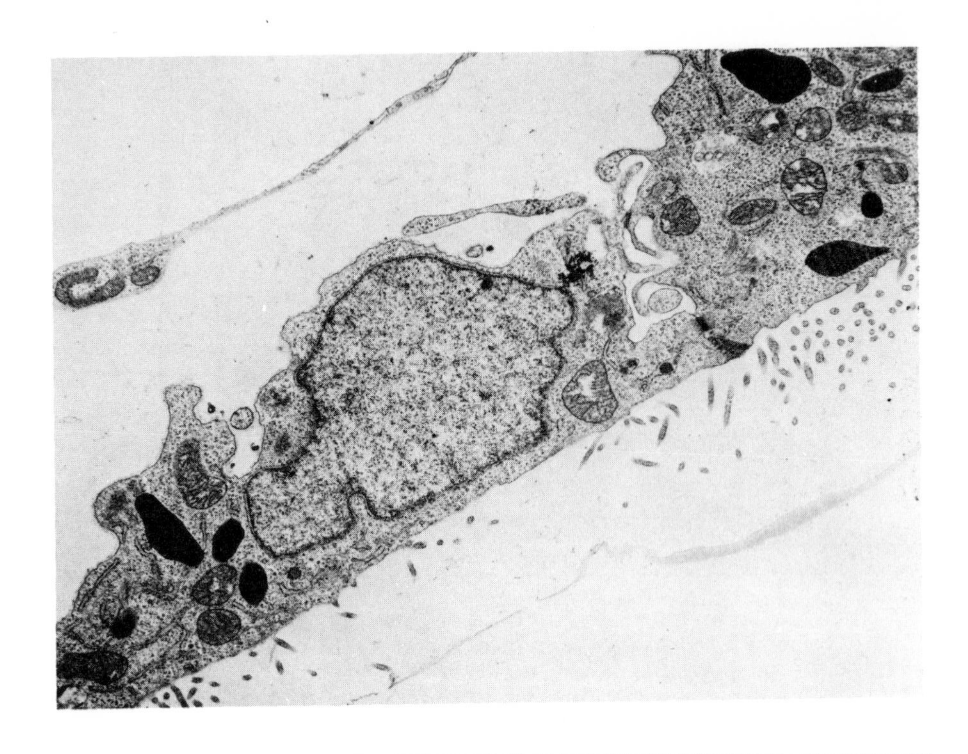

Fig. 13. Electron micrograph of the wall of a rabbit blastocyst after intrauterine administration of 6 mg of antipain. There is an extremely thin remnant of blastocyst coverings (below) retained at the surface of the trophoblast. The microvilli do not reach into it as they do in regular implantation stages. Trophoblast and entoderm (upper left corner) appear rather healthy. x 5700.

in initiation of implantation, together with certain glycosidases as postulated above. This hypothesis seems very attractive and its validity will be investigated further.

The physiological regulation of implantation-associated proteinase activity and of implantation

Much experimental work has been done by many authors on the physiological regulation of implantation via maternal hormones (for review, see Psychoyos, 1973). The question whether or not embryo-derived hormones may also play a role in the attachment phase of implantation is still a matter of speculation because too little hard facts have been accumulated on this subject so far.

It is generally assumed and in many details also substantiated that sex steroid hormones regulate reproduction via an alteration of certain biosynthetic or secretory activities (see Beier, this volume). The development of implantation-associated blastocyst proteinase activity as well as dissolution of the blastocyst coverings and onset of implantation have been shown to be dependent on maternal progesterone in the rabbit (Denker, 1972). There is no indication for a functionally significant estrogen surge in this species. The hormonal regulation of the proteinase system could be achieved, at the molecular level, via either an alteration of the biosynthesis or activation of the enzyme(s), or changes in the activity of proteinase inhibitors present in the uterus and the blastocyst. In fact, there is evidence for the presence of considerable quantities of various proteinase inhibitors in these tissues. Rabbit trophoblast shows little gelatinolytic activity when cryostat sections or homogenates are assayed (on the exceptionally high activity of trophoblastic knobs after experimental application of low doses of inhibitors see above), but high activity becomes demonstrable after electrophoresis (Denker and Petzoldt, 1977; Denker, 1976 c). Recently, uteroglobin, the dominant protein of rabbit uterine secretion at these stages, was found to have trypsin inhibiting activity (see Beier, this volume).

On the basis of the shown possibility of blocking implantation through the application of proteinase inhibitors it seems fascinating to undertake a detailed study of interactions between trophoblast-dependent and uterine proteinases, activation of proenzymes possibly present, and regulation by natural proteinase inhibitors present at these sites.

Inhibition of implantation by administration of proteinase inhibitors seems to be applicable as a new concept for contraception, as shown in the described animal experiments. However, the fact that an increased incidence of dystopic implantation was observed (Denker,

1976 c) shows that specific interference with embryo implantation may cause peculiar types of side-effects which will have to be investigated further in detail.

Acknowledgments

The author wishes to express his gratitude to Miss Gerda Bohr and Miss Edith Höricht for excellent technical assistance. Sincere thanks are due to Prof. Dr. H. Fritz (München, Germany) for valuable suggestions and for samples of inhibitors. Trasylol R and antipain were kindly provided by Dr. E. Truscheit and Dr. W. Wingender, Bayer AG (Wuppertal-Elberfeld, Germany). These investigations were supported by Deutsche Forschungsgemeinschaft grants No. De 181/1-4 and 7.

References

Böving, B.G. (1962). Anatomical analysis of rabbit trophoblast invasion. Carnegie Inst. Washington, Publ. No. 621, Contrib. Embryol. 37, 33-55.

Böving, B.G. (1963). Implantation mechanisms. In: Conference on Physiological Mechanisms Concerned with Conception, pp. 321-396. Ed. C.G. Hartman. Pergamon Press, Oxford, London, New York and Paris.

Böving, B.G. and Larsen, J.F. (1973). Implantation. In: Human Reproduction: Conception and Contraception, pp. 133-156. Eds. E.S.E. Hafez and T.N. Evans, Harper & Row, Hagerstown (Maryland).

Dabich, D. and Andary, T.J. (1974). Prevention of blastocyst implantation in mice with proteinase inhibitors. Fertil. Steril. 25, 954-957.

Dabich, D. and Andary, T.J. (1976). Tryptic- and chymotryptic-like proteinases in early and late preimplantation mouse blastocysts. Biochim.Biophys. Acta 444, 147-153.

Denker, H.-W. (1969 a). Zur Enzym-Topochemie von Frühentwicklung und Implantation des Kaninchens. Thesis, Med. Fak. University of Marburg (Germany). (See also Denker, 1971 a-c).

Denker, H.-W. (1969 b). Substratfilmtest für den Proteasennachweis. XIII. Sympos. Gesellsch. f. Histochemie, Graz 1969. (Acta histochem. (Jena) Suppl. X, 303-305 (1971)).

Denker, H.-W. (1970 a). Topochemie hochmolekularer Kohlenhydratsubstanzen in Frühentwicklung und Implantation des Kaninchens. I. Allgemeine Lokalisierung und Charakterisierung hochmolekularer Kohlenhydratsubstanzen in frühen Embryonalstadien. Zool. Jahrb., Abt. allgem. Zool. Physiol., 75, 141-245.

Denker, H.-W. (1970 b). Topochemie hochmolekularer Kohlenhydratsubstanzen in Frühentwicklung und Implantation des Kaninchens. II. Beiträge zu entwicklungsphysiologischen Fragestellungen. Zool. Jahrb., Abt. allgem. Zool. Physiol., 75, 246-308.

Denker, H.-W. (1971 a) Enzym-Topochemie von Frühentwicklung und Implantation des Kaninchens. I. Glykogenstoffwechsel. Histochemie 25, 256-267.

Denker, H.-W. (1971 b). Enzym-Topochemie von Frühentwicklung und Implantation des Kaninchens. II. Glykosidasen. Histochemie 25, 268-285.

Denker, H.-W. (1971 c). Enzym-Topochemie von Frühentwicklung und Implantation des Kaninchens. III. Proteasen. Histochemie 25, 344-360.

Denker, H.-W. (1972). Blastocyst protease and implantation: effect of ovariectomy and progesterone substitution in the rabbit. Acta endocr. (Kbh.) 70, 591-602.

Denker, H.-W. (1973). Blastocyst enzymes and implantation in the rabbit. Abstract, VIth Ann. Meet. Soc. Study of Reprod., Athens (Georgia), 1973. Biol. Reprod. 9, 102-103.

Denker, H.-W. (1974 a). Protease substrate film test. Histochemistry 38, 331-338, and 39, 193.

Denker, H.-W. (1974 b). Trophoblastic factors involved in lysis of the blastocyst coverings and in implantation in the rabbit: observations on inversely orientated blastocysts. J. Embryol. exp. Morphol. 32, 739-748.

Denker, H.-W. (1976 a). Interaction of proteinase inhibitors with blastocyst proteinases involved in implantation. In: Protides of the Biological Fluids. Proceedings of the XXIIIrd Colloquium, Brugge 1975, pp. 63-68. Ed. H. Peeters. Pergamon Press, Oxford etc.

Denker, H.-W. (1976 b). Wechselbeziehungen zwischen Blastozyste und Endometrium bei der Implantation: Beeinflussung der endometrialen Aminosäure-Arylamidase-Aktivität durch die Blastozyste. Verh. Anat. Ges. 70; Anat. Anz. Suppl. 140, 839-847.

Denker, H.-W. (1976 c). Mechanismen der Implantation des Säugetier-embryos und ihre experimentelle Beeinflussung. Habilitationsschrift, Med. Fak. RWTH Aachen (Germany). (English translation: Implantation: the role of proteinases, and blockage of implantation through proteinase inhibitors. Adv. Anat. Embryol. Cell Biol., in press).

Denker, H.-W. (1977 a). Zur Spezifität und Empfindlichkeit des Proteasen-Substratfilmtests auf Gelatinebasis. XVIII. Sympos. Ges. f. Histochemie, Bozen 1975. Acta histochem. (Jena) Suppl. XVIII, 153-158.

Denker, H.-W. (1977 b). Role of trophoblast-dependent and of uterine proteases in initiation of implantation. In: Human Fertilization. Proceedings of an International Workshop, Essen 1976. Eds.H. Ludwig and P.F. Tauber. Thieme-Verlag, Stuttgart (in press).

Denker, H.-W. and Hafez, E.S.E. (1975). Proteases and implantation in the rabbit: role of trophoblast vs. uterine secretion. Cytobiologie 11, 101-109.

Denker, H.-W. and van Hoorn, G. (1974). Peptidases related to implantation in the rabbit: local stimulation of arylamidase secretion by the blastocyst immediately preceding implantation. Abstract, VIIth Ann. Meet. Soc. Study of Reproduct., Ottawa(Canada).

Denker, H.-W. and Petzoldt, U. (1977). Proteinases involved in implantation initiation in the rabbit: micro disc electrophoretic studies. Cytobiologie (in press).

Denker, H.-W. and Stangl, R. (1976). Versuche zur Lokalisierung und Abgrenzung verschiedener Aminosäure-Arylamidasen in Uterus und Blastozyste des Kaninchens. XVII. Sympos. Ges. f. Histochemie, Bozen 1974. Acta histochem. (Jena) Suppl. XVI, 249-257.

Enders, A.C. (1976). Anatomical aspects of implantation. J. Reprod. Fert. Suppl. 25, 1-15.

Enders, A.C. and Schlafke, S. (1971). Penetration of the uterine epithelium during implantation in the rabbit. Am. J. Anat. 132, 219-240.

Finn, C.A. (1974). The induction of implantation in mice by actinomycin D.J. Endocr. 60, 199-200.

Hoorn, G. van, and H.-W. Denker (1975). Effect of the blastocyst on a uterine anino acid arylamidase in the rabbit. J. Reprod. Fert. 45, 359-362.

Kirby, D.R.S. (1970). The extra-uterine mouse egg as an experimental model. In: Schering Symposium on Mechanisms Involved in Conception, Berlin 1969. Adv. Biosciences 4, pp. 255-273. Ed. G. Raspé. Pergamon Press / Vieweg, Oxford etc.

Kirby, D.R.S., Potts, D.M. and Wilson, I.B. (1967). On the orientation of the implanting blastocyst. J. Embryol. exp. Morph. 17, 527-532.

Kirchner, C. (1972). Uterine protease activities and lysis of the blastocyst covering in the rabbit. J. Embryol. exp. Morphol. 28, 177-183.

Kirchner, C., Hirschhäuser, C. and Kionke, M. (1971). Protease activity in rabbit uterine secretion 24 hours before implantation. J. Reprod. Fert. 27, 259-260.

McLaren, A. (1973). Blastocyst activation. In: The Regulation of Mammalian Reproduction, pp. 321-334. Eds. S.J. Segal et al. Charles C. Thomas, Publ., Springfield (Illinois).

Mintz, B. (1971). Control of embryo implantation and survival.In: Schering Sympos. on Intrinsic and Extrinsic Factors in Early Mammalian Development, Venice 1970. Adv. Biosciences 6, pp.317-342. Ed. G. Raspé. Pergamon Press / Vieweg, Oxford etc.

Petzoldt, U. (1971). Untersuchung über das anorganische Milieu in Uterus und Blastozyste des Kaninchens. Zool. Jahrb., Abt. allg. Zool. Physiol. 75, 547-593.

Pinsker, M.C., Sacco, A.G. and Mintz, B. (1974). Implantation - associated proteinase in mouse uterine fluid. Develop. Biol. 38, 285-290.

Psychoyos, A. (1973). Endocrine control of egg implantation. In: Handbook of Physiology, Sect. 7 (Endocrinology) Vol. II (Female Reproductive System) Part 2, pp. 187-215. Ed. R.O. Greep. American Physiological Society, Washington, D.C.

Rifkin, D.B. and Pollack, R. (1976). Proteases produced by normal and malignant cells in culture. In: Proteolysis and Physiological Regulation. Miami Winter Sympos. Vol. 11, pp. 263-285. Eds. D.W. Ribbons and K. Brew. Academic Press, New York - San Francisco - London.

Schlafke, S. and Enders, A.C. (1975). Cellular basis of interaction between trophoblast and uterus at implantation. Biol. Reprod. 12, 41-65.

Steer, H.W. (1970). The trophoblastic knobs of the preimplanted rabbit blastocyst: a light and electron microscopic study. J. Anat. (Lond.) 107, 315-325.

Steer, H.W. (1971). Implantation of the rabbit blastocyst: the adhesive phase of implantation. J. Anat. (Lond.) 109, 215-227.

Zimmermann, W. (1961). Untersuchungen am Hauskaninchen. III. Die Wasserstoffionenkonzentration im weiblichen Genitaltrakt und in der Keimblase vor, während und nach der Nidation. Verh. Deutsch. Zool. Ges., Bonn 1960. Zool. Anz. Suppl. 24, 143-149.

DISCUSSION

Enders: The trophoblast at day 7 is highly pinocytotic and has a fairly good lysosomal system to handle these phagosomes. In your system the blastocyst is placed in a lytic environment, namely on the gelatin. Is it possible that one of the things you are observing is the internal lysosomal system rather than an external system involving an endopeptidase?

Denker: In principle it would be possible, but there is no indication that the proteinase is of lysosomal origin. We were not able to match the known biochemical properties of our enzyme with any known properties of lysosomal proteinases. Experiments on this line are in progress. So far it appears that our proteinase has a slightly alkaline pH optimum, and also the substrate specificity is different. Several lysosomal proteinase substrates are not affected by our proteinase. According to inhibitor experiments it is neither a SH-proteinase nor a metallo-enzyme but a member of the trypsin family of enzymes.

Feigelson: If you don't think this proteinase is emanating from the lysosome, do you then believe that perhaps it is activated in some manner or there is de novo protein synthesis? For either of these possibilities, do you envision a hormonal control?

Denker: First, I think it is a very good point that activation of a proenzyme might be involved. We are trying to find that out but so far I can't give you an answer. There could be several possibilities; the proenzyme could be provided by the endometrium or by the trophoblast and then activated. One proteinase which I have mentioned is trophoblast-dependent and the other one which I mentioned is mainly endometrium-dependent; therefore, at least two proteinases could interact, activating one another, but possibly there will be more proteinases involved. As far as hormonal control is concerned, a series of ovariectomy and hormone replacement therapy experiments has been performed (Denker, Acta Endocrin. 70:591, 1972). You can prevent implantation by ovariectomizing rabbits shortly before implantation at six days post coitum and, in that case, the normal increase in proteinase activity does not occur. Replacement therapy with progesterone will normalize the situation completely, i.e. blastocyst proteinase activity is restored, and the dissolution of blastocyst coverings and attachment proceed normally. So, proteinase activity is dependent on maternal progesterone, probably indirectly, but I have not done any adrenalectomy studies and I cannot account for any estrogens which might be coming from other sources. Anyway, there is no indication for a physiologically important estrogen surge in the rabbit.

Strickland: I would like to make a couple of comments and ask a question. We don't think that the plasminogen activator produced by mouse trophoblast has any function in hatching of the zona pellucida. In fact, the initiation of protease synthesis by the mouse trophoblast is really too late to account for the shedding of the zona pellucida. The second point in terms of inhibitor studies is that the rabbit enzyme seems to be different from the mouse enzyme, because the plasminogen activator that we get from the mouse trophoblast is not inhibited by macromolecular inhibitors of proteases. For example, soybean trypsin inhibitor doesn't inhibit it, whereas the low molecular weight inhibitors do. What is inhibited in our system by the macromolecular protease inhibitors is plasmin, the derivative molecule. If the shedding of the blastocyst covering in the rabbit is due solely to the trophoblast, then the rabbit is different from the mouse which requires a combination of trophoblastic factors and uterine factors. One experiment that would settle that question directly would be the _in vitro_ culture of a blastocyst that was covered. If you flush a blastocyst at six days and put it _in vitro_, does it shed the covering?

Denker: There is certainly a concerted action of trophoblast- and endometrium-dependent enzymes and they are certainly not only proteinases, but also other enzymes, for example glycosidases. This concerted action might involve a proenzyme activation or sequential breakdown of different parts of glycoproteins. Also, we have to envisage a role for proteinase inhibitors because we know that several proteinase inhibitors are present in the uterus and the uterine secretions. We could also think of a splitting off of enzyme inhibitor complexes by other enzymes. Local activation of a proenzyme is a very attractive hypothesis in respect to the very high activity of proteinase inhibitors in these sites. We have started _in vitro_ culture studies and we hope these experiments will yield important information.

Strickland: The embryo associated aspect of hatching is shown by the fact that if you take the 2 cell embryo, which has never seen the uterus, out of the oviduct and culture it _in vitro_, it hatches.

Denker: Hatching _in vitro_ in the mouse is completely different from the hatching _in vivo_; there is much indication that _in vivo_ there is a dissolution of the zona pellucida and _in vitro_ you have a rupturing, so I wouldn't compare these things. But the fact that there is this difference between these two situations does not mean that the trophoblast does not contribute anything in the mouse, but there also could be a combination of endometrial enzymes and trophoblastic enzymes which might activate each other. I think we have to study this complicated system very much in detail; I hope it is not as complicated as the blood clotting or the complement system.

Boving: In some sections where a trophoblastic knob is apposed to the lemmas of the rabbit, I notice that the inner one which is derived from the oviduct is dissolved whereas the more frothy uterine contribution on the outside is not. Until I saw your picture of the inhibited blastocyst with the vast amount of uterine secretion accumulated outside, I had the simple happy idea that this was clear proof that the dissolving agency was from the knob and was working out. But now I've started to worry that since the lemmas of the rabbit are of multiple origins and possibly multiple compositions, perhaps the inner one is more susceptible to the enzymes coming from the knob and the outer one just less susceptible and perhaps interacting with some external factor. Do you have any evidence that discriminates activity on the two different kinds of lemmas?

Denker: I was very much interested in this point and we have tried to look at it carefully, but the results were a bit disappointing. I have not seen very clear evidence of erosion from inside or from outside. There were a few cases in which after partial inhibition by low doses of inhibitors the inner layer disappeared earlier than the other layers; however, in other cases it simply looked like all these different layers dissolved at the same time from inside and outside. I wouldn't rule that out; enzymes from both the trophoblast and the uterine secretions might contribute.

Enders: I can't leave the comments on the extracellular coats alone. You can get rabbit blastocysts to expand *in vitro* to the point where they will break through those extracellular coats but they don't resemble in any way the normal situation. They don't develop trophoblastic knobs, they just get large and split the external material. What I want to ask, however, is something different. Your illustrations of the dissolution of gelatin do not seem to focus around the knobs, but along all of the anti-mesometrial trophoblast. Would you care to comment on that observation?

Denker: I was very much puzzled by this for quite a while and I think we now have some clues. In this histochemical system we don't find much activity of the trophoblast. But, when you take the trophoblast, isolate it, and do the gelatin test after electrophoresis, you get very much activity of apparently the same enzyme with the same electrophoretic mobility as you find in the disintegrating blastocyst coverings. If you try the whole homogenate without the electrophoresis, you don't find much activity. So, I simply imagine that there are inhibitors, or maybe activation of a proenzyme. I think a good possibility would be that there are proteinase inhibitors in the trophoblast, and this wouldn't be surprising because we often find high activities of proteinase inhibitors where proteinases are being produced, such as in the pancreas.

7

Studies on the Role of Plasminogen Activator in Embryo Implantation

SIDNEY STRICKLAND

The process of embryo implantation in many mammals is an invasive event in which the embryo penetrates through the uterine epithelium and its underlying basement membrane and finally broaches maternal blood vessels before advancement ceases. The trophoblast cells of the blastocyst have long been recognized as the cells responsible for this invasion, and pathologists often draw attention to the degradative nature of these cells by pointing out their "pseudomalignant" properties (Novak and Woodruff, 1974). In view of these interesting characteristics, it is perhaps surprising that little is known about the biochemical basis for trophoblast invasion.

The tissue destruction that accompanies implantation can be regarded as analogous in many respects to that associated with follicle rupture at ovulation. Recently, evidence has been presented that the protease, plasminogen activator, is involved in follicle rupture (Beers, Strickland and Reich, 1975). For this reason, and since several investigators have reported a proteolytic activity associated with implanting embryos (Blackwood et al., 1968; Owers and Blandau, 1971; Denker, 1972), it was of interest to study the trophoblast for the possible synthesis of plasminogen activator.

Much of the recent interest in the formation of plasminogen activator by cultured cells stems from the observation that transformed cells produce much larger amounts of this enzyme than their normal counterparts (Unkeless et al., 1973; Ossowski et al., 1973). The importance of these results was that they suggested a function other than blood clot homeostasis for the plasminogen activator:plasminogen system.

The proposed role for these proteases, based on the ability of malignant cells to invade surrounding tissue, was that plasminogen activator participates in processes that involve tissue destruction and/or turnover of the extracellular matrix.

As with all seminal observations, the experimental results described above were significant not only in themselves, but also for the insight they provided into the earlier literature on proteolysis. For example, Albrechtsen had detected high levels of plasminogen activator in many normal tissues (Albrechtsen, 1957), but his findings could not be related clearly to any known physiological processes. Secondly, due to the recent advances in understanding of lysosome functions, many observations of cell mediated proteolysis were ascribed to lysosomal proteases, despite the acid pH optimum of lysosomal enzymes, and the consistent failure to demonstrate the secretion of these enzymes under conditions that might be plausible for normal function in extracellular proteolysis. In short, many previous results deserved reconsideration in light of a possibly expanded role for plasminogen activator.

The suitability of plasminogen activator as a general extracellular protease in normal processes is readily apparent: *(1)* The enzyme has a pH optimum around neutrality, the pH of most extracellular fluids. *(2)* Plasminogen activator is secreted. *(3)* The proteolytic activity of the activator can be amplified many-fold by its catalytic conversion of extracellular plasminogen to the active serum protease plasmin. Thus, it is possible for a relatively few cells to have an impact on the surrounding tissue. *(4)* The resultant proteolytic activity of plasmin is probably localized *in vivo* due to the high serum concentrations of plasmin inhibitors, thereby limiting the diffusion of active protease and restraining its action in time and space.

This chapter summarizes briefly our findings on embryo plasminogen activation; experimental verification and proper bibliographic documentation are not included, but can be found in the original manuscript (Strickland, Reich and Sherman, 1976).

Plasminogen Activator and Implantation

We have surveyed the early stages in the development and differentiation of cultured mouse embryos for plasminogen activator production. The results can be summarized as follows: *(1)* Blastocysts collected from mice on the fourth day of pregnancy were cultured *in vitro* for four days. The resultant blastocyst outgrowth produced fibrinolytic zones when assayed by a fibrin-agar overlay technique. *(2)* The fibrinolytic activity was strictly plasminogen-dependent and hence due to plasminogen activator in the conditioned medium. Thus, the activity was abolished when the assays were performed in the absence of plasminogen, and full activity was then restored by addition of purified plasminogen. *(3)* The level of activity was proportional to the number of blastocysts in the culture. *(4)* The time course of enzyme production in cultures of second, third, fourth, or fifth day embryos was a function of equivalent gestation age and not of time in culture. *(5)* The onset of enzyme secretion was not directly related to the emergence of embryos from the zona pellucida. Blastocysts (fourth day), which normally require about 30 hours to hatch spontaneously *in vitro*, can be artificially freed from their zona pellucidae by treatment with pronase (Mintz, 1962). When blastocysts hatched with pronase on the fourth day were cultured and assayed for plasminogen activator, the time course of enzyme production was indistinguishable from that of untreated cultures. *(6)* Elevated levels of enzyme were present intracellularly as well as in conditioned medium. The enzyme was undetectable in cell lysates before equivalent gestation day 6; subsequently, the increase in intracellular enzyme levels roughly paralleled that assayed in conditioned medium. *(7)* The temporal pattern of plasminogen activator production by the blastocyst cultures was complex. Initial enzyme activity appeared on about the sixth day, rising to a temporary maximum at the eighth or ninth day; this was followed by a transient decline with a second and continuing increase starting at the eleventh day. This second phase of enzyme secretion was maintained until at least the fifteenth day, by which time the level of activity was 5-fold higher than the plateau value at the eighth day. This pattern suggested the participation of two types of cells, with the first initiating enzyme production by the sixth day and declining after reaching a maximum by the ninth day, and the second beginning to secrete detectable amounts at a later time with progressive increases thereafter. *(8)* The pattern of enzyme production *in vitro* by blastocysts that had been delayed from implanting by maternal ovariectomy was indistinguishable from control blastocysts. *(9)* Fractionation of the blastocyst into its constituent cell types has identified the trophoblast as the cell responsible for the first phase of enzyme synthesis. Beginning on the seventh day, pure trophoblast cultures *in vitro* produce increasing amounts of plasminogen activator until the ninth day, after which enzyme secretion gradually decreases.

Conclusion

The trophoblast is one of the most invasive cells known. It can invade and destroy virtually all tissues in addition to those normally encountered in the uterus, but its invasiveness is expressed only during a limited period of its life cycle. The invasive period has been determined primarily from transplantation of early embryos to ectopic sites, and it extends from the sixth to the tenth equivalent day of gestation in the mouse (Kirby, 1965). This interval corresponds closely with the period during which cultured trophoblast is secreting plasminogen activator, and this suggests that the enzyme is a contributing factor in implantation. Since, as noted, the tissue destruction that occurs during implantation can be considered analogous to that accompanying follicle rupture at ovulation, it is noteworthy that the production of plasminogen activator is also similar in the two processes. Thus, the trophoblast produces enzyme at the time of maximum tissue disruption in implantation, just as the ovarian granulosa cells do in ovulation.

In addition, the production of plasminogen activator by several cell types has provided a versatile assay for studying the hormonal control of cellular functions. The synthesis of enzyme by ovarian granulosa cells is modulated *in vitro* by gonadotropins, cyclic nucleotides, and prostaglandins, and this has permitted an examination of the process of ovulation by way of the cellular response to these substances (Strickland and Beers, 1976). Similarly, macrophage plasminogen activator production, which is inhibited by low levels of glucocorticoids, has provided insights about some of the anti-inflammatory activity of these hormones (Vassalli, Hamilton and Reich, 1976). By analogy, enzyme formation by early embryonic cells suggests that plasminogen activator may also be an attractive probe for monitoring regulatory processes during development.

Acknowledgments

This work was performed in collaboration with Drs. E. Reich and M.I. Sherman, and was supported by grants from the American Cancer Society and The Rockefeller Foundation.

References

Albrechtsen, O.K. (1957) "The Fibrinolytic Activity of Animal Tissues," *Acta Physiol. Scand. 39,* 284-290.

Beers, W.H., Strickland, S. and Reich, E. (1975) "Ovarian Plasminogen Activator: Relationship to Ovulation and Hormonal Regulation," *Cell 6,* 387-394.

Blackwood, C.E., Hosannah, Y. and Mandl, I. (1968) "Proteolytic Enzyme Studies in Developing Rat Tissues," *J. Reprod. Fertil. 17,* 19-33.

Denker, H.W. (1972) "Blastocyst Protease and Implantation: Effect of Ovariectomy and Progesterone Substitution in the Rabbit," *Acta Endocrinol. 70,* 591-602.

Kirby, D.R.S. (1965) "The Invasiveness of the Trophoblast," in *The Early Conceptus, Normal and Abnormal,* W.W. Park, ec., University of St. Andrews Press, Edinburgh, p. 68-73.

Mintz, B. (1962) "Experimental Studies of the Developing Egg: Removal of the Zona Pellucida," *Science 138,* 594-595.

Novak, E.R. and Woodruff, J.A. (1974) "Hydatidiform Mole and Choriocarcinoma," in *Novak's Gynecologic and Obstetric Pathology,* W.B. Saunders, Philadelphia, p. 600.

Ossowski, L., Unkeless, J.C., Tobia, A., Quigley, J.P., Rifkin, D.B., and Reich, E. (1973) "An Enzymatic Function Associated with Transformation of Fibroblasts by Oncogenic Viruses. II. Mammalian Fibroblast Cultures Transformed by DNA and RNA Tumor Viruses," *J. Exp. Med. 137,* 112-126.

Owers, N.O. and Blandau, R.J. (1971), "Proteolytic Activity of the Rat and Guinea Pig Blastocyst *in Vitro,*" in *Biology of the Blastocyst,* R.J. Blandau, ed., University of Chicago Press, pp. 207-223.

Strickland, S. and Beers, W.H. (1976) "Studies on the Role of Plasminogen Activator in Ovulation: *in Vitro* Response of Granulosa Cells to Gonadotropins, Cyclic Nucleotides and Prostaglandins," *J. Biol. Chem. 251.* 5694-5702.

Strickland, S., Reich, E. and Sherman, M.I. (1976) "Plasminogen Activator in Early Embryogenesis: Enzyme Production by Trophoblast and Parietal Endoderm," *Cell 9,* 231-240.

Unkeless, J.C., Tobia, A., Ossowski, L., Quigley, J.P., Rifkin, D.B., and Reich, E. (1973) "An Enzymatic Function Associated with Transformation of Fibroblasts by Oncogenic Viruses. I. Chick Embryo Fibroblast Cultures Transformed by Avian RNA Tumor Viruses," *J. Exp. Med. 137*, 85-111.

Vassalli, J.-D., Hamilton, J. and Reich, E. (1976) "Macrophage Plasminogen Activator: Modulation of Enzyme Production by Anti-Inflammatory Steroids, Mitotic Inhibitors, and Cyclic Nucleotides," *Cell 8*, 271-281.

8

Physiology of Uteroglobin

HENNING M. BEIER

The rabbit uterus has recently become an increasingly popular system for studying embryonic-maternal relationships prior to implantation. The main advantages are that the endocrinology, the biochemistry and morphology of uterine secretion is well documented, and that the blastocyst has proved amenable to experimental manipulation, as are the techniques of egg transfer and in-vitro-culture. In particular, it has been useful for the studies on uterine secretion that informations have been accumulated from several laboratories on uteroglobin, the predominant rabbit endometrial secretory protein. The main disadvantage is of course the still unknown function of uteroglobin. The organizers of this symposium have been extremly polite and diplomatic, when they invited me to report on the "physiology" rather than on the "function" of this enigmatic protein. Since the first description of uteroglobin as the identical major protein component of rabbit uterine secretion and blastocyst fluid (Beier, 1966) it has been the permanent question, whether uteroglobin regulates blastocyst development, implantation or any other pre- or peri-implantational process. A protein which is released into the developing blastocyst's environment in such abundant amounts, comprising about half of the protein content of the endometrial secretion, and which permeates into the blastocyst fluid, seems to represent more than just another luxurious production of Mother Nature. It is the purpose of this review to present available data in order to develop a possible concept on uteroglobin function.

Where does uteroglobin come from?

In a series of light microscopical and electron microscopical studies it has been demonstrated that uteroglobin is a secretory product of the epithelial cells of the endometrium. It is released after its biosynthesis in the endometrial cells, in consequence of progestational stimuli. As a matter of fact, uteroglobin has proved to be an excellent model for investigating the progesterone control of uterine secretion. The synthesis and release of uteroglobin is accompanied by characteristic morphological transformations of the epithelial cells, the features of which have been well known to endocrinologists earlier than the biochemical events within the nucleus and the cytoplasm of the cells. Practical implications of the tissue and cell transformation have been used by means of the Clauberg-test (1930) or the McPhail-index. Recently, higher resolutions by scanning and transmission electron microscopy are demonstrating more clearly a significant and rapid cell transformation induced by progesterone or synthetic progestins. The scanning electron micrographs (Fig.3 - 6) beautifully document the surface transformation of the endometrial cavum epithelium. These transformations occur, regardless whether the endocrine status of the rabbit is a physiologically normal one (pregnant,pseudopregnant)or whether the endocrine status is artificially built up by hormone supplementation in ovariectomized animals (Beier and Mootz, 1976; Beier, Mootz and Kühnel, 1977).

Protein patterns of uterine secretion have been analyzed after flushing the uterine cornua with physiological saline, concentrating these wash fluids to a small volume by ultrafiltration, and subjecting these samples to acrylamide gel electrophoresis (slab gels, disc gel columns). The uterine fluid pattern from oestrous animals is not very different from the patterns of oviducal fluid or even blood serum: albumin and transferrin appear by far as the strongest fractions. The usually prominent fractions of serum macroglobulins and lipoproteins cannot be detected after electrophoretical resolutions of oestrous uterine fluid. However, during pregnancy and pseudopregnancy, specific protein patterns develop. The patterns from Day 4 to Day 7 post coitum show considerable changes from day to day: a sharp prealbumin-peak forms, albumin ratios decrease, pregnancy-related

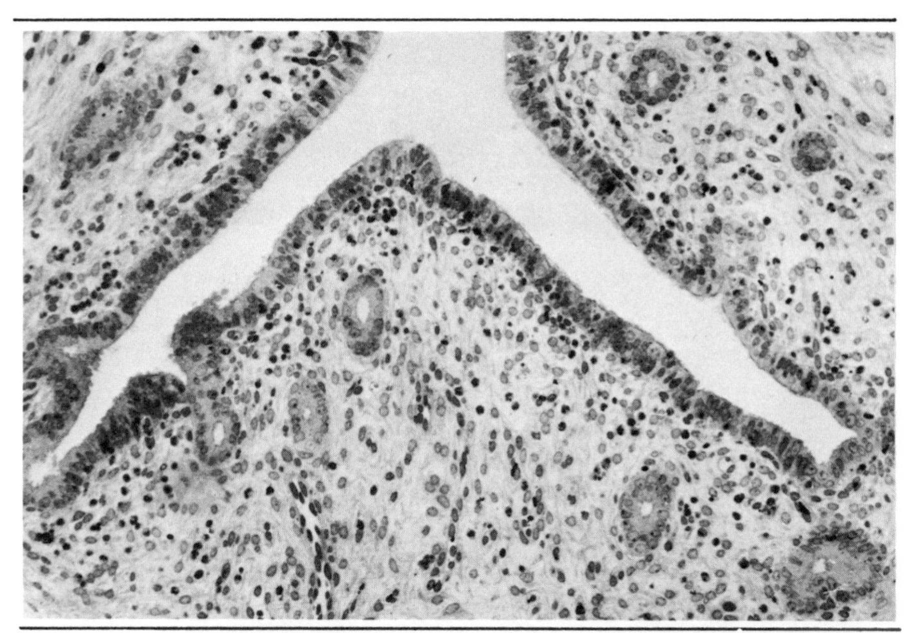

Fig. 1 Rabbit endometrium from an oestrous female. The cavum epithelium shows a relatively smooth surface, and only a few glands are visible. Semithin section, toluidine blue staining / x210.

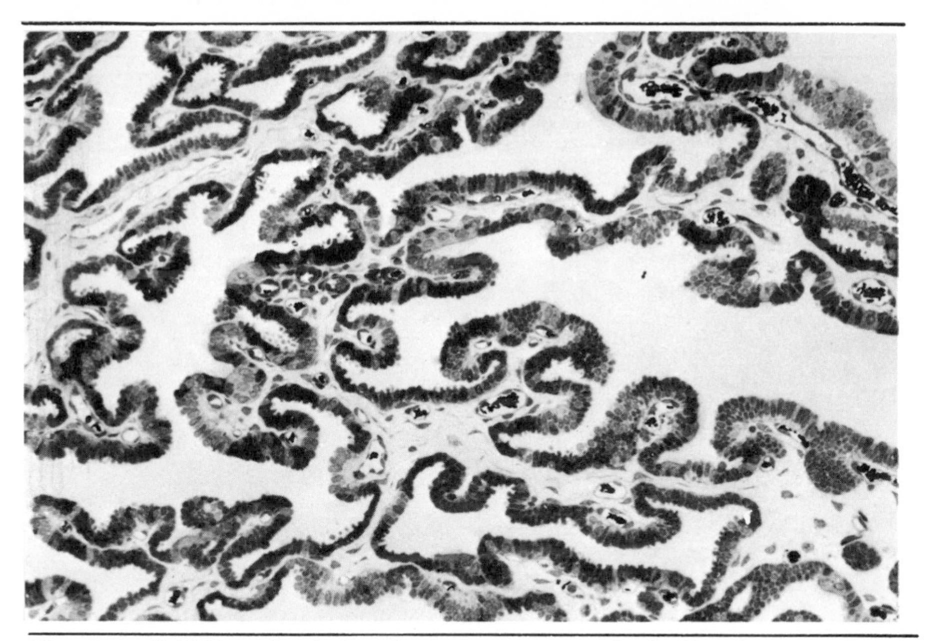

Fig. 2 Rabbit endometrium from a pregnant animal (Day 6 post coitum). There is a remarkable proliferative and secretory transformation visible, compared to the oestrous stage (see above). Semithin section, toluidine blue staining / x210.

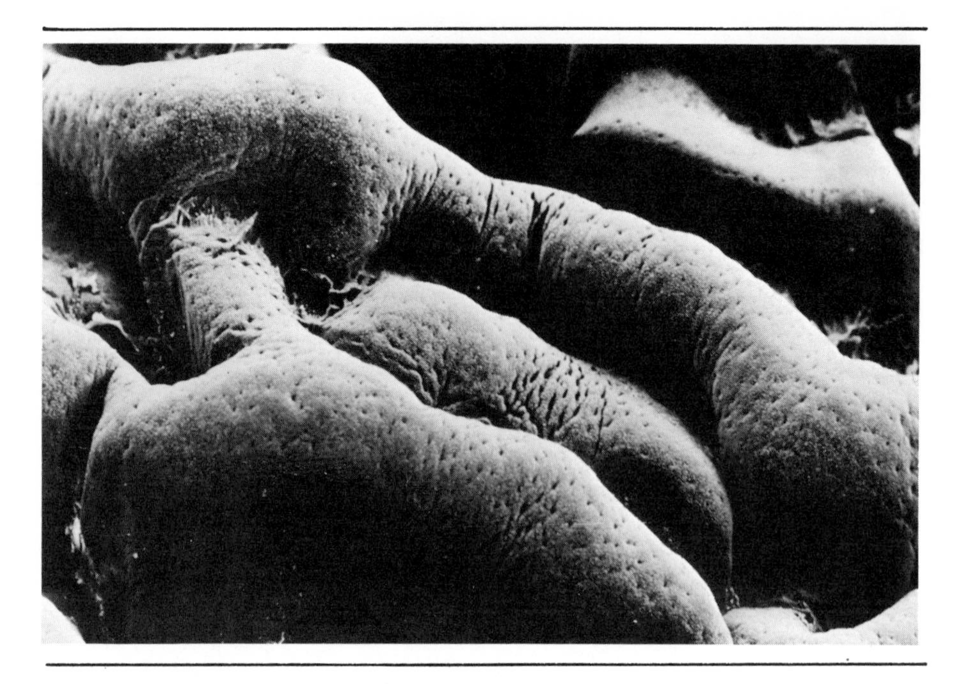

Fig. 3 Endometrial surface as it is apparent in oestrous rabbits or in oestrogen-treated castrated females. Scanning EM, x 80.

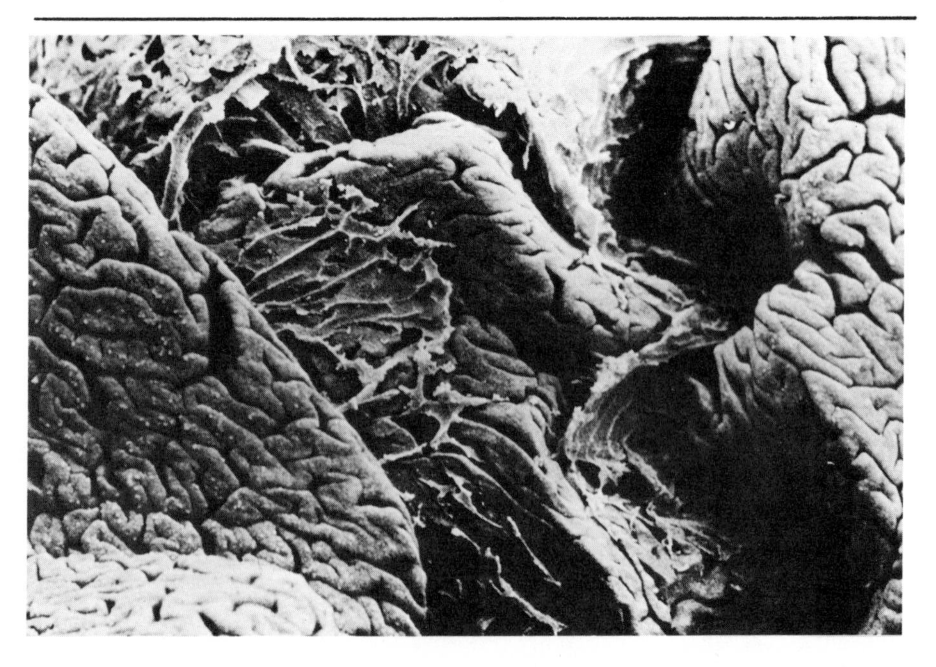

Fig. 4 Endometrial surface after progestagen stimulated transformation as it appears during preimplantation. Scanning EM, x 80.

Fig. 5 Cavum epithelial cells at the beginning of the secretory transformation. Cells regularly covered with microvilli and typical ciliated cells. Scanning EM x 4,000.

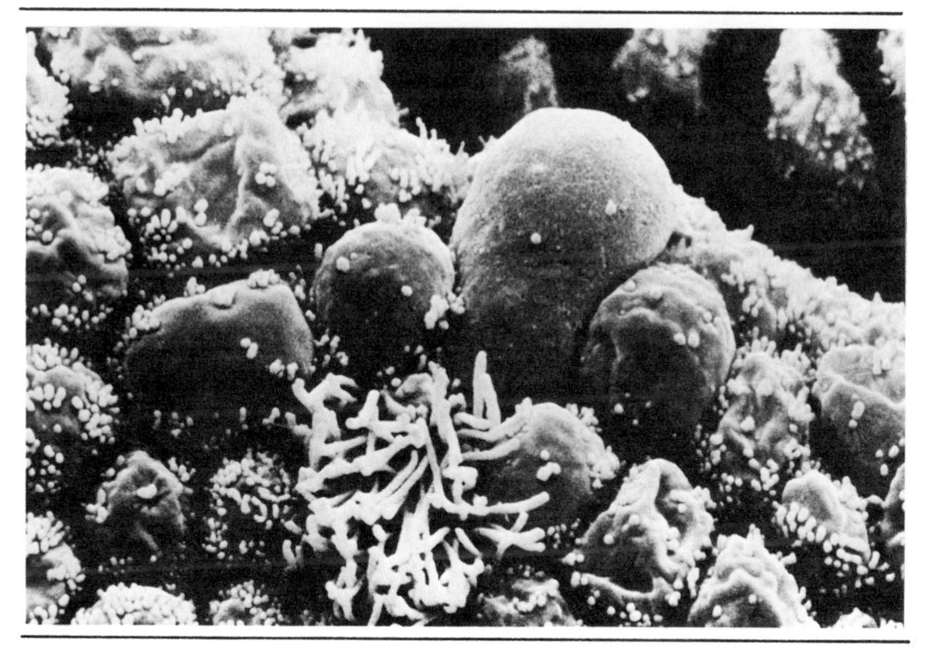

Fig. 6 Cavum epithelial cells as they appear in secretory shape. Apical protrusions and reduced microvilli. Scanning EM, x 8,000.

postalbumin appears, the predominant uteroglobin fraction increases, and a broad spectrum of glycoproteins appears, among which one peak develops towards Day 7 a maximum (ß-glycoprotein). Interestingly, in pseudopregnant uteri, a comparable protein pattern develops, however, detailed quantitative and chronological differences have been analyzed (Beier and Kühnel, 1973).

As knowledge of the biochemistry of uterine secretion components accumulates, new advances are attained concerning the various ways in which these components are released into the lumen of the genital tract, and about the origin of these molecules. Analytical evidence from qualitative and from quantitative studies, employing acrylamide gel electrophoresis, Ouchterlony double immunodiffusion, and other immuno-electrophoretical combinations, suggests that, whereas many genital tract fluid proteins are identical to those of the blood serum, several are only found in uterine secretion. As shown earlier (Beier, 1967), there exists a selective transport of serum proteins into the uterine luminal fluid. Serum-identical proteins are not lined up within a certain category of small molecular sizes, but there are, on the contrary, rather diverse molecular weights, disregarding any preference for small protein components that would suggest a simple sieve effect by the blood-uterine secretion barrier. The partial disparity in the relative amounts of these few identified serum proteins in uterine secretion, compared to their serum proportions, further emphasizes the selectivity of transudation process. The albumin/globulin ratio has been calculated for blood serum as 1.32, however, for Day 6 uterine secretion only as 0.15, indicating a significant relative albumin decrease in uterine fluid. More than obvious is this feature, in addition, with the amounts of immunoglobulins in blood serum and uterine fluid. Whereas serum contains normally 15 - 17 rel.% of immunoglobulins, uterine pre-implantational secretion has not more than 2 rel.%.

Endocrine regulation of uteroglobin synthesis and release

It has been an interesting question since the first observations on progesterone effects on uteroglobin appearance in uterine secretion, whether this steroid acts stimulating directly on the synthesis or only on the release, or even on both processes together. By now, evidence has been attained that progesterone and also several other progestagens are capable to stimulate synthesis largely, however release totally. Uteroglobin synthesis is mainly activated by progesterone, although this protein is produced in detectable amounts by the uterus of oestrogen treated castrated does. Under such experimental conditions uteroglobin only is present in the epithelial cells, but not extruded into the cavum uteri in amounts which could be detected by immunohistochemical means, as work by Kirchner (1976) indicates. However, within the uterus we never could obtain any uteroglobin release under the influence of oestrogens, even if considerable dosage of 17ß-oestradiol was used (Beier, 1974). The release response to progesterone is to a certain extent dose-dependent, the most effective dosage being in the range of 0.6 - 3.0 mg/day/animal/average weight of 2.5 - 3.0 kg. All higher dosages have been without a remarkable effect on increase of uteroglobin concentrations in uterine secretion.

Doubtless, progesterone is required to establish the characteristic secretion pattern of uteroglobin during preimplantation in vivo. But moreover, Beato and Arnemann (1975) have shown that this secretion even is intact when the uterus is removed from the mother, isolated, and perfused under defined experimental conditions.

Production of uteroglobin as the major endometrial secretion protein logically asked for identification of true synthesis of this protein by demonstration of the messenger RNA (mRNA). The endometrium of progesterone stimulated rabbits has been used as source for the isolation of uteroglobin mRNA . The translation of poly(A)-containing mRNA for uteroglobin has been demonstrated in different systems (Beato and Rungger, 1975; Bullock, Woo and O'Malley, 1976). Identification of newly in vitro synthesized uteroglobin was presented by means of monospecific antibodies. These results definitely point out that uteroglobin is synthesized de novo in the uterus and that progesterone stimulates this synthesis. Bullock et al. (1976) have demonstrated that the specific mRNA for uteroglobin accounts for an increasing proportion of the poly(A)-rich endometrial mRNA during the preimplantation period, reaching a maximum at Day 4 post coitum in normal pregnancy. The pattern of change in uteroglobin mRNA is similar to the pattern of secretion of uteroglobin during normal

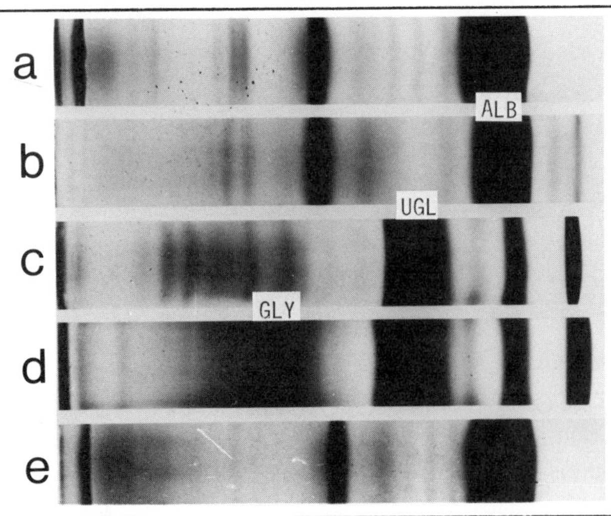

Fig. 7 Disc electrophoretic protein patterns of the rabbit uterine secretion during early pregnancy, compared to the usual protein pattern of rabbit blood serum. (a) Blood serum, (b) oestrous uterine fluid, (c) uterine secretion, represented by concentrated flushings, at Day 6 post coitum, (d) at Day 7 p.c., (e) at Day 9 p.c. Migration of the proteins was from the left to the right, the fastest fractions are prealbumins, followed by albumin (ALB), within the postalbumin region uteroglobin (UGL) predominates remarkably at Days 6 and 7 p.c. The ß-glycoprotein fraction peaks at Day 7 (GLY).
Disc electrophoresis, tris-glycine buffer, pH 9.0

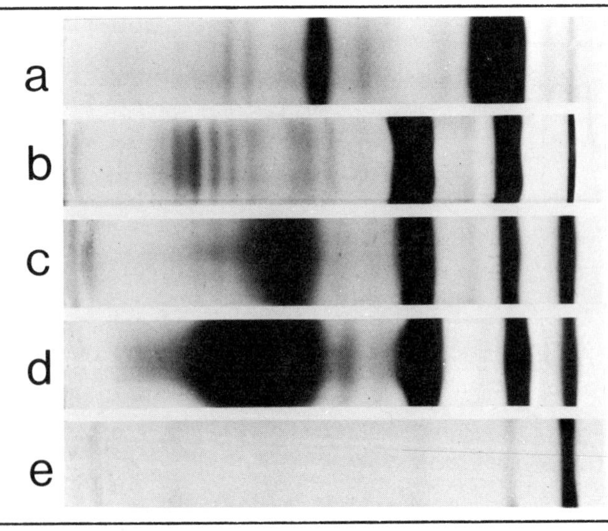

Fig. 8 Disc electrophoretic protein patterns of the rabbit uterine secretion during pseudopregnancy. (a) oestrous uterine fluid, (b) uterine secretion, represented by concentrated flushings, at Day 4 p.c., (c) At Day 6 p.c., (d) at Day 7 p.c., (e) at Day 11 p.c. Migration of the proteins was from the left to the right.
Disc electrophoresis, tris-glycine buffer, pH 9.0.

preimplantation and reflects the changing endocrine control mechanisms of the maternal system.

So far, it is unclear what process in detail is terminating uteroglobin presence in the secretion. It may well be that the progesterone/ oestradiol ratio is more important than the level of one of both steroids alone. Since the protein pattern observed during pseudopregnancy suggests that implantation itself terminates uteroglobin release or synthesis, there may be a specific "message" delivered from the blastocyst or from the decidual tissue, as has been proposed by Johnson (1974). The perfusion experiments on isolated uteri and their uteroglobin production (Beato and Arnemann, 1975) indicate that the "switch-off" in vivo is a termination of the synthesis rather than only the release, since these studies show no long-term accumulation of uteroglobin within the cytosol of the endometrial cells.

Approaches to study uteroglobin function

Some biochemical and biological parameters of uteroglobin, so far known, are compiled in tables 1 and 2. Approaches to analyze uteroglobin physiology concentrate on the studies of its biochemical properties, including binding and transport functions, as well as enzyme-inhibitor activities. Additionally, a differential analysis of uteroglobin and its homologous molecules of the male reproductive tract and the lung is necessary and will be useful (Fig.9).

The uteroglobin-like protein from the seminal plasma and from the bronchial secretion are immunological identical antigens, as could be confirmed by several independent investigators (Beier, Bohn and Müller, 1975; Beier, 1977; Beier, Kirchner and Mootz, 1978; Noske and Feigelson, 1976; Bullock and Bhatt, 1976). The most challenging questions arise from comparative studies on uteroglobin and the uteroglobin-like antigens. Does this "ectopic uteroglobin" show similar or identical properties, particularly with regard to the most intriguing features of steroid binding and trypsin inhibition? Purified proteins have been used in recent studies on progesterone binding to authentic uteroglobin and the uteroglobin-like antigen

TABLE 1

BIOCHEMICAL PROPERTIES OF UTEROGLOBIN

(a) Separation from other uterine secretion proteins by acrylamide gel electrophoresis and identification as postalbumin fraction (Beier, 1966, 1967)

(b) Identification of uteroglobin, the postalbumin fraction as identical antigen to the β_1-U-globulin from classical agar-immunoelectrophoresis performed by Schwick (1965), by means of a direct acrylamide-agar-immunoelectrophoretical combination (Beier, 1966, 1967)

(c) Sedimentation in analytical ultracentrifugation (Svedberg units analysis) $S_{c=1.0\%} = 1.38$ (Beier, 1968 a)

(d) Molecular weight around 14,000-15,000 (Murray et al., 1972; Bullock and Connell, 1973)

(e) Composition of two identical subunits of approximately 7,000 - 8,000 MW, linked by S-S-bonds and composed of 75 amino acids (McGaughey and Murray, 1972; Beato and Baier, 1975, Beato, 1977)

(f) Isoelectrofocussing reveals an isoelectric point of 5.4 (McGaughey and Murray, 1972)

(g) Globular "glycoprotein", only small amount of carbohydrate components attached; sialic acid is missing among the identified carbohydrate components (Beier, 1967, 1968 b, Krishnan and Daniel, 1968; Kirchner, 1969)

(h) Binding of progesterone with high affinity, in a reaction which does not involve the sulfhydryl groups, requires the opening of the disulfide bridges (Beato, 1977). Affinity for progesterone has been reported with different association constants:
$5.95 \times 10^8\ M^{-1}$ (Arthur, Cowan and Daniel, 1972)
$3.00 \times 10^6\ M^{-1}$ (Beato and Baier, 1975; Beato, 1977)
very much lower than these (Rahman et al., 1975)
not calculated, but estimated as low (Urzua et al., 1970)

(i) Inhibition of trypsin, tested in fibrin-agar-electrophoresis (Beier, 1970; Johnson, 1974; Beier, 1976)

(j) Uteroglobin is synthesized by endometrial cells, as indicated by isolation of mRNA for uteroglobin from this tissue, and translation of this mRNA in Xenopus laevis oocytes (Beato and Rungger, 1975), mRNA isolation by Bullock, Woo and O'Malley, 1976.

TABLE 2

BIOLOGICAL FEATURES OF UTEROGLOBIN

(a) Significant and predominant protein in rabbit uterine secretion during preimplantation in normal pregnancy and in equivalent stages of pseudopregnancy (Beier, 1966, 1967, 1968b; Krishnan and Daniel, 1967; Kirchner, 1969; Urzua et al., 1970; Beier et al., 1970, 1971; Daniel, 1971)

(b) Predominant protein of blastocyst fluid (Beier, 1966, 1967; Hamana and Hafez, 1970; Petzoldt, 1974)

(c) Uteroglobin is not detectable in cultured blastocysts, when development is accomplished from the 2-celled stage up to the large expanded blastocyst stage in BSA supplemented in vitro culture media (Beier and Maurer, 1975)

(d) Release from uterine epithelial cells is stimulated and controlled by progesterone (Beier, 1968 a; Beier et al., 1970; Urzua et al., 1970; Arthur and Daniel, 1972)

(e) Release can be stimulated by Chlormadinone acetate and Norgestrel (Beier and Beier-Hellwig, 1973)

(f) Release from uterine epithelial cells is delayed by post-coital oestrogen injections (17β-oestradiol benzoate) (Beier et al., 1971)

(g) Release into uterine lumen is terminated mainly by oestrogen after implantation (Bullock and Willen, 1974)

(h) Uteroglobin is antigenic in mice (Beier, 1966), rats (Schwick, 1965) guinea pigs (Beier, 1966, 1968 b; Kirchner, 1972; Johnson et al., 1972), goats (Noske and Feigelson, 1976), sheep (Beier et al.,1975)

(i) Antigenic cross-reactions can be attained under certain conditions by means of all highly potent antiserum preparations with antigens from seminal vesicle secretion, seminal plasma (Beier et al.,1975), from lung tissue extracts (Noske and Feigelson, 1976), and from lung lavage (Beier, Kirchner and Mootz, 1978).

Additional studies on rabbit esophagus tissue and flushings, jejunal flushings, and also human endometrium, human oviduct, and human seminal plasma indicate uteroglobin cross-reacting antigens to be present there (Feigelson, Noske, Goswami and Kay, 1977)

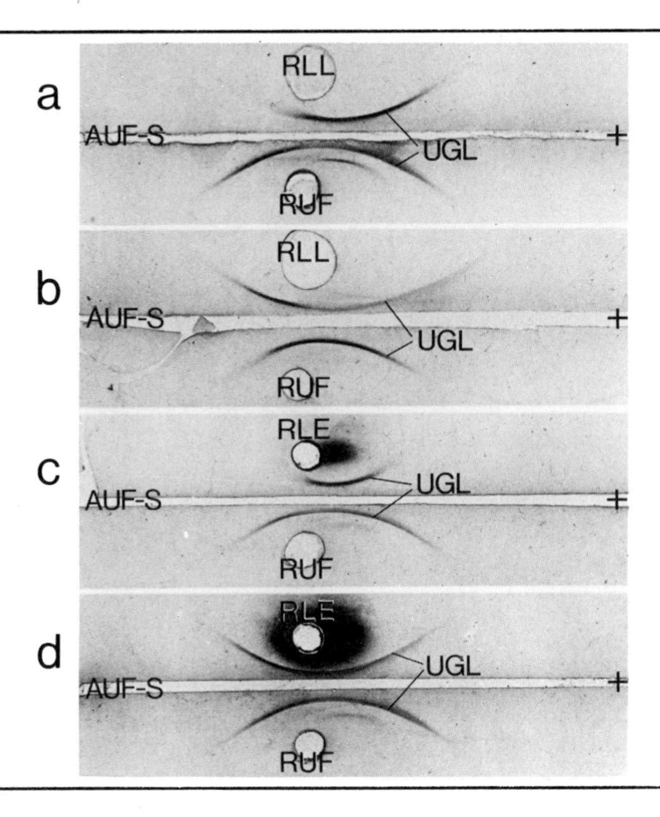

Fig. 9 Immunologic reaction of anti-rabbit uterine fluid-serum from the sheep (AUF-S) against different secretions and tissue extracts from rabbit uterus and lung. The antigen preparations are: rabbit lung lavage (RLL), protein concentration 0.015% (a) 0.15% (b); rabbit lung tissue extract (RLE), protein concentration 1 % (c) and 4 % (d);rabbit uterine fluid (RUF), protein concentration 1 % (a, c and d) and 0.5%(b). Uteroglobin precipitates (UGL) are developed differently according to various protein concentrations.
Immunoelectrophoresis, diethylbarbiturate-acetate buffer, pH 8.2.
(From Beier, Kirchner and Mootz, 1978).

Fig. 10 Binding of progesterone to endometrial uteroglobin and to the uteroglobin-like protein of the lung. Purified endometrial uteroglobin (24 µg/ml) and the uteroglobin-like protein of the lung (26 µg/ml) were incubated at 4°C for 2 hrs with increasing concentrations of (^{3}H)progesterone in the presence of ovalbumin (2 mg/ml). The amount of the protein bound radioactivity was determined by the charcoal absorption technique and is plotted as a function of the concentration of free progesterone. The values represent the average of two determinations performed in duplicate. (From Beato and Beier, 1978).

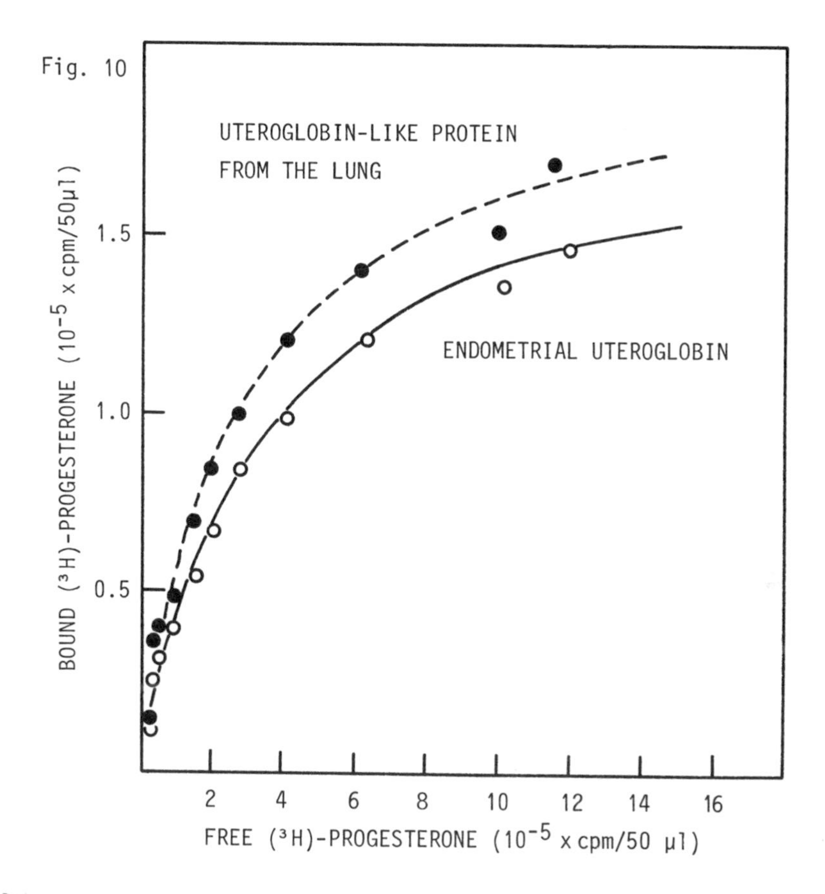

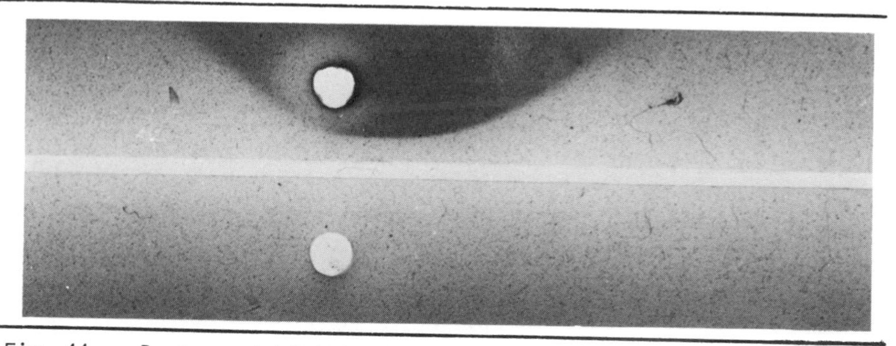

Fig. 11 Protease inhibition by the uteroglobin fraction in fibrin-agar-electrophoresis. The samples of this reaction are rabbit uterine secretion from preimplantational stage Day 6 (protein concentration in total 1 %), and rabbit uteroglobin-like protein from the lung, which was isolated by immunabsorption from rabbit lung tissue extract (protein concentration of the isolated material 1 %). The well was filled with trypsin (0.005 %), and a broad zone of inhibition attained. Inhibition is demonstrated by the black area at the typical site of uteroglobin precipitation in classical agar-immunoelectrophoresis (Beier, 1970).The isolated protein from the lung does not inhibit trypsin digestion in detectable amounts in this test.

from the lung (Beato and Beier, 1978). There is an identical progesterone binding to both proteins (Fig.10). However, synthesis and secretion are only controlled by ovarian steroids within the uterus, since all investigated lungs, even from ovariectomized animals, male and newborn rabbits contained the uteroglobin-like protein.

Rather surprisingly, we have found quite different results by means of fibrin-agar-electrophoresis to test the trypsin inhibitory activity of uteroglobin and the homologous antigen from bronchial secretion. Preliminary results, which need further confirmation, show that the uteroglobin fraction of endometrial secretion does inhibit trypsin whereas the uteroglobin-like protein fraction of the lung tissue does not (Beier, 1977). It remains to be clarified whether an unknown factor (molecule) is involved in this proteinase inhibition, since the highest activities are always found, in our system, when total uterine flushings are tested and compared to purified or isolated fractions (Fig.11). Another possible difficulty may arise from the chemical purification procedure itself by alteration of the biological activity of the uteroglobin molecule. Regardless of further studies and results on the comparison of uteroglobin and the uteroglobin-like antigen, it should be pointed out here, that we cannot accept any longer the dogma, that immunological identity is proof for the total identity of two antigen molecules. Biological activities of a protein may not be localized or dependent on the immunologically determinant parts of the molecule.

Studies on biological systems, where uteroglobin may act: the in vivo and the in vitro growing blastocyst

Injections of 17ß-oestradiol benzoate at 6 hrs and 30 hrs after mating reveal a significant delay in the secretory pattern sequence of the rabbit uterus (Fig. 12). Compared to the normal preimplantation patterns, there is a delay of 2 - 5 days, dependent on the stages compared. At earlier stages, the delay is less extended (2 - 3 days) than in later developmental stages (4 - 5 days delay). This feature is not only true for the protein patterns, but also for the histology of the endometrium and the enzyme histochemistry of the endometrial epithelia. We have claimed, that these asynchronous protein patterns

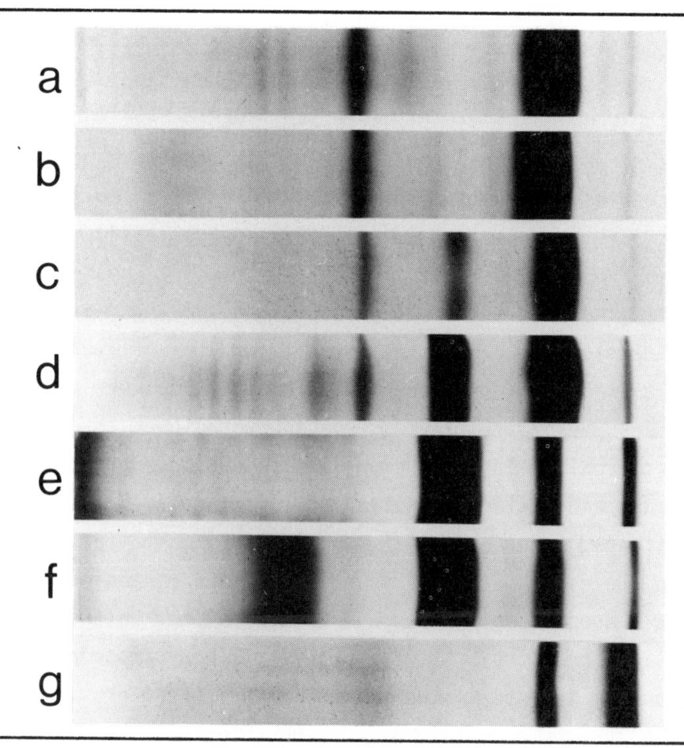

Fig. 12 Disc electrophoretic protein patterns of rabbit uterine fluid from delayed secretion. (a) Oestrous pattern, (b) oestrous pattern after two injections of 100 µg and 150 µg, respectively, oestradiol-17ß benzoate, (c) delayed secretion pattern from Day 4 p.c., (d) from Day 6 p.c. (e) from Day 8 p.c., (f) from Day 12 p.c., and (g) from Day 16 p.c.

TABLE 3

UTEROGLOBIN IN NORMAL PREIMPLANTATIONAL UTERINE SECRETION AND IN DELAYED SECRETION

Day	0	4	6	7/8	9/10	12	14
Normal	-	38.9	39.5	38.4	4.4	-	-
Delay	-	15.0	25.5	44.2	43.2	42.3	5.0

Values expressed as relative percentage of total uterine secretion protein. Mean values of at least three determinations of four animals.

contribute particularly to an unfavourable uterine environment for blastocyst development (Beier, 1970; Beier, Kühnel and Petry, 1971; Beier, 1974). Subsequent egg transfer experiments have shown clearly the fact, that normally developed blastocysts require a normally developed uterine environment to accomplish implantation and further development (Beier, Mootz and Kühnel, 1972; Adams, 1974). We transferred normal Day 4 blastocysts into Day 8 pregnant, post-coitally oestrogen treated, and such delayed secretion showing recipient uteri. These transferred blastocysts implanted around Day 12 (recipient's reproductive stage) and developed in an experimental series ca. 40% normal fetuses. Several of these fetuses were allowed to develop to viable young rabbits. In conclusion, we are convinced that the clearly synchronized uterine environment, particularly the secretion protein patterns, is an essential part in early mammalian embryogenesis. The model of Delayed Secretion has not been merely studied to demonstrate a questionable growth-inhibition effect on the native blastocysts. One particular item, however, is very important: the asynchronous egg transfer in delayed secretory uteri provides biological evidence for the necessity of a proper uterine environment to blastocyst development. Consequently, the essential protein environment for normal blastocyst development is composed of a considerable number of macromolecules. Within this,uteroglobin obviously is the major component, moreover so characteristic that it represents a specific marker molecule for the ovarian hormonal status of the animal.

Interestingly, the experimentally induced delay of uterine secretion is not the only possibility to desynchronize the maternal and embryonic systems. Prefertilization treatment of oestrous rabbits with progesterone (up to 2 mg/day/animal) for 8 days (Day-6 to +1) and induction of ovulation (by HCG-injection on Day 0) with consecutive artificial insemination results in normal egg development during oviducal passage, however, degeneration of blastocysts appears after their arrival in the uterus, mostly within 2 days being exposed to the uterine secretion environment (De Visser, 1976). If this treatment is applied, the intrauterine milieu is changed in so far, as the protein patterns are advanced compared to the normal preimplantation. The exogenous progesterone induces uteroglobin synthesis and release before ovulation and fertilization. There is a maximum of uteroglobin secretion from Day-1 until Day +2 , designated consequently as an Advanced Secretion. Comparable observations are reported by Kendle and Telford (1970) and by McCarthy, Foote and Maurer (1977). This phenomenon leads as clearly to failure of implantation as it appears during Delayed Secretion.

We have paid particular attention to the presence of uterine secretion proteins in blastocyst fluid. In a recent investigation on the protein patterns of rabbit blastocyst fluid and blastocyst homogenates after development in vivo and in vitro, we have tried to present evidence for the origin of the blastocyst fluid proteins (Beier and Maurer, 1975). Special emphasis was directed to the protein patterns of in vitro grown blastocysts, since these embryos developed from the 2- and 4-celled stages to expanded blastocysts without any rabbit protein in the culture medium. There, bovine serum albumin was used as the only protein source. Patterns from in vivo and in vitro development differ significantly, as judged by means of acrylamide gel electrophoresis and by several immunochemical test methods. These results demonstrate, that blastocysts grown in vitro do not contain uteroglobin or ß-glycoprotein in detectable amounts. Compared to the in vivo developed blastocysts, this is a striking difference, because uteroglobin and ß-glycoprotein have been demonstrated in fluids of in vivo expanded blastocysts in considerable amounts, at least in such quantities that our routinely applied immunochemical tests are applicable with positive results. However, our study indicates, that the in vitro developing blastocyst cannot synthesize uteroglobin. It may well be true, that in vivo growing blastocysts also do not synthesize molecules identical to the uterine secretion proteins. All evidence now agrees with the concept that uterine secretion proteins permeate into the blastocyst fluid (Beier, 1967; Kulangara and Crutchfield, 1973; Schlafke and Enders, 1973), into the blastocyst cells, particularly trophoblast cells on the 7th day p.c. (Kirchner, 1976), and into the blastocyst coverings (Kirchner, 1972). It seems that the blastocyst utilizes these environmental proteins. Our experiments suggest, that under the conditions of in vitro culture, other proteins supplemented to the culture medium pass into the blastocyst compartments in comparable quantities. This in turn indicates, that the environmental proteins under in vitro conditions seem not to enfold a specific embryotropic activity, but more likely

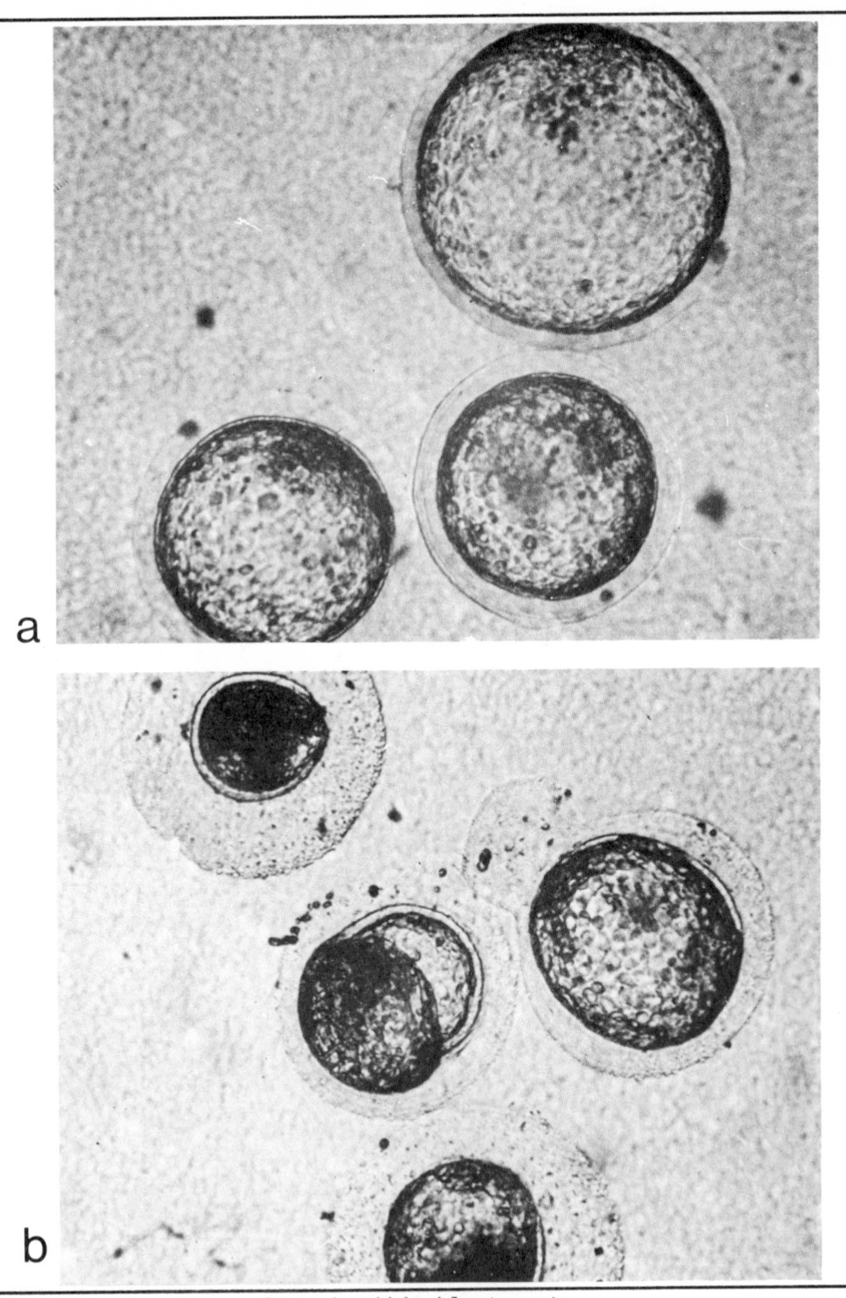

Fig. 13 In vitro cultured rabbit blastocysts.
(a) Expanded blastocysts,showing overall radial expansion and extension of thinned mucin coat;these embryos were kept 24 hrs in Maurer's medium (1.5% BSA) from Day 4 to Day 5 p.c.
(b) Herniated blastocysts, showing broken zonae and different stages of trophoblast herniation. The rigid mucin coat prevents hatching. Embryos kept in Maurer's medium (1.5% BSA) from Day 3 to Day 5 p.c.

general physico-chemical or biochemical effects. If we extrapolate these conclusions from the in vitro situation to the normal physiological situation in utero, it appears conceivable that the uterine secretion proteins act as integrated parts of a maternally-embryonically linked molecular system, the function of which is more a protective activity for the genetically individual embryo (Beier, 1974), than a direct embryotropic role as suggested by Krishnan and Daniel (1967).

In addition, uteroglobin may play an important role in the biochemical reactions and the "metabolism" of the blastocyst coverings, particularly, when the blastocyst enters the uterine cavity on Day 4. Striking evidence for the involvement of uteroglobin in the physico-chemical conditioning of the blastocyst coverings, the zona pellucida and the mucin coat (mucoprotein layer), has been obtained from the comparison of two experiments on rabbit blastocysts. Blastocysts flushed from the isthmic part of the oviduct or from the upper uterine segment (approximately 70 hrs p.c.) during Delayed Secretion show herniations of the trophoblast. The same pictures of trophoblast herniation can be easily obtained by culturing early cleavage stages up to the blastocyst stage in vitro, using a defined medium (Maurer and Beier,1976) with or without BSA as the only protein source. Both blastocysts are faced to an unfavourable environment, on the one hand to a desynchronized uterine secretion and on the other hand to a non-natural medium, also "desynchronized" by lack of uteroglobin or any other uterine protein.

Likely in consequence of this disproportion of uterine proteins, in particular of uteroglobin, the mucin coat does not show the elasticity and flexibility as usually on Day 4, when the normally developing blastocysts start expansion. This expansion cannot take place when the rigid coverings under desynchronized conditions act as a strait-jacket for the blastocyst. We have presented evidence for the influence of the uterine proteins on the coverings and their physico-chemical alteration by supplementing in vitro protein-free culture media with uterine secretion proteins (Maurer and Beier, 1976). Expansion does appear frequently, however, the overall development of the embryonic system occurs more slowly than normal. Since uteroglobin acts like a protease-inhibitor (Beier, 1976, 1977), it may be regulating proteases which control the structural metabolism of the blastocyst coverings. These proteases can be of uterine or embryonic (trophoblast) origin,

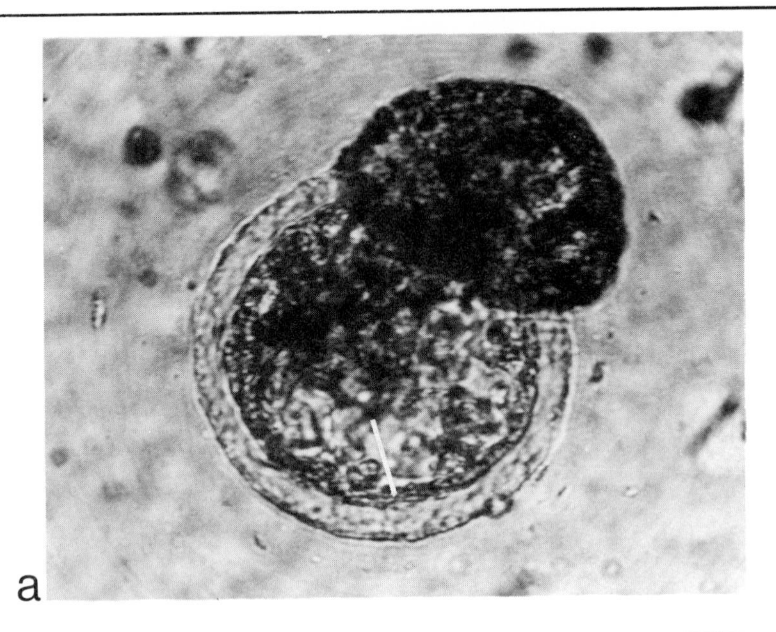

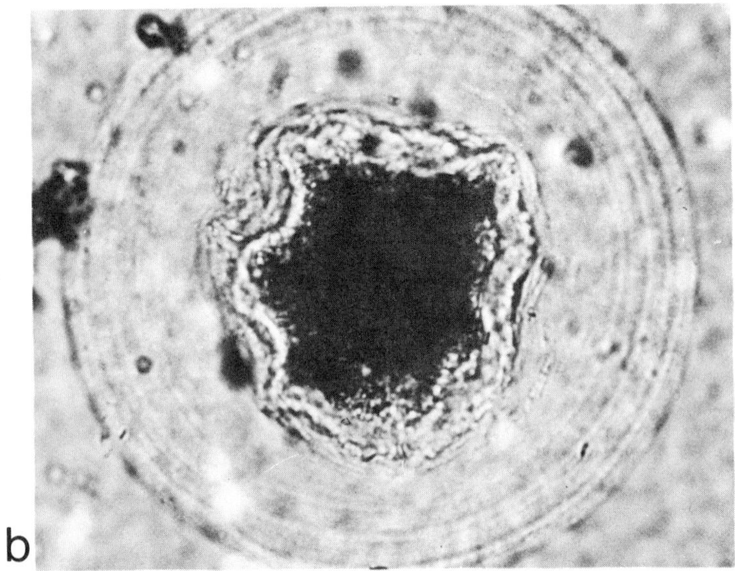

Fig. 14 In vitro cultured rabbit blastocysts.
(a) Herniated blastocyst with single trophoblast herniation. Trophoblast cells are unable to break the rigid mucin layer.
(b) Dead blastocyst after 56 hrs of culture in Maurer's medium (1.5% BSA). Culture started with Day 3 blastocysts, which expanded in culture, however died after collapse on Day 5. Mucin coat seems to enfold high degree of elasticity, gaining back earlier thickness. Zona does not show this physico-chemical property, because the thinned remnants are forming a wave-like figure inside the mucoprotein layer.

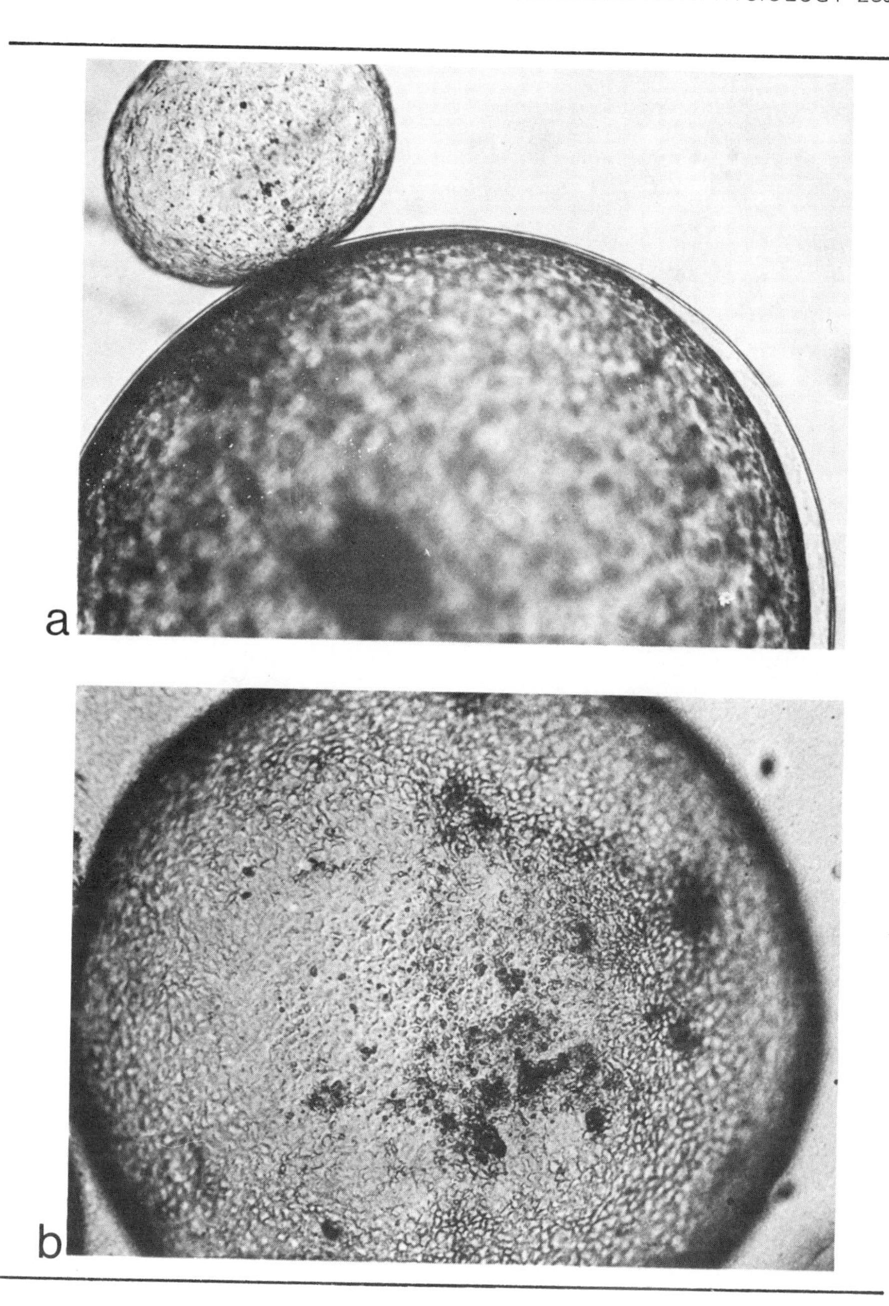

Fig. 15 In vitro cultured rabbit blastocysts.
(a) Large blastocyst just hatching through the very thin blastocyst coverings. This hatching does not occur in utero, however, frequently in vitro. This blastocyst was kept for 3 days in culture (Maurer's medium, 1.5% BSA), from Day 5 p.c. to Day 8.
(b) Large blastocyst after hatching. This blastocyst was kept 4 days in culture, like the above described blastocyst. Development is not normal with regard to the embryonic disc, which is retarded and defective.

and may be controlled by uteroglobin and other uterine secretion components (cf. Denker, this symposium).

The development of a normal blastocyst requires a certain number of essential environmental factors, which mainly are under maternal endocrine control. Only one single feature, however, is apparently clear after several series of in vitro culture studies: the pattern of uterine environmental factors has to develop in synchrony with the demands of the growing blastocyst. We have no conclusive answer to the question, what role uteroglobin plays in this biological concert. There is yet no proof that uteroglobin acts as a single "blastokinetic" component or as the only inducer or regulator of blastocyst expansion and differentiation. There shall be several opportunities to present new reviews until this simple question about uteroglobin function receives an adequate answer.

Acknowledgements

The author acknowledges with sincere thanks the continuous fruitful cooperation with Dr. Ursula Mootz, Dr. Ralph Maurer, Maria Petuelli and Sabine Hembeck. The support from the Physiology and Pathology of Reproduction Program of the Deutsche Forschungsgemeinschaft, Bonn, is gratefully acknowledged (Grants Be 524/5 and Ku 210/8) as is the support received from the National Institutes of Health, Bethesda, Maryland, during a research visit to the National Institute of Environmental Health Sciences, Research Triangle Park/North Carolina (Exchange Visitor Program G-05-111).

References

Adams, C.E. (1974) Asynchronous egg transfer in the rabbit. J. Reprod. Fert. 35, 613-614

Arthur, A.T., Cowan, B.D. and Daniel, J.C. Jr. (1972) Steroid binding to blastokinin. Fert. Steril. 23, 85-92

Beato, M. (1977) Physico-chemical characterization of uteroglobin and its interaction with progesterone. In: Development in Mammals (Ed.: M.H. Johnson) Vol. 2, 173-198, North-Holland Publ.Comp. Amsterdam, New York, Oxford

Beato, M. and Arnemann, J. (1975) Hormone-dependent synthesis and secretion of uteroglobin in isolated rabbit uterus. FEBS Letters 58, 126-129

Beato, M. and Baier, R. (1975) Binding of progesterone to the proteins of the uterine luminal fluid. Identification of uteroglobin as the binding protein. Biochim. Biophys. Acta 392, 346-356

Beato, M. and Beier, H.M. (1978) Binding of progesterone to the purified uteroglobin-like protein of the lung compared to uteroglobin. J. Reprod. Fert (MS submitted)

Beato, M. and Rungger, D. (1975) Translation of the messenger RNA for rabbit uteroglobin in Xenopus oocytes. FEBS Letters 59, 305-309

Beier, H.M. (1966) Das Proteinmilieu in Serum, Uterus und Blastocysten des Kaninchens vor der Nidation. In: Biochemie der Morphogenese, Chairman: W. Beermann, Deutsche Forschungsgemeinschaft, Konstanz

Beier, H.M. (1967) Veränderungen am Proteinmuster des Uterus bei dessen Ernährungsfunktion für die Blastocyste des Kaninchens. Verh. dt. zool. Ges. (Heidelberg 1967) 31, 139-148

Beier, H.M. (1968a) Uteroglobin: a hormone-sensitive endometrial protein involved in blastocyst development. Biochim. Biophys. Acta 160, 289-291

Beier, H.M. (1968b) Biochemisch-entwicklungsphysiologische Untersuchungen am Proteinmilieu für die Blastocystenentwicklung des Kaninchens (Oryctolagus cuniculus). Zool. Jb. Anat. 85, 72-190

Beier, H.M. (1970) Hormonal stimulation of protease inhibitor activity in endometrial secretion during early pregnancy. Acta endocr. Copenh. 63, 141-149

Beier, H.M. (1974) Oviducal and uterine fluids. J. Reprod.Fert. 37, 221-237

Beier, H.M. (1976) Uteroglobin and related biochemical changes in the reproductive tract during early pregnancy in the rabbit. J. Reprod. Fertil., Suppl. 25, 53-69

Beier, H.M. (1977) Immunologische und biochemische Analysen am Uteroglobin und dem Uteroglobin-ähnlichen Antigen der Lunge. Med. Welt 28, (N.F.), 788-792

Beier, H.M. and Beier-Hellwig, K. (1973) Specific secretory protein of the female genital tract. Acta endocr. Copenh. Suppl. 180, 404-425

Beier, H.M., Kirchner, C., and Mootz, U. (1978) Uteroglobin-like antigen in the pulmonary epithelium and secretion of the lung. Cell Tiss. Res. (in press)

Beier, H.M. and Kühnel, W. (1973) Pseudopregnancy in the rabbit after stimulation by human chorionic gonadotropin. Hormone Res. 4, 1-27

Beier, H.M., Kühnel, W.,and Petry, G. (1971) Uterine secretion proteins as extrinsic factors in preimplantation development. Adv. Biosci. 6, 165-189

Beier, H.M. and Maurer R.R. (1975) Uteroglobin and other proteins in rabbit blastocyst fluid after development in vivo and in vitro. Cell Tiss. Res. 159, 1-10

Beier, H.M., Bohn, H., and Müller, W. (1975) Uteroglobin-like antigen in the male genital tract secretions. Cell Tiss. Res. 165, 1-11

Beier, H.M. and Mootz, U. (1976) Oestrogen-mediated delayed secretion of uteroglobin in ovariectomized rabbits. Fifth Intern. Congr. Endocrinology Abstr. Vol. 182-183, Hamburg

Beier, H.M., Mootz, U., and Kühnel, W. (1972) Asynchrone Eitransplantationen während der verzögerten Uterussekretion beim Kaninchen. The 7th Int. Congr. Anim. Reprod. Artif. Insem. München, 3, 1891-1896

Beier, H.M., Mootz, U., and Kühnel W. (1977) Endokrinologische Studien an der östrogeninduzierten verzögerten Transformation und Sekretion des Kaninchenendometriums. Acta anatom. 99, 250

Beier, H.M., Petry, G., and Kühnel, W. (1970) Endometrial secretion and early mammalian development. In: 21st Coll. Ges. Biol. Chem. Mosbach 1970, Mammalian Reproduction. p. 264-285. Eds.: H. Gibian and E.J. Plotz. Springer, Berlin - Heidelberg - New York

Bullock, D.W. and Bhatt, B.M. (1976) Studies on uteroglobin messenger RNA.ICE-Satellite Symp. "Proteins and Steroids in Early Mammalian Development", Aachen 1976, in: Reproductive Endocrinology (Eds. H. M. Beier and P. Karlson) Springer-Verlag, Berlin - Heidelberg - New York (in press)

Bullock, D.W. and Connell, K.M. (1973) Occurrance and molecular weight of rabbit uterine "blastokinin". Biol. Reprod. 9, 125-132

Bullock, D.W. and Willen, G.F. (1974) Regulation of a specific uterine protein by estrogen and progesterone in ovariectomized rabbits. Proc. Soc. experim. Biol. Med. 146, 294-298

Bullock, D.W., Woo, S.L.C., and O'Malley, B.W. (1976) Uteroglobin messenger RNA. Translation in vitro. Biol. Reprod. 15, 435-443

Clauberg, C. (1930) Der biologische Test für das Corpus-Luteum-Hormon. Klin. Wschr. 9, 2004-2005

Daniel, J.C. Jr. (1971) Uterine proteins and embryonic development. Adv. Biosci. 6, 191-203

Feigelson, M., Noske, I.G., Goswami, A.K., and Kay, E. (1977) Reproductive tract fluid proteins and their hormonal control. Ann. N.Y. Acad. Sci. 286, 273-286

Hamana, K. and Hafez, E.S.E. (1970) Disc electrophoretic patterns of uteroglobin and serum proteins in rabbit blastocoelic fluid. J. Reprod. Fert. 21, 557-560

Johnson, M.H. (1974) Studies using antibodies to the macromolecular secretions of the early pregnant uterus. In: Immunology in Obstetrics and Gynecology. Proc. 1st Intern. Congr. Padua 1973. p. 123-133. Eds.: A. Centaro and N. Carretti. Excerpta Medica, Amsterdam

Johnson, M.H., Cowan, B.D., and Daniel, J.C. Jr. (1972) An immunologic assay for blastokinin. Fert. Steril. 23, 93-100

Kendle, K.E. and Telford, J.M. (1970) Investigations into the mechanism of the antifertility action of minimal doses of megestrol acetate in the rabbit. Brit. J. Pharmacol. 40, 759-774

Kirchner, C. (1969) Untersuchungen an uterusspezifischen Glycoproteinen während der frühen Gravidität des Kaninchens (Oryctolagus cuniculus). Wilhelm Roux Archiv Entw.Mech. Org. 164, 97-133

Kirchner, C. (1972) Immune histologic studies on the synthesis of a uterine-specific protein in the rabbit and its passage through the blastocyst coverings. Fert. Steril. 23, 131-136

Kirchner, C. (1976) Uteroglobin in the rabbit: I. Intracellular localization in the oviduct, uterus, and the preimplantation blastocyst. Cell Tiss. Res. 170, 415-424

Krishnan, R.S. and Daniel J.C. Jr. (1967) "Blastokinin": Inducer and regulator of blastocyst development in the rabbit uterus. Science 158, 490-492

Krishnan, R.S. and Daniel, J.C. Jr. (1968) Composition of "blastokinin" from rabbit uterus. Biochim. Biophys. Acta (Amst.) 168, 579-582

Kulangara, A.C. and Crutchfield, F.L. (1973) Passage of bovine serum albumin from the mother to rabbit blastocysts. II. Passage from uterine lumen to blastocyst fluid. J. Embryol. exp. Morphol. 30, 471-482

Maurer, R.R. and Beier, H.M. (1976) Uterine proteins and development in vitro of rabbit preimplantation embryos. J. Reprod. Fert. 48, 33-41

McCarthy, S.M., Foote, R.H., and Maurer, R.R. (1977) Embryo mortality and altered uterine luminal proteins in progesterone-treated rabbits. Fertil. Steril. 28, 101-107

McGaughey, R.W. and Murray, F.A. (1972) Properties of blastokinin: amino acid composition, evidence for subunits, and estimation of isoelectric point. Fert. Steril. 23, 399-404

Murray, F.A., McGaughey, R.W., and Yarus, M.J. (1972) Blastokinin: Its size and shape, and an indication of the existence of subunits. Fert. Steril. 23, 69-77

Noske, I.G. and Feigelson, M. (1976) Immunological evidence of uteroglobin (blastokinin) in the male reproductive tract and in nonreproductive ductal tissues and their secretions. Biol. Reprod. 15, 704-713

Petzoldt, U. (1974) Micro-disc electrophoresis of soluble proteins in rabbit blastocysts. J. Embryol. exp.Morphol.31, 479-487

Rahman, S.S.U., Velayo, N., Domres, P., and Billiar,R.B. (1975) Evaluation of progesterone binding to uteroglobin. Fert. Steril. 26, 991-995

Schlafke, S. and Enders, A.C. (1973) Protein uptake by rat preimplantation stages. Anat. Rec. 175, 539-560

Schwick, H.G. (1965) Chemisch-entwicklungsphysiologische Beziehungen von Uterus zu Blastocyste des Kaninchens (Oryctolagus cuniculus). Wilhelm Roux Archiv Entw.Mech.Org. 156, 283-343

Urzua, M.A., Stambaugh, R., Flickinger, G., and Mastroianni, L. Jr. (1970) Uterine and oviduct fluid protein patterns in the rabbit before and after ovulation. Fert. Steril. 21, 860-865

de Visser, J. (1976) Degeneration of rabbit ova by prefertilization progesterone treatment: effect on endometrium and uteroglobin secretion. ICE-Satellite Symp. "Proteins and Steroids in Early Mammalian Development", Aachen 1976, in: Reproductive Endocrinology (Eds.: H.M. Beier and P. Karlson) Springer-Verlag Berlin - Heidelberg - New York (in press)

DISCUSSION

Spilman: What was the dose of estradiol used in experiments designed to study the delayed formation of uteroglobin?

Beier: This was 100 µg/animal and a second injection of 100 or 150 µg/animal. That was the initial approach. Now we are able to reduce this injection to about 10-50 µg. We haven't the conclusive answer yet for a dose-response, but we are approaching a relatively low level.

Spilman: Post-coital treatment with estrogen could have an effect on the rate of egg transport as well as causing delayed secretion of uteroglobin. This might also have an effect on the number of implants.

Beier: This is a very important point. It is important to think about any involvement of egg transport, as well as other things, interferring with normal development. However, the point I want to focus your attention on is that the egg transfer experiments are giving more conclusive results, and that the morphology of the blastocysts in a desynchronized environment is very similar whether the rate of egg transport was greatly altered or not.

Spilman: I think the point you made about the immunological activity of uteroglobin *vs.* its biological activity is very important. Have you ever tested the uteroglobin from lung tissue in an *in vitro* embryo system? This may show you something that the trypsin inhibition test didn't show with regard to its "biological" activity.

Beier: That is something that must be done also for the uteroglobin-like antigens from the seminal plasma and oviduct. At the present time we can only say there is crossreactivity in the immunological test, and very similar steroid binding properties of the lung uteroglobin-like antigen.

Gadsby: Has anyone looked at the endocrine control of the secretion of uteroglobin from lung tissue?

Feigelson: Our group has looked at the endocrine control of the lung uteroglobin antigen. We do not see any alteration in the amount of antigen in the lung during pregnancy, nor do we see induction of the antigen in the lung after progesterone. We do, however, see a modest induction of the lung uteroglobin antigen in ovariectomized animals following cortisol administration. The lung of course is a known target tissue for cortisol. It seems that there is a differential hormonal control of this protein at each of its various sites. In fact in the oviduct uteroglobin antigen is not induced by progesterone; it is induced to some degree by estrogen in the oviduct. The uterus seems to be the only site where it responds to progesterone.

Beier: It seems to me that the reason why there is no significant progesterone effect in the oviduct is because the oviduct tissue doesn't show a remarkable amount of progesterone receptors. Again this fits in the whole feature that we have found on the receptor content of the delayed secretion uteri.

Feigelson: The differential hormonal control may indeed be due to the differential receptors for the various hormones at these sites. It is hard to believe that there would be many, or any, progesterone receptors in the lung.

Bullock: The only information we have on the lung message is that the proportion of total message activity accounted for by the uteroglobin message activity in the lung does not change between nonpregnant rabbits and day 4 of pregnancy, whereas over the same period the uterine mRNA for uteroglobin increases 11-fold.

I would like to mention some data which illustrate the incredible complexities of trying to look at what uteroglobin does by means of passive immunization. Rabbits were injected on Day 5 of pregnancy with anti-uteroglobin antiserum and various nonspecific antisera into the uterine lumen. The anti-uteroglobin was given in one horn and the control antiserum in the opposite horn. There are two reports in the literature, one by Krishnan and one by Johnson, indicating that this treatment will prevent implantation. These workers used nonimmune sera as controls, and it seemed desirable to reinvestigate this phenomenon using immune sera containing nonspecific antibodies as controls.

We found that monospecific goat anti-uteroglobin antiserum did prevent implantation in a roughly dose related manner. The first control serum we used was a goat anti-avidin antiserum, which also prevented implantation. Rabbit anti-aviden antiserum did not seem to inhibit implantation to quite as great an extent as did the goat anti-aviden. We then injected another rabbit antiserum, this time against goat IgG, which did not prevent implantation compared to control horns which received no treatment. Therefore, we began to suspect that it was the serum that the rabbits did not like rather than the antibodies that it contained. Next we injected normal goat serum, and this prevented implantation in a dose dependent manner. It seemed that if we could partially purify the antibodies from the goat sera, we might get rid of the noxious factors that were preventing implantation. We made an ammonium sulfate cut and repeated the experiment. The IgG fraction of the goat anti-uteroglobin still had an effect on implantation, although the results were quite variable. I suspect that these erratic results may be haphazard ones due to the proximity of the blastocyst to the injection site at the time of treatment. An IgG preparation of the goat anti-avidin had no effect on implantation at low doses, and the previous inhibition of the implantation was substantially reversed at higher doses. The gamma globulin fraction of the goat serum also showed the same reversal of effect. Although these data are preliminary, they illustrate the extremely rigorous controls that one must have to make any sense out of this sort of experiment; they do give some suggestion that anti-uteroglobin antibodies have some selective effect on suppressing implantation. If that is so, how can it work if we are only injecting very small amounts of antibody? We calculate that one would need to inject something like 800 mg of purified antibody protein to neutralize all of the uteroglobin in the uterus at this stage of pregnancy. We thought that perhaps the antibodies that we were injecting might bind to the blastocyst and thus prevent implantation. We tested this possibility by labelling the gamma globulin fractions of the antisera with fluorescein isothiocyanate, injected equivalent amounts of the antiserum into the uterine lumen on Day 6 of pregnancy, and then flushed out the blastocysts 6 hours later. We found that the uteroglobin antibodies bound to the blastocyst, and there was marked fluorescence around the zona and also around the surface of the trophoblast. Unfortunately, fluorescein-labelled anti-aviden IgG also bound quite nicely and showed marked fluorescence around the zona pellucida and to some extent around the trophoblast. The dose of antiuteroglobin that showed fluorescent binding prevented implantation, whereas the dose of anti-avidin that showed fluorescent binding did not prevent implantation. I don't think these experiments indicate much of a critical role for uteroglobin in implantation. I think there are possibly two things going on. At higher doses of anti-avidin there might be nonspecific binding which interferes with implantation in a physical way; at lower doses, anti-uteroglobin may bind specifically and interfere with some specific action of uteroglobin. I think that the whole business is incredibly complicated, and I quite agree with Dr. Beier that we are a long way from knowing what uteroglobin does.

Boving: I am reminded of a paper by A. C. Kulangara that may give a clue as to why there is some difficulty. In general, his conclusion was that the reports of various proteins within the rabbit blastocyst fluid must be treated with skepticism since the tendency is for proteins to enter into the space between the blastocyst and the lemmas very quickly. Unless he got his material out within just a few minutes, shrinkage occurred and he got spurious results. In addition, he found that many workers may have fallen into a technical trap in that if the blastocysts are recovered by flushing or rinsed to remove surface contamination there is a tendency for the fluids to change the permeability of the blastocyst. However, if the materials inside are sampled by withdrawal through needles, the chances are pretty good that contamination will be introduced by the needle. His evidence for this was that large needles always gave larger protein values for the fluid contents of blastocysts than did very small needles. I think that some of the problems that have come up may have a rather mechanical explanation.

9

Uteroglobin: Distribution in Tissues and Secretions of the Rabbit and Human and Endocrine Control

MURIEL FEIGELSON

INTRODUCTION

Oviducal and uterine secretions of the mammal provide the milieu in which critical reproductive and early developmental processes take place, *e.g.*, sperm capacitation and acrosome reaction, cumulus cell dispersion, fertilization, early embryogenesis, and implantation. Proteins secreted into the lumen of the female genital tract, especially those regulated by reproductive hormones, may indeed participate specifically and possibly crucially in one or more of these reproductive and early developmental events.

The goals of the studies of our laboratory were to identify such female reproductive tract proteins and explore their molecular nature, specificity, hormonal control, and ultimately their possible reproductive and/or developmental functions. We initially undertook an analysis of the protein patterns of rabbit oviducal and uterine fluids, in search of such reproductively relevant proteins (Feigelson and Kay, 1972). Our efforts were soon focused on the study of one protein, which seemed at that time to be unique to female genital tract secretions and responsive to ovarian hormones. We examined the molecular properties of this protein; inquired into its possible identity with uteroglobin, a protein known to be a constituent of uterine fluid of the pregnant rabbit; searched for its possible presence in the male reproductive tract and in non-reproductive tissues and secretions; and explored hormonal influences on this protein in the female genital tract and at extra-genital sites.

"CONE PROTEIN": MOLECULAR PROPERTIES AND IDENTITY WITH UTEROGLOBIN

The protein of interest was originally observed in oviducal fluid (Feigelson and Kay, 1972; Kay and Feigelson, 1972) and subsequently in uterine fluid (Goswami and Feigelson, 1974) of the estrous rabbit. It was found to be of low molecular weight (16,000 daltons), non-detectable in rabbit serum, and to migrate to the post-albumin region upon acrylamide gel electrophoresis under non-denaturing conditions. It could be seen to assume a cone-like configuration on such gels, if denatured and stained very briefly with Amido Black in 7% acetic acid and immediately destained electrophoretically (Kay and Feigelson, 1972) (Fig. 1). This protein was accordingly designated as "cone protein". Cone formation on acrylamide gels has been related to its low molecular weight and has served as a convenient, sensitive and specific marker for further studies on this protein.

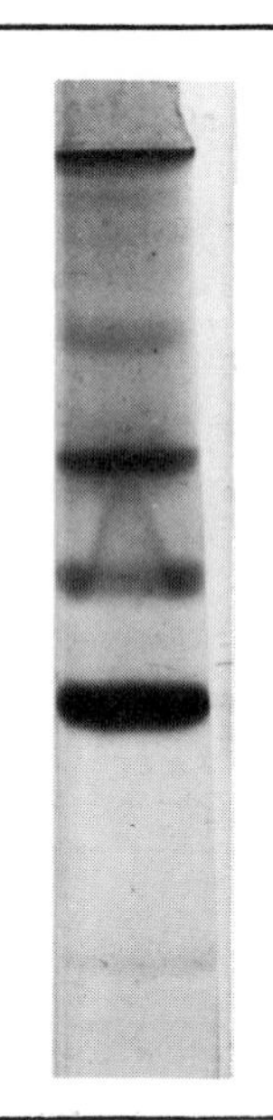

Fig. 1. Acrylamide gels electrophorogram of rabbit oviducal fluid, illustrating cone-formation of the post-albumin band.

During the early phases of this investigation, the possibility that the female genital tract "cone protein" was identical with the uterine fluid protein of the pregnant rabbit, known as uteroglobin (UG) (Beier, 1968) or blastokinin (Krishnan and Daniel, 1967) was raised. Although some of the properties of "cone protein" were in common with those reported for UG, e.g., electrophoretic mobility on acrylamide gels and inducibility in the uterus by progesterone (see below), a good many characteristics seemed dissimilar at that time. A molecular weight of 27,000-30,000 (Krishnan and Daniel, 1967; Beier, 1968; Urzua et al, 1970) had been assigned to UG and there was no evidence of its presence in uterine fluid of non-pregnant rabbits (Krishnan and Daniel, 1967) nor in oviducal fluid of any endocrine state. However, many of these properties originally attributed to UG have been subsequently modified. The molecular weight of UG has since been revised to 15,000 (Murray et al, 1972) and UG was observed in low quantities in estrous rabbit uterine (Bullock and Connell, 1973) and oviducal (Petzoldt, et al., 1972) fluids. Furthermore, binding of UG to estrogen and progesterone, as originally described by Urzua et al. (1970) and Arthur, et al. (1972), was not demonstrable in our laboratory in the case of the "cone protein", following rigorous separation from contaminating albumin (Goswami and Feigelson, 1974). However, subsequent studies in other laboratories have reported extremely weak steroid binding (Rahman et al., 1975) and the requirement for a strongly reducing environment in order to elicit significant progesterone binding (Beato and Baier, 1975); under non-reducing conditions, as in our experiments, steroid binding would not be expected. Thus, the apparent inconsistancies between the properties of UG and those of the "cone protein" were gradually reconciled.

Direct evidence of the immunological identity between "cone protein" and UG was forthcoming with the preparation, in our laboratory, of a monospecific antibody raised in the goat against "cone protein" purified to homogeneity from estrous rabbit uterine fluid (Noske and Feigelson, 1976). An antiserum prepared against pregnant rabbit uterine fluid UG was kindly provided by Dr. David Bullock of Baylor University,

Houston, Texas and absorbed to monospecificity in our laboratory. Precipitin bands of identity in the Ouchterlony double immunodiffusional system and precipitin arcs of identical mobility upon immunophoresis were observed when uterine fluid from pregnant rabbits diffused versus both of these antisera. At this time, therefore, all evidence points to the identity between the "cone protein" and UG. With view towards economy of terminology, this protein henceforth will be designated as UG in this communication.

TISSUE DISTRIBUTION OF UG IN THE RABBIT AND HUMAN

Employing both antisera, described above, in double immunodiffusion, immunoelectrophoretic, radial diffusion and "rocket" electroimmunophoretic systems, we have explored the distribution of the UG amongst various tissues and their secretions in both sexes of the rabbit and human; these results are summarized in Table 1.

UG has been immunologically detected in soluble extracts of tissues surrounding the female genital tract of the estrous and pregnant rabbit, *viz*., ovary, oviduct, endometrium, myometrium and vagina, tissues which may serve to elaborate and secrete this protein into the female genital tract lumen, where UG is found in significant amounts. Ovariectomy results in depletion of the UG antigen from each of these sites.

Of particular interest has been the finding that UG antigen is not confined to the female reproductive tract, where it has been presumed to play a significant role during early pregnancy (Krishnan and Daniel, 1967). Employing the two antisera in the immunological test systems cited above, a protein with immunological identity to UG in female tract tissue extracts and secretions and to purified uterine fluid UG and the capability of cone formation on acrylamide gels has been detected in the fluids and tissues of the male genital

Table 1. Distribution of uteroglobin antigen in tissues and secretions of the rabbit and human.

	Rabbit		Human	
	Tissue	Secretion	Tissue	Secretion
Reproductive Tract				
Female				
Ovary	+			
Oviduct	+	+		+
Endometrium	+	+	+	+
Myometrium	+			
Vagina	+			
Male				
Testis	-			
Cauda epididymis	+			
Vas deferens	+	+		
Seminal plasma		+		+
Non-Reproductive Tissues				
Male and Female				
Lung	+	+	+	
Bronchus	+	+		
Trachea	+	+		
Esophagus	+	+		
Jejunum		+		
Heart	-			
Spleen	-			
Liver	-			
Kidney	-			
Serum		-		-

(+) indicates the presence, (-), the absence, of UG, employing Ouchterlony double immunodiffusional, immunoelectrophoretic, radial immunodiffusional (Mancini et al., 1965) and "rocket" electro-immunophoresis (Laurell, 1966) test systems, utilizing goat antisera to UG, purified from uterine fluids from normal estrous rabbits and from rabbits during the first week of pregnancy.

tract, viz., epididymis, vas deferens tissue and secretions and seminal plasma (Noske and Feigelson, 1976). Beier et al., 1975, and Kirchner and Schroer, 1976, have likewise reported the presence of UG antigen in certain portions of the male genital tract.

It was of some interest to inquire as to whether UG was indeed exclusively relegated to tissues and secretions of reproductive tract of both sexes. UG has been undetectable in highly concentrated pre-preparations of serum and tissue extracts of all rabbit non-reproductive tissues in which it was sought (see Table 1), with the exception however of tissues and lumenal secretions of two organ systems comprised of ductal structures, the respiratory and digestive tracts (Noske and Feigelson, 1976). It was found that tracheal, bronchial and lung tissue extracts and secretions, as well as esophageal tissue extracts and fluids and jejunal secretions of rabbits of either sex all contain UG antigens which cross-react with identity and comparable electrophoretic mobility to female genital tract preparations and purified uterine fluid UG. Furthermore, when diffusing against either antiserum to uterine fluid UG, lung extracts incorporated into the agarose matrix of Ouchterlony diffusion plates were capable of complete absorbtion of the precipitin bands formed by UG from female tract fluids. Moreover, fractionation of lung tissue extracts by gel filtration yielded an immunologically reactive lung protein which co-eluted with its counterparts from estrous rabbit uterine and oviducal fluids. Gel electrophoresis of UG-rich fractions of lung extracts, formed the same cone-like structures emerging from the post albumin band, as did uterine fluid, when stained and destained for cone formation. These observations, taken together, provide evidence of the presence of a protein in rabbit respiratory and digestive tract tissues and lumenal secretions, particularly lung, with similar antigenic sites and molecular characteristics as those of female genital tract UG.

The possibility that a protein comparable to rabbit UG may be present in analagous human tissues and secretions has been explored. We have detected a protein in components of the reproductive tract of the human male and progestational female, viz., seminal plasma, endometrial extracts and uterine and oviducal secretions (Noske and Feigelson, 1976) and in human lung extracts (unpublished preliminary observations) which cross-react with anti-rabbit UG, (Table 1). However, preparations of human materials must be much more highly concentrated than their counterparts from rabbit to produce visible immunoprecipitin bands. Furthermore, human preparations form immunophoretic precipitin arcs which manifest shoulders and extensions not observed in analagous rabbit preparations. We thus conclude that a protein similar, although not identical, to rabbit UG is present in the same ductal tissues and secretions of the human as in the rabbit.

HORMONAL REGULATION OF UG

Exploration of the endocrine regulation of UG within the female reproductive tract, by means of densitometric evaluation of stained acrylamide gels, has revealed differential regulation of this protein within the oviduct and uterus (Goswami and Feigelson, 1974). Table 2 illustrates that UG is depleted to non-detectable values in the secretions of both ducts. A course of estradiol-17β treatment in such ovariectomized does induces the appearance of this protein in oviducal fluid and to an even greater magnitude in uterine fluid. Progesterone treatment, on the other hand, elicits only negligible amounts of UG in the oviducal fluid of these animals, however, markedly induces the protein in uterine secretions. UG levels in fluids retrieved from

Table 2. Differential responses of uteroglobin in oviducal and uterine fluids to ovarian hormones.

Endocrine State		Uteroglobin/Albumin x 10^3 * Oviducal Fluid	Uterine Fluid
Intact estrous	(5)	42 ± 1.3	33 ± 0.8
Ovariectomized	(5)	0	0
Ovariectomized + estradiol-17β	(4)	37 ± 1.2	363 ± 13.0
Ovariectomized + progesterone	(4)	3 ± 0.4	2975 ± 231.0

*Means ± SE of ratios of uteroglobin/albumin. Five days following ovariectomy rabbits received 10 daily doses of propylene glycol, 5 μg/kg estradiol-17β or 1 mg/kg progesterone. Fluids were collected by double ligation, electrophoresed on acrylamide gels and stained with Amido black. The post albumin band which forms a conical configuration was quantitated densitometrically. ( ), indicate number of rabbits. (Goswami and Feigelson, 1974).

oviducts and uteri during 3 to 5 days of gestation were consistent with the endocrine state of the animals, viz., UG levels were low in oviducal fluid and very marked in uterine fluid, due to progesterone domination of these animals.

Studies on the hormonal regulation of UG were extended to the lung, employing quantitative immunologic techniques. Levels of UG antigen in uterine fluids and lung extracts prior to and during the first week of pregnancy, following ovariectomy, and subsequent to steroid hormone administration were quantitated by radial immunodiffusion and "rocket" electroimmunodiffusion techniques. Table 3 illustrates that, whereas uterine fluid UG antigen levels of the 6-day pregnant rabbit are elevated approximately 30-fold above those of estrous does, the level of this antigen in lung is not significantly altered, (Feigelson, 1976; Feigelson et al., 1977). Table 3 further indicates that no UG is detectable in the uterine fluid of ovariectomized rabbits. A course

Table 3. Differential responses of uteroglobin antigen from rabbit uterine flushings and lung to pregnancy and steroid hormones.

Hormonal State	Uteroglobin Concentrations* (antigen units/μg protein)			
	Uterine Flushings	p-value	Lung	p-value
Estrous (Control)	5.5 ± 1.8 (6)		1.00 ± 0.12 (7)	
Pregnant (6 days)	178.0 ± 23.0 (12)	$< .01$	0.74 ± 0.10 (8)	$> .05$
Ovariectomized + diluent (Control)	0 (4)		1.06 ± 0.15 (4)	
Ovariectomized + estradiol-17β	5.7 ± 3.1 (4)		1.04 ± 0.24 (4)	$> .05$
Ovariectomized + progesterone	77.3 ± 9.6 (4)		0.97 ± 0.06 (4)	$> .05$
Ovariectomized + testosterone	76.9 ± 17.8 (4)		1.36 ± 0.36 (4)	$> .05$
Ovariectomized + cortisol	0 (3)		1.68 ± 0.10 (3)	$< .02$

* Mean levels of uteroglobin antigen ± SE, as quantitatively estimated employing radial immunodiffusion and Laurell "rocket" electroimmunophoresis (Laurell, 1966) utilizing goat antiserum prepared against uteroglobin isolated from estrous rabbit uterine fluid. Five days following ovariectomy, rabbits received 10 daily doses of 5 μg/kg estradiol-17β, 1 mg/kg progesterone, 25 mg/kg testosterone propionate or 20 mg/kg cortisol acetate (Feigelson, 1976; Feigelson, et al., 1977).

of estradiol or progesterone treatment in these animals induces this protein in uterine fluid to a modest and marked extent, respectively. These results corroborate findings employing non-immunologic quantitative procedures (Table 2). In contrast, Table 3 illustrates that lung UG antigen is not significantly altered following ovariectomy or administration of either ovarian steroid (Feigelson, et al., 1977). Thus, UG is differentially regulated by ovarian steroids in oviduct, uterus and lung. Neither estrogen nor progesterone regulate UG in lung; estrogen, but not progesterone, induces this protein in oviduct;

both ovarian steroids induce UG in the uterus, progesterone being the major inducer at this site.

The presence of UG antigen in male genital tract prompted exploration of the effects of androgen on UG titers. Table 3 illustrates that testosterone administration to ovariectomized does significantly induces the UG antigen in uterine fluid, with little effect of this protein in the lung (Feigelson et al., 1977).

Since lung is known to be a target tissue for glucocorticoids, (Giannopoulos et al., 1973; Kotas and Avery, 1971), the effects of cortisol on UG in lung and uterus were investigated. Table 3 illustrates that, whereas cortisol is ineffective in inducing uterine fluid UG antigen levels of ovariectomized rabbits, this steroid significantly (63%) induces lung UG antigen (Feigelson et al., 1977). Thus, all four steroid hormones tested demonstrate differential regulation of UG antigen in uterus and lung. Gonadal hormones of both sexes induce this protein in the uterus, but are inactive as regulators of the lung protein. Conversely, glucocorticoids are inactive in regulating uterine UG, but induce this antigen in the lung.

SUMMARY AND CONCLUSIONS

The low molecular weight glycoprotein, known as uteroglobin or blastokinin, as implied by the two terms used to designate it, has been thought to be localized in the female genital tract, where it has been presumed to play a role in pre-implantation processes during early pregnancy. Studies undertaken in this laboratory have established that this protein, or a protein with identical antigenic and other molecular properties, is present, as well, in the tissues and secretions of the male genital tract and in two non-reproductive organ systems comprised of ductal structures, the respiratory and digestive tracts of the rabbit. Furthermore, a protein with similar, although not identical, immunologic properties is observed in the human genital tract and lung of both sexes.

Considerable evidence points to the site of synthesis of UG as being the secretory epithelium common to each of these tissues, from which it is secreted into the respective lumena. In the uterus, radioactively labeled amino acids instilled into the lumen *in vivo* (Murray and Daniel, 1973; Bullock and Willen, 1974), perfused *in vitro* (Beato and Arneman, 1975), or added *in vitro* to endometrial cultures (Whitson and Murray, 1974) are, in each instance, preferentially incorporated into UG. Furthermore, a translatable mRNA coding for UG has been found to be present in endometrium (Beato and Rungger, 1975; Levey and Daniel, 1976 and Bullock et al., 1976). Both uteroglobin biosynthetic rates and mRNA levels are increased in uteri of pregnant or progesterone treated animals. A translatable mRNA coding for uteroglobin, whose levels are unaltered during pregnancy, and which presumably acts to direct UG synthesis *in vivo*, has been reported to be present in the lung (Bullock, 1977).

We have found that UG is subject to differential hormone regulation in at least three of its tissue sites: estrogen weakly induces this protein in the oviduct and uterus, but not in the lung; progesterone strongly induces UG in the uterus, where it effects high levels of this protein during the first week of gestation, but is inactive in the oviduct and lung; testosterone induces it in uterus, but not lung; glucocorticoid enhances UG in lung, but is inactive in the uterus.

The biological function of this hormonally sensitive protein, is, of course, of critical concern. Its role in promoting blastocyst development (Krishnan and Daniel, 1967) has not been confirmed (Maurer and Beier, 1976). Since progesterone binding by UG requires a strongly reducing local environment (Beato and Baier, 1975), its role as a steroid binding protein would seem to be physiologically problematic. Moreover, *a priori*, neither of these actions would seem to be its primary, or at least sole, biological function, if its distribution in the male genital tract and in non-reproductive tissues is considered. Thus, the underlying physiological action of this hormonally responsive secretory protein present in several reproductive and non-reproductive ductal structures remains enigmatic.

ACKNOWLEDGEMENTS

This study has been supported by NIH Grant, HD-05116. The author gratefully acknowledges the skillful technical assistance of Emily Scott.

REFERENCES

Arthur, A. T., Cowan, R.D. and Daniel, J. C. (1972). Steroid binding to blastokinin. Fertil. Steril. 23, 85-92.

Beato, M. and Arnemann, J. (1975). Hormone dependent synthesis and secretion of uteroglobin in isolated rabbit uterus. FEBS Letters 58, 126-129.

Beato, M. and Rungger, D. (1975). Translation of messenger RNA for rabbit uteroglobin in xenopus oocytes. FEBS Letters 59, 305-309.

Beier, H.M. (1968). Uteroglobin - A hormone sensitive endometrial protein involved in blastocyst development. Biochem. Biophys. Acta 160, 289-291.

Beier, H. M., Bohn, H. and Muller, W. (1975). Uteroglobin-like antigen in the male genital tract secretions. Cell Tiss. Res. 165, 1-12.

Bullock, D. W. (1977). *In vitro* translation of messenger RNA for a uteroglobin-like protein from rabbit lung. Biol. Reprod. 17, 104-107.

Bullock, D.W. and Connell, K. M. (1973). Occurrence and molecular weight of rabbit uterine "blastokinin". Biol. Reprod. 9, 125-132.

Bullock, D. W. and Willen, G. F. (1974). Regulation of a specific uterine protein by estrogen and progesterone in ovariectomized rabbits. Proc. Soc. Exp. Biol. Med. 146, 294-298.

Bullock, D. W., Woo, S. L. and O'Malley, B. W. (1976). Uteroglobin messenger RNA: Translation *in vitro*. Biol. Reprod. 15, 435-443.

Feigelson, M. (1976). Reproductive tract fluid proteins. In: Protides of the Biological Fluids, Vol. 24, pp. 143-146, H. Peeters, ed., Pergamon Press, Oxford.

Feigelson, M. and Kay, E. (1972). Protein patterns of rabbit oviducal fluid. Biol. Reprod. 6, 244-252.

Feigelson, M., Noske, I. G., Goswami, A.K. and Kay, E. (1977). Reproductive tract fluid proteins and their hormonal control. Ann. N. Y. Acad. Sci. 286, 273-286.

Giannopoulas, G., Shree, M. and Solomon, S. (1973). Glucocorticoid receptors in lung. II. Specific binding of glucocorticoids to nuclear receptor components of rabbit fetal lung. J. Biol. Chem. 248, 5016-5023.

Goswami, A. and Feigelson, M. (1974). Differential regulation of a low molecular weight protein in oviductal and uterine fluids by ovarian hormones. Endocrinology 95, 669-675.

Kay, E. and Feigelson, M. (1972). Estrogen modulated protein in rabbit oviducal fluid. Biochim. Biophys. Acta 271, 436-441.

Kirchner, C. and Schroer, H. G. (1976). Uterine secretion-like proteins in the seminal plasma of the rabbit. J. Reprod. Fertil. 47, 325-330.

Kotas, R. V. and Avery, M. E. (1971). Accelerated appearance of pulmonary surfactant in the fetal rabbit. J. Appl. Physiol. 30, 358-361.

Krishnan, R. S. and Daniel, J. C. (1967). "Blastokinin": Inducer and regulator of blastocyst development in the rabbit uterus. Science 158, 490-492.

Levey, I. L. and Daniel, J. C. (1976). Isolation and translation of blastokinin mRNA. Biol. Reprod. 14, 163-174.

Maurer, R.R. and Beier, H. M. (1976). Uterine proteins and development in vitro of rabbit pre-implantation embryos. J. Reprod. Fertil. 48, 33-41.

Murray, F. A. and Daniel, J. C. (1973). Synthetic pattern of proteins in rabbit uterine flushings. Fertil. Steril. 24, 692-697.

Murray, F. A., McGaughey, R. M. and Yarus, M. J. (1972). Blastokinin: Its size and shape and an indication of the existence of subunits. Fertil. Steril. 23, 69-80.

Noske, I. G. and Feigelson, M. (1976). Immunological evidence of uteroglobin (blastokinin) in the male reproductive tract and in nonreproductive ductal tissues and their secretions. Biol. Reprod. 15, 704-713.

Petzoldt, U., Dames, W., Gottschewski, G. H. M. and Neuhoff, V. (1972). Protein patterns in early developmental stages of the rabbit. Cytobiologie 5, 272-280.

Rahman, S. S. U., Velayo, N., Domres, P. and Billiar, R. B. (1975). Evaluation of progesterone binding to uteroglobin. Fertil. Steril. 26, 991-995.

Urzua, M. A., Stambaugh, R., Flickinger, C. and Mastroianni, L. (1970). Uterine and oviduct fluid protein patterns in the rabbit before and after ovulation. Fertil. Steril. 21, 860-865.

Whitson, G. L. and Murray, F. A. (1974). Cell culture of mammalian endometrium and synthesis of blastokinin _in vitro._ Science 183, 668-670.

10

Steroid Hormones and Their Synthesis in the Early Embryo

JOHN E. GADSBY and R. B. HEAP

Introduction

Ideas about the synthesis of active substances by the blastocyst that are involved in the processes of attachment and implantation, or in the maternal recognition of pregnancy, have been reviewed recently (Bullock, 1977; Heap et al., 1977). Experiments in rats and mice have shown that after ovariectomy a sequential treatment consisting of progesterone followed by a minute amount of oestradiol will allow normal implantation. In the rat, implantation is associated with a surge in oestrogen secretion which is ovarian in origin (see Heald, 1976). However, ovarian oestrogen is not essential in all species as implantation and normal gestation will occur in some animals treated only with progesterone after ovariectomy. These animals include rabbit, guinea-pig, hamster, monkey, sheep and pig (see Heap et al., 1977 for references).

If preimplantation oestrogen secretion is obligatory for the establishment of normal pregnancy, an extra-ovarian source of synthesis must be sought in these species. Adrenal synthesis of oestrogens seems an unlikely site of synthesis in early pregnancy. In normal circumstances, aromatization of C_{19} and C_{21} steroid precursors by adrenal tissue is a marginal activity which has been demonstrated in rodents and primates (Engel, 1973). In non-pregnant sheep, adrenal secretion of oestrogens in the form of oestrone (but not oestradiol-17β) has been found in animals with an autotransplanted adrenal (McCracken and Baird, 1969). The work reported by Harrison et al. (1972), however, shows that normal pregnancy can be established in the adrenalectomized ewe without the administration of exogenous hormones at the time of blastocyst attachment. The evidence in sheep indicates that ovarian progesterone is the only hormone required for normal implantation after removal of the ovaries or adrenals.

An alternative source of preimplantation oestrogen synthesis may reside in the blastocyst itself. Steroids have been detected in the blastocyst and blastocoel fluid of the rabbit. Seamark and Lutwak-Mann (1972) reported the presence of progesterone and smaller amounts of 17α-hydroxyprogesterone and 20α-dihydroprogesterone in blastocoel fluid. These authors considered that the steroids were not necessarily synthesized by the embryo, but may be transported to the blastocyst by way of the uterine secretions. A similar view was expressed as a result of the studies of Fuchs and Beling (1974) and Borland et al. (1977).

The determination of oestrogens in blastocysts recovered from rabbits has proved inconclusive. Dickmann et al. (1975) reported a variable concentration of oestradiol-17β, ranging from 0.3 - 2.7 pg/blastocyst, but the absence from their paper of radioimmunoassay validation precludes an assessment of the significance of these low values. In a more extensive study Borland et al. (1977) found evidence of oestradiol-17β in 6 of 17 groups of blastocysts recovered between 110 - 159h after mating. Positive values in blastocysts were frequently correlated with positive values in uterine flushings, indicating that the steroid may have been absorbed from maternal sources rather than synthesized by embryonic tissue.

A different approach to the problem of whether the blastocyst is a steroid-synthesizing tissue has been adopted by Dickmann and co-workers (see Dickmann et al., 1976 for references). These workers have claimed on histochemical grounds the existence of enzymes involved in the pathway of steroid synthesis; Δ^5-3β-hydroxysteroid dehydrogenase (3β HSD; 3(or 17)β-hydroxysteroid : NAD(P) oxidoreductase, EC 1.1.1.51) using dehydroepiandrosterone (DHA) as substrate; and 17β-hydroxysteroid dehydrogenase (17βHSD, oestradiol-17β; NAD 17-oxidoreductase, EC 1.1.1.62) with oestradiol-17β as substrate. These enzymes were found in preimplantation blastocysts of rats, mice, hamsters and rabbits. The enzyme, 3βHSD was also found in pig blastocysts up to the 8-cell stage. Using a similar procedure Flood (1974) found no hydroxysteroid dehydrogenase activity in pig blastocysts at Day 11 of pregnancy (Day 0 is the day of first service) but described the histochemical localization of an enzyme which utilized 3β-hydroxy-5β-androstan-17-one as substrate in blastocyst tissue taken at Day 13 of pregnancy. When DHA was used as substrate, the reaction was less marked. By Day 20, the intensity of the histochemical response for both substrates had increased to levels normally found in the fully developed placenta. Dickmann and co-workers (1976) have deduced from their histochemical and biological studies that the preimplantation blastocyst has steroidogenic properties and that their synthesis of oestrogens is of functional significance in the process of implantation.

The term, steroidogenesis, implies steroid synthesis from non-steroidal precursors such as acetate. In this respect, the histochemical evidence adduced by Dickmann and co-workers is inconclusive, as it relates to only two of numerous enzyme systems that are involved in the synthesis of oestrogens, and fails to demonstrate the presence of a crucial step, the aromatization of neutral steroids. A further reservation derives from the biochemical studies of Sherman and Atienza (1977) which did not substantiate these histochemical observations. Mouse preimplantation embryos failed to convert [^{3}H]pregnenolone to progesterone, or to metabolize [^{3}H]DHA to androstenedione in culture. An active 3βHSD was demonstrated in culture only after trophoblast outgrowth and at a stage that corresponded with the post-implantation period in vivo. A similar finding was obtained in the rat (Marcal et al., 1975). However, in view of the fact that pre-implantation oestrogen in rats and mice is ovarian in origin, these results do not disprove the occurrence of blastocyst oestrogen synthesis in the rabbit, a species, as we have already noted, in which ovarian oestrogen secretion is not essential for implantation.

Huff and Eik-Nes (1966) were the first to provide biochemical proof of steroidogenesis in rabbit blastocysts. Six-day old rabbit blastocysts were incubated in a defined medium with labelled steroid precursors. Acetate was converted into pregnenolone. Pregnenolone, 17α-hydroxypregnenolone, progesterone and androstenedione were converted into labelled products. There was no evidence for the conversion of pregnenolone to progesterone, though small amounts of compounds with a chromatographic mobility similar to 17α-hydroxypregnenolone, testosterone and androstenedione were obtained. When blastocysts were incubated with 17α-hydroxypregnenolone, compounds resembling androstenedione and dehydroepiandrosterone (DHA) were formed but their identity was not established by recrystallization. This work provided no evidence for the synthesis of hormonal steroids such as progesterone and oestrogens by rabbit blastocysts.

The studies described in the present paper provide biochemical proof of the in vitro synthesis of hormonal steroids by blastocyst tissue in the pig. Evidence has been obtained of oestrogen synthesis before the time that the embryonic tissue becomes attached to the uterine epithelium and we shall consider the probable role of these steroids in the establishment of pregnancy.

Morphology of pig blastocyst

The pig blastocyst is a valuable experimental model for the study of the ontogeny of steroidogenesis in embryonic tissue because it provides a large amount of cellular material for biochemical study, and the phases of development are relatively prolonged compared with those in rodents. The zona pellucida is lost at about Day 6 - 8 when its diameter is about 2 mm, but definitive attachment associated with the formation of interlocking microvilli does not take place until Day 18. The trophoblast does not invade maternal tissue in this species and placentation is diffuse. Placentation differs from that of the ewe and cow in which the uterine epithelium becomes symplasmic.

The pig blastocyst at Day 10 consists of a flaccid sac with a diameter of 5 - 10 mm. After the next 48h the blastocyst undergoes rapid elongation to a length of approximately 1 m. Points of loose attachment occur on Day 14 and trophoblast cells become closely apposed to the uterine epithelial cells which show apical protruberances (Crombie, 1972). After blastocyst elongation the embryo undergoes marked changes with the formation of the amnion, the extension of the mesoderm between trophoblast and endoderm, the formation of the exocoel, and the differentiation of a transient yolk-sac. By Day 18 the allantoic vesicle has started to grow out into the exocoel, and subsequently develops rapidly as the yolk-sac regresses.

Hormone of early pregnancy in the sow

Figure 1 summarizes changes in the plasma concentrations of progesterone and prostaglandin $F_{2\alpha}$ and in the urinary concentration of oestrone sulphate during the oestrous cycle and pregnancy.

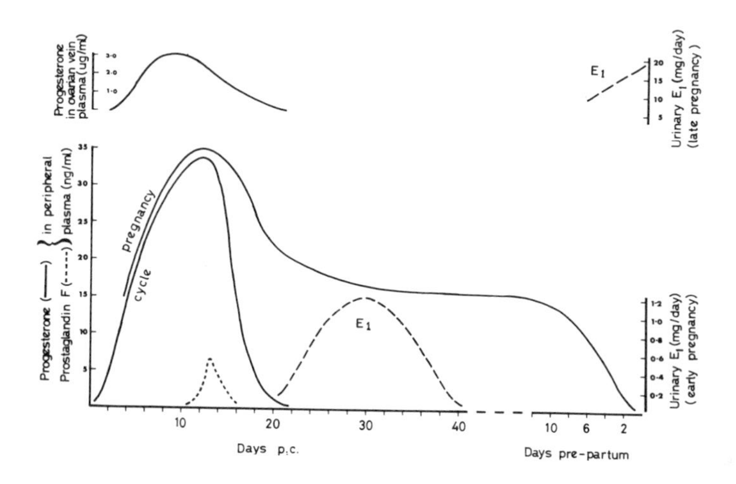

Fig. 1. The pattern of hormone changes throughout pregnancy in the sow. Diagram shows progesterone concentration in ovarian vein during the oestrous cycle (21 days), and in peripheral blood during the oestrous cycle and pregnancy (114 days); prostaglandin $F_{2\alpha}$ concentration during the oestrous cycle in the uterine vein; and urinary oestrone (mainly oestrone sulphate) excretion in early and late gestation. Data compiled from the work of Masuda et al. (1967); Guthrie et al. (1972); Ash and Heap (1975); Gleeson et al. (1974), and Fèvre et al. (1968).

In the luteal phase of the oestrous cycle, the plasma progesterone concentration in the anterior vena cava increases to about 35ng/ml by Day 12 after oestrus, and declines rapidly thereafter. The frequency and amplitude of pulses of prostaglandin $F_{2\alpha}$ ($PGF_{2\alpha}$) release measured in the utero-ovarian vein increase after about Day 12, coincident in some instances with the sharp decline in progesterone concentration. The onset of $PGF_{2\alpha}$ release is associated with an increased sensitivity of the corpus luteum to the luteolytic action of this uterine factor (Guthrie and Polge, 1976). In pregnancy the decline in plasma progesterone concentration is arrested soon after Day 12, and by Day 24 it has decreased only to 15 - 20ng/ml. Plasma levels of about

10ng/ml are maintained for the remainder of gestation until just before the onset of parturition when a rapid fall is observed (Ash and Heap, 1975). Progesterone secretion during gestation in this species is ovarian in origin since removal of the ovaries at any stage results in abortion. A small contribution, however, is derived from the gravid uterus in late pregnancy as the progesterone concentration in umbilical venous blood is greater than that in the maternal artery (Barnes et al., 1974).

Oestrogen production in pregnancy is biphasic (Figs. 1 and 2).

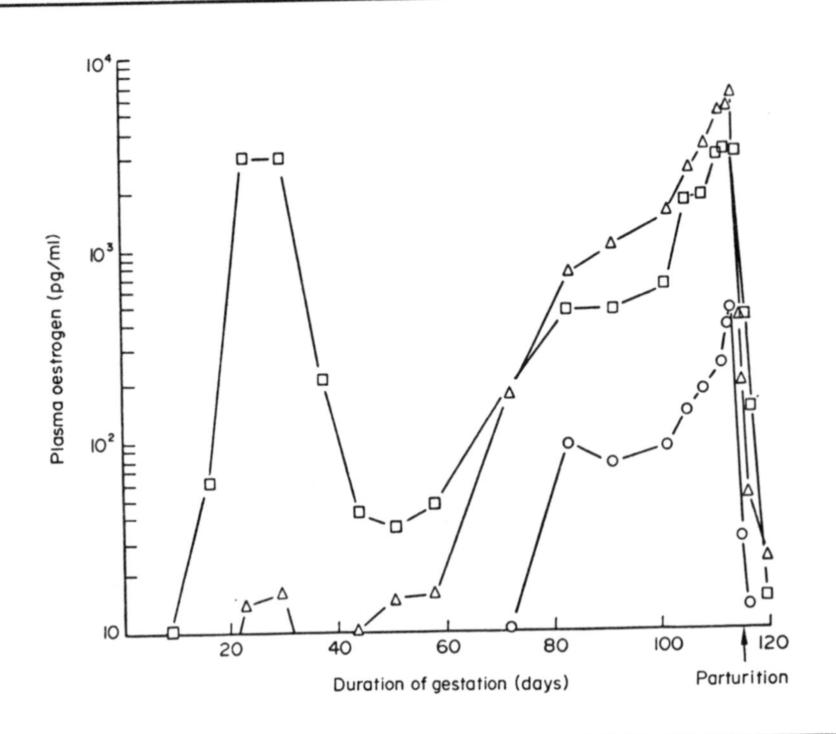

Fig. 2. The concentration of unconjugated oestrone (Δ), oestradiol-17β (O), and oestrone sulphate (□) in the plasma of a gilt at implantation, throughout gestation, and at parturition (Robertson & King, 1974, by kind permission).

Urinary oestrogen excretion shows a peak at about 25 - 30 days of pregnancy, followed by a second peak which reaches maximum values near term. Fèvre et al., (1968) showed that the conjugated oestrone excreted was uterine in origin as ovariectomy or maternal hypophysectomy did not reduce the rate of production. The measurement of conjugated plasma oestrogens in the anterior vena cava (Fig. 2) follows a similar biphasic pattern of secretion. Oestrone sulphate is the main conjugated oestrogen and it is first detected on about Day 16 and reaches maximum values between Days 23 and 30 (Robertson and King, 1974). This concentration then declines and begins to rise a second time after about Day 60 to peak values just before delivery. During the first peak of oestrone sulphate production, unconjugated oestrone is undectable before Day 20, and then reaches a value of only 15pg/ml compared with a concentration of oestrone sulphate of 3ng/ml. During the second phase of oestrogen production, the concentration of unconjugated oestrone is very high and similar to that of oestrone sulphate. High concentrations of unconjugated oestradiol are also present during the same period, contrasting with the absence of oestradiol in circulation during the first phase of oestrogen production. It is notable from the studies of Robertson and King (1974) that neither oestradiol nor oestrone could be detected in an unconjugated form in circulation before, or during the time of definitive attachment at Day 18. Guthrie et al. (1972), however, reported significant values of total unconjugated oestrogens of about 20pg/ml during the first 24 days of pregnancy, though these were no higher at the time of attachment than those found during the oestrous cycle. The determination of plasma oestrogens, therefore, provides no evidence of a preimplantation surge in the ovarian secretion of unconjugated oestrogens in this species, and, as we have already mentioned, progesterone is the only exogenous hormone required for normal attachment after ovariectomy in early pregnancy.

Blastocyst steroid hormones

Blastocyst tissue taken from 14 and 17 day pregnant sows was found to contain 2.23 and 0.86ng/g wet weight of total unconjugated oestrogen, respectively (Perry et al., 1973). This finding has subsequently been confirmed and extended by an investigation of the identity of blastocyst oestrogens (Fig. 3). Blastocyst extracts were subjected to

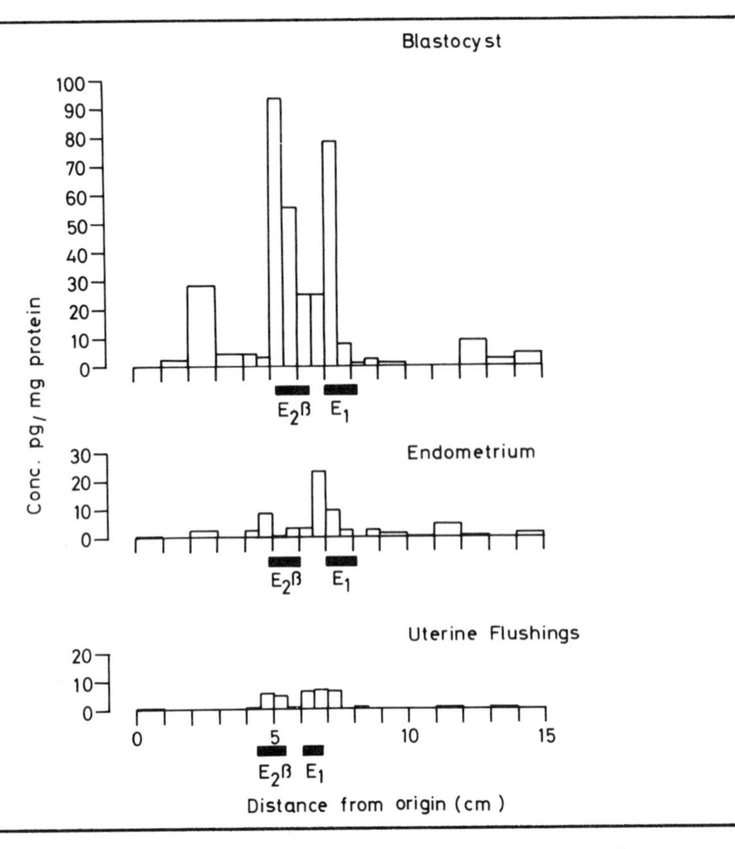

Fig. 3. Oestrogen concentration in the blastocyst, endometrium and uterine flushings of a 17-day pregnant sow. Ether extracts of the total blastocyst and uterine flushings and of a sample of endometrium were subjected to thin layer chromatography (TLC) in cyclohexane:ethylacetate, 1:1. The sample lane and an adjacent blank lane were eluted in 1cm bands or in 0.5cm bands in the region corresponding to the positions of oestrone (E_1) and oestradiol-17β ($E_2\beta$). Aliquots of the eluates were assayed for oestrogen by radioimmunoassay using an antiserum (BF 461/6) with a high cross-reaction against oestradiol-17β and oestrone. The results shown above are expressed as the concentration of oestradiol-17β equivalents in each band eluted from the TLC plate, corrected for the protein of the sample. The black bars represent the relative positions of authentic E_1 and $E_2\beta$ markers run on the same plate (Gadsby, 1977).

thin-layer chromatography and radioimmunoassay. For comparison, a similar study was carried out on the uterine flushings and on endometrial tissue taken from a pregnant sow on Day 17 after mating (Fig. 3). This study showed that blastocyst tissue contained immunoreactive oestrogens corresponding in chromatographic mobility to oestrone and oestradiol-17β. The concentration of these oestrogens in blastocyst tissue was appreciably greater than in either the endometrium or the uterine flushings, when expressed per unit weight of protein.

An extensive study of endogenous steroids has been carried out using a celite micro-column technique for the chromatographic separation of steroids, prior to radioimmunoassay (Fig. 4). This investigation was designed to observe the occurrence of progesterone and oestrogens in the blastocyst and embryonic tissues in relation to the establishment of pregnancy. In addition these hormones were assayed in the endometrium and uterine flushings of the same animals. Figure 4 shows the endogenous steroid concentrations of blastocyst and embryonic tissue removed from the uteri of pregnant sows at different stages of early pregnancy.

Blastocysts recovered on Day 12 or 13 were still spherical in shape or just beginning to elongate. They contained a high concentration of progesterone which had decreased in elongated blastocysts at 14 to 16 days of gestation. The progesterone concentration of embryonic tissues (18 to 21 days) taken after definitive attachment increased to a value similar to that found before blastocyst elongation. The concentration of oestrone and oestradiol-17β showed a similar pattern, with peak values at 12 to 13 days, which decreased to about 70 or 80pg/mg protein at Day 16. The concentration of oestrone increased slightly as attachment proceeded.

Endometrial tissue taken from the same animals was found to contain a progesterone concentration similar to that in blastocyst and embryonic tissue varying between 80 and 270pg/mg, but the concentrations of oestrone and oestradiol-17β were found to be very much lower (less than 10pg/mg). Endometrial progesterone concentration reached a maximum value at Day 16, while oestrone showed a peak at Day 20 - 21 (25pg/mg) and oestradiol-17β remained low throughout.

Figure 5 shows the concentrations of progestagens, oestrone and oestradiol-17β in uterine flushings taken in the same series.

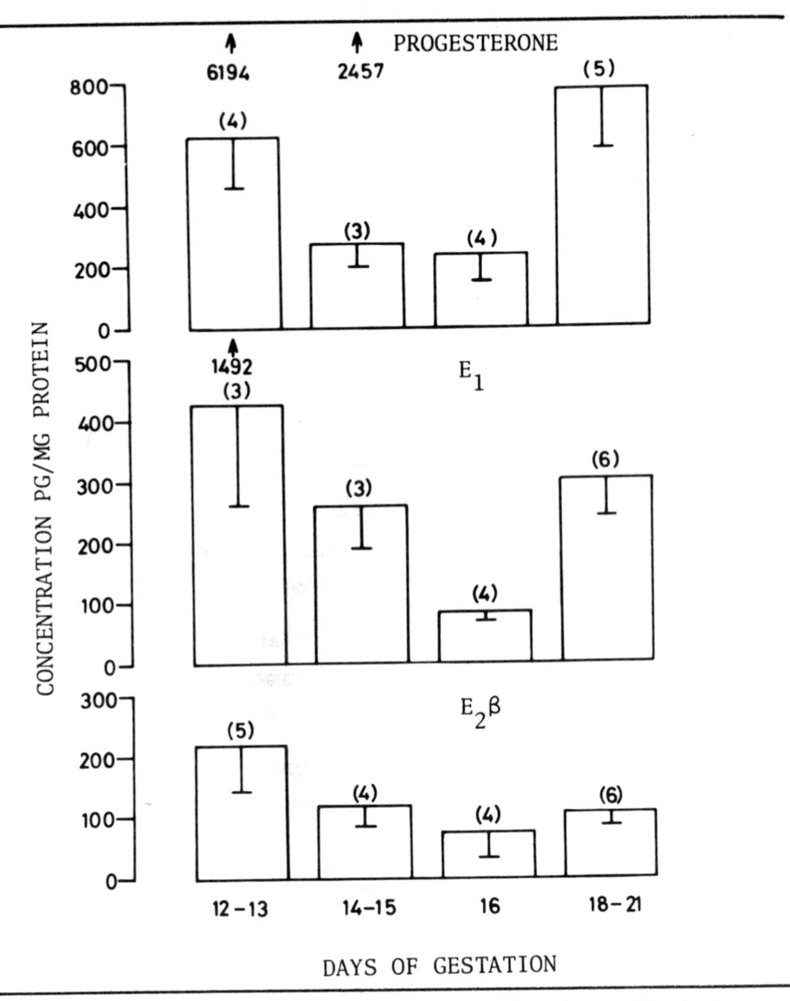

Fig. 4. Steroids in pig blastocysts and embryonic tissue flushed (12-18 days) or dissected (20-21 days) from the uterus. Tissues were homogenized, extracted with ether, and ether extracts were applied to celite micro-columns in iso-octane. Fractions were eluted, containing progesterone, oestrone and oestradiol-17β. Aliquots were taken for radioimmunoassay using antiserum BF 465/6 for progesterone (Heap et al. 1973), BF 461/6 for oestrone, and BF 510/5, specific for oestradiol-17β. The results corrected for procedural losses and protein content are given as mean ± S.E.M. (numbers of animals in brackets). Three exceptionally high values have been omitted from the calculation of the mean and are shown below small vertical arrows.

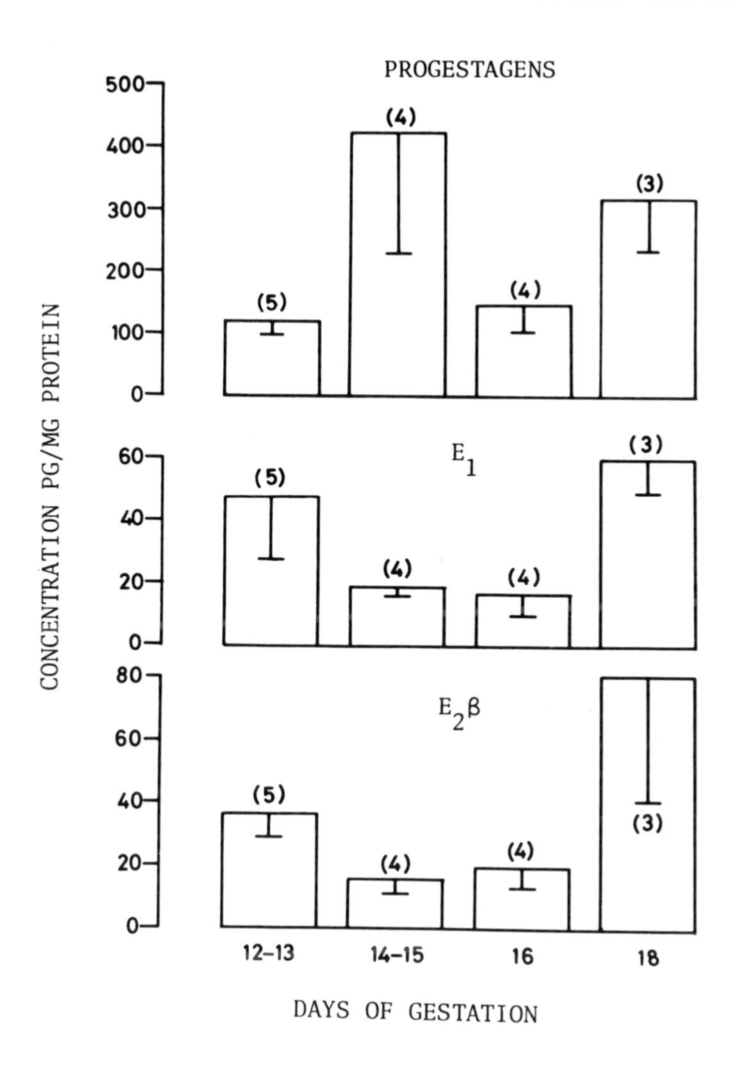

Fig. 5. Steroid concentrations in uterine flushings of pregnant sows. After removal of blastocyst tissue by centrifugation at 600g, uterine flushings were lyophilised. The residue was dissolved in distilled water and extracted with diethyl ether. Chromatographic separation on celite columns and subsequent radioimmunoassay was carried out as described before (see legend Figure 4).

During the period of study the progestagen values fluctuated between 120 and 425pg/mg and were similar in magnitude to those found in endometrial tissue. The term 'progestagen' is used in this context because it was found in an earlier experiment that uterine flushings recovered at 17 days of pregnancy contained very little progesterone but a substantial amount of a related compound which had a chromatographic mobility similar to a 5-pregnanedione reduced at C_5. This compound showed an appreciable cross-reaction with the antiserum, BF 465/6 (Gadsby, 1977). The concentrations of oestrone and oestradiol-17β in uterine flushings were intermediate between those found in embryonic and maternal tissues. This suggests the existence of a diffusion gradient from embryo to endometrium. At 12 to 13 days, for example, the oestradiol-17β concentrations were:-

blastocyst, 220pg/mg
uterine flushings, 36pg/mg
endometrium, 0.3pg/mg.

Considering the low plasma concentration of oestrogen present in the maternal circulation at this time the results provide strong evidence of the embryonic synthesis of oestrogens in vivo, and that the capacity for this synthesis is present as early as Day 12 to 13. However no such firm conclusion can be drawn regarding the origin of progesterone in the blastocyst. Progesterone also appears in the blastocyst at Day 12 to 13 when plasma progesterone levels are at their maximum (about 35 to 40ng/ml, Fig. 1), and this seems the most probable source. But these findings do not exclude the possibility that the blastocyst is capable of synthesising its own progesterone. Nevertheless, progesterone, whether embryonic or maternal in origin, may be an important precursor of oestrogen synthesis in vivo.

Oestrogen synthesis by the blastocyst

Previous studies carried out in our laboratory have shown that elongated pig blastocyst tissue, incubated in vitro with neutral steroid precursors such as [^{3}H]androstenedione or [^{3}H]DHA, is capable of producing phenolic compounds, which can be resolved into [^{3}H]oestrone and [^{3}H]oestradiol-17β by repeated chromatography (Perry et al., 1973; Perry et al., 1976). Definitive evidence for the synthesis of oestrone and oestradiol-17β, from these labelled precursors, has been recently obtained using the technique of recrystallisation to constant specific activity, after the addition of authentic unlabelled steroid. In addition, this tissue was also shown to be able to synthesize oestrone and oestradiol-17β from progesterone and oestradiol-17β from pregnenolone. These conversions were only demonstrated when the incubation was carried out in the presence of exogenous cofactors (Gadsby et al., 1976). Thus it is apparent that the elongated blastocyst is capable of synthesizing oestrogens in vitro from pregnenolone, progesterone, DHA and androstenedione probably according to the pathway shown in Figure 6.

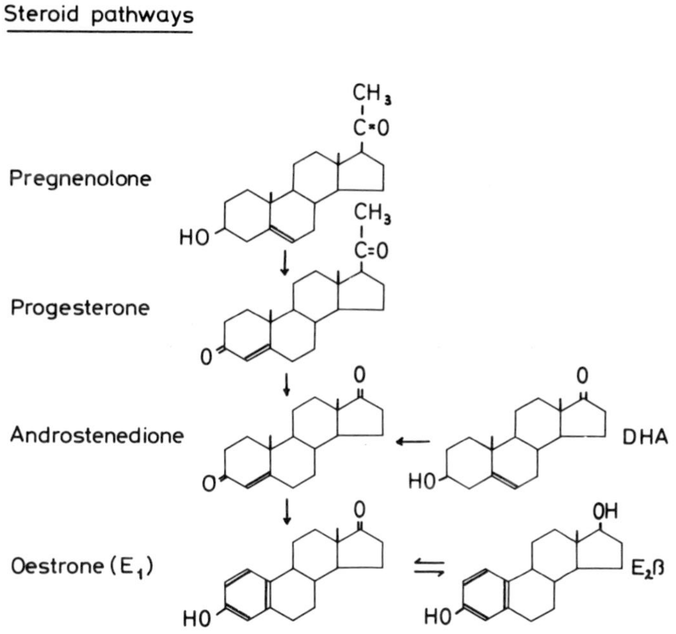

Steroid pathways

Pregnenolone

Progesterone

Androstenedione

DHA

Oestrone (E_1)

$E_2\beta$

Fig. 6. Pathways of steroid metabolism and oestrogen synthesis in the pig blastocyst. Definitive evidence has been obtained for the synthesis in vitro of oestrone and oestradiol-17β from pregnenolone, progesterone, androstenedione and dehydroepiandrosterone (DHA), and for the conversion of oestrone to oestradiol-17β.

With regard to the physiological role of oestrogen synthetic capacity, it became important to establish the time of onset of this synthesis. Incubations in vitro were carried out as described earlier with blastocyst and embryonic tissue at different stages of gestation. The results of this study, with [^{3}H]DHA as the precursor, are shown in Figure 7.

The ability to synthesize oestrogens is first apparent at Day 12, although this activity is variable, suggesting that the relevant enzymes are produced about this time. Detection of these enzymes at Day 10, may have been precluded by the small amount of tissue which is available at this stage. The appearance of oestrogen synthetic activity in vitro corresponds with the occurrence of significant quantities of oestrone and oestradiol-17β in blastocysts in vivo (Fig. 4). After attachment on Day 18 the incorporation of precursor into oestrogens based on wet weight declined. This decline has been associated with the development of the allantochorion (Perry et al., 1976). However, as the blastocyst enlarges and the total amount of tissue increases, it is probable that the total capacity for oestrogen synthesis would be very high indeed. A similar pattern of change was observed in the endogenous concentrations of steroids before definitive attachment.

Oestrogens in circulation at this time of early pregnancy are in a conjugated form. There is an apparent discrepancy between this finding and our present results, which show that oestrogens synthesized by the blastocyst consist of the unconjugated form of oestrone and oestradiol-17β. This discrepancy can be explained by the presence of an active sulphotransferase enzyme system which converts oestradiol-17β to oestrone (Pack and Brooks, 1974; Heap et al., 1977). Thus, it has been proposed that the oestrogen secreted by the blastocyst is conjugated in the endometrium and enters the plasma in this form as depicted in Figure 8.

Steroid metabolism by the blastocyst

An analysis by thin-layer chromatography of the radioactivity that remains after extraction of the oestrogens, demonstrated that pig blastocyst tissue is also able to metabolise pregnenolone, progesterone, DHA and androstenedione, almost completely. So far the

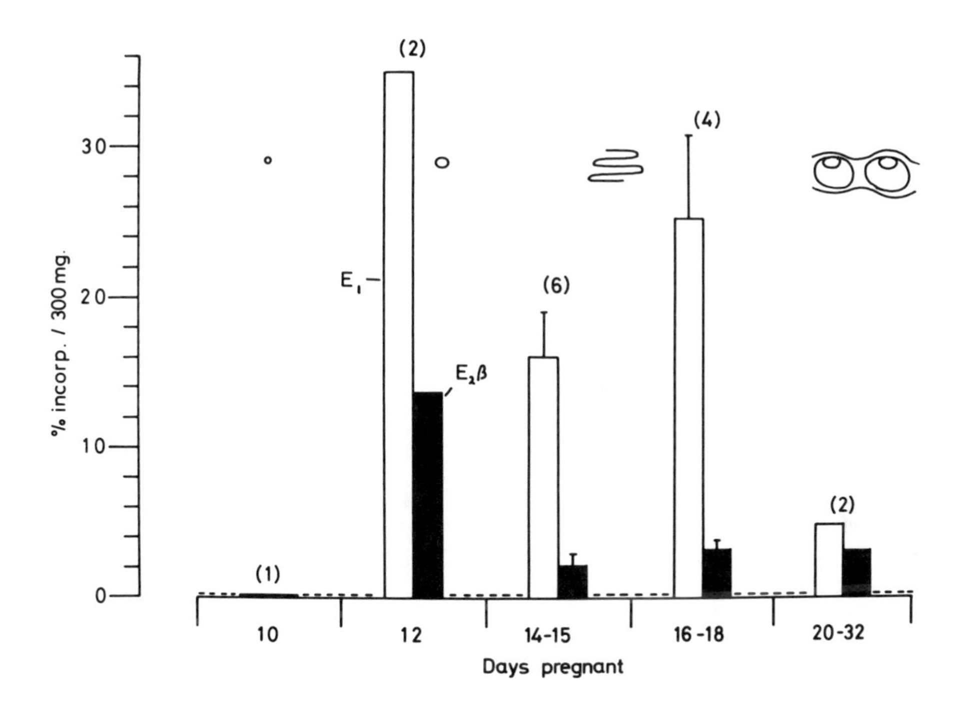

Fig. 7. Oestrogen synthesis from [^{3}H]dehydroepiandrosterone (DHA) in vitro by early embryonic tissue in the pig, between Day 10 and Days 20 - 32 of pregnancy. The percentage incorporation of DHA per 300 mg fresh tissue (except at Days 10 and 12 when 7 and 3 blastocysts per flask were used, respectively) into oestrone (E_1, open columns) and oestradiol-17β ($E_2\beta$, black columns) was measured. The horizontal dotted line indicates the level of incorporation found in control incubations in the absence of trophoblast tissue, or in the presence of endometrial or myometrial tissue. Symbols above the columns indicate the stages of embryo development from a blastocyst of 2mm diameter at Day 10 post coitum (p.c.) to an elongated bilaminar blastocyst of 1m length at Days 14 and 15 p.c., and to a distended chorionic sac at Days 20 to 32 p.c. Numbers of animals are given in brackets.

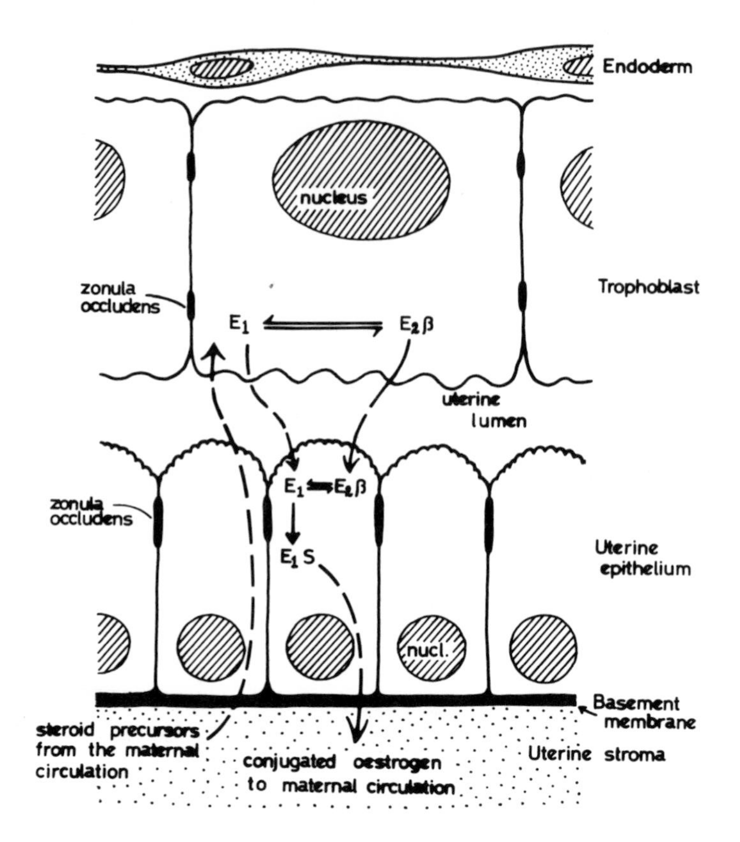

Fig. 8. Postulated interaction between trophoblast and uterine endometrium of the pig, early in pregnancy, based on in vitro experiments using isotopically labelled steroid precursors. The embryonic tissue synthesizes the hormones oestrone (E_1) and oestradiol ($E_2\beta$), whereas the endometrium converts oestrone to oestrone sulphate (E_1S), which is hormonally inactive.

compounds obtained are unidentified, but a majority of them are chromatographically more polar than the added precursor, suggesting that enzymes involved in hydroxylation and reduction also exist in this tissue. This active metabolism may explain why attempts to obtain biochemical evidence of progesterone synthesis from pregnenolone, have not been successful. However, the existence of the enzyme 3βHSD, is implied by the synthesis of oestrogens from DHA (Fig. 6).

The detection of large concentrations of progesterone in the blastocyst may be attributable to the existence of some mechanism, possibly a binding protein, which protects the steroid in vivo from metabolism. In rabbit blastocysts, more than 90% of the progesterone was found to be in the blastocoel fluid (Borland et al., 1977). If this is true of the pig blastocyst, progesterone may be compartmentalized and protected from metabolizing enzymes within the trophoblast cells.

Further proof of the steroid metabolizing properties of the pig blastocyst is to be found in the metabolism (as well as synthesis) of oestrogens. Evidence has been obtained for the formation of a small amount of oestradiol-17α and of an unknown phenolic compound from oestrone. This unknown compound is produced in an appreciable amount and it is identical, after thin-layer chromatography and after acetylation, to oestrone. However, it can be distinguished from oestrone after recrystallization to constant specific radioactivity.

CONCLUDING REMARKS

We have described experiments concerned with the question of whether the blastocyst synthesizes steroid hormones, such as progesterone and oestrogens, or absorbs them from maternal circulation. The findings show that the pig blastocyst contains a high concentration of oestrone and oestradiol-17β, and that it has the capacity to synthesize these steroids in vitro from as early as Day 12. A decreasing gradient exists in endogenous oestrogen concentrations within the gravid uterus, with a high concentration in blastocyst tissue and a low concentration in the endometrium and peripheral plasma.

The precursors required for blastocyst oestrogen synthesis in vivo have not been conclusively determined, but studies in vitro indicate that progesterone as well as C_{19}-neutral steroids may be utilized.

High concentrations of endogenous progesterone have been found in blastocyst tissue as observed previously in the rabbit. The high concentrations of progesterone in plasma and endometrial tissue during early pregnancy in the pig may provide a source of blastocyst oestrogen precursor.

In parallel with the acquisition of oestrogen synthetic activity in blastocyst tissue, the endometrium develops enzymes that modify the action of oestrogens by the oxidation of oestradiol-17β to oestrone, and by the sulphoconjugation of oestrone. The hypothesis has been proposed that a product of these blastocyst-endometrial interactions is the secretion of oestrone sulphate in early pregnancy which may undergo hydrolysis and metabolism in the hypothalamus and in corpora lutea.

Oestrogens in early pregnancy have been implicated in the mechanism of implantation in several species. Blastocyst oestrogen synthesis in the pig may also be involved in the mechanism of attachment as this event occurs within a few days after the onset of steroid synthesis by blastocyst tissue. A possible luteotrophic effect of blastocyst oestrogen synthesis and its involvement in the maternal recognition of pregnancy have been discussed (Perry et al., 1976; Heap et al., 1977).

Acknowledgments

J.E.G. received an A.R.C. Research Scholarship. We thank Dr B.J.A. Furr for gifts of antisera, Mr R.W. Ash, Mr J. Smith, and Dr C. Polge for experimental animals, Mr R.B. Burton for technical assistance and Mrs D. George for typing the manuscript.

References

Ash, R.W. and Heap, R.B. (1975). Oestrogen, progesterone and corticosteroid concentrations in peripheral plasma of sows during pregnancy, parturition, lactation and after weaning. J. Endocr. 64, 141-154.

Barnes, R.J., Comline, R.S. and Silver, M. (1974). Foetal and maternal plasma progesterone concentrations in the sow. J. Endocr. 62, 419-420.

Borland, R.M., Erikson, G.F. and Ducibella, T. (1977). Accumulation of steroids in rabbit preimplantation blastocysts. J. Reprod. Fert. 49, 219-224.

Bullock, D.W. (1977). Steroids from the preimplantation blastocyst. In: Development in Mammals, vol. 2. Ed. M.H. Johnson. Elsevier: North Holland Biomed. Press, Amsterdam.

Crombie, P.R. (1972). Ultrastructure of the pig's placenta throughout pregnancy. Ph.D Thesis, University of Cambridge.

Dickmann, Z., Dey, S.K. and Gupta, J.S. (1975). Steroidogenesis in rabbit preimplantation embryos. Proc. natn. Acad. Sci., U.S.A. 72, 298-300.

Dickmann, Z., Dey, S.K. and Gupta, J.S. (1976). A new concept : control of early pregnancy by steroid hormones originating in the preimplantation embryo. Vitams. Horm. 34, 215-242.

Engel, L.L. (1973). The biosynthesis of estrogens. In: Endocrinology, Handb. Physiol. Sect. 7, vol. 7, Female Reproductive System, part 1, pp. 467-484, American Physiological Society, Washington, D. C.

Fèvre, J., Lèglise, P.-C. and Rombauts, P. (1968). Du rôle de l'hypophyse et des ovaires dans la biosynthèse des oestrogènes au cours de la gestation chez la truie. Ann. Biol. anim. Bioch. Biophys. 8, 225-233.

Flood, P.F. (1974). Steroid-metabolizing enzymes in the early pig conceptus and in the related endometrium. J. Endocr. 63, 413-414.

Fuchs, A.R. and Beling, C. (1974). Evidence for early ovarian recognition of blastocysts in rabbits. Endocrinology 95, 1054-1058.

Gadsby, J.E. (1977). Comparative studies of steroid hormones in blastocysts, uterine flushings and endometrium of pig, sheep and cow. In: Monographs on Endocrinology. Ed. H.M.Beier and P. Karlson. Springer Verlag : Berlin, Heidelberg and New York (in press).

Gadsby, J.E., Burton, R.D., Heap, R.B. and Perry, J.S. (1976). Steroid metabolism and synthesis in early embryonic tissue of pig, sheep and cow. J. Endocr. 71, 45-46P.

Gleeson, A.R., Thorburn, G.D. and Cox, R.I. (1974). Prostaglandin F concentrations in the utero-ovarian venous plasma of the sow during the late luteal phase of the oestrous cycle. Prostaglandins 5, 521-529.

Guthrie, H.D., Henricks, D.M. and Handlin, D.L. (1972). Plasma estrogen, progesterone and luteinizing hormone prior to estrus and during early pregnancy in pigs. Endocrinology 91, 675-679.

Guthrie, H.D. and Polge, C. (1976). Luteal function and oestrus in gilts treated with a synthetic analogue of prostaglandin-$F_{2\alpha}$ (ICI 79,939) at various times during the oestrous cycle. J. Reprod. Fert. 48, 423-425.

Harrison, F.A., Heap, R.B. and Paterson, J.Y.F. (1972). Pregnancy in adrenalectomized ewes : glucose production and steroid changes in late pregnancy. J. Physiol., Lond. 227, 21-22P.

Heald, P.J. (1976). Biochemical aspects of implantation. J. Reprod Fert., Suppl. 25, 29-52.

Heap, R.B., Gadsby, J.E., Wyatt, C. and Perry, J.S. (1977). The synthesis of steroids and proteins in the pig blastocyst. In : Monographs on Endocrinology. Ed. H.M. Beier and P. Karlson. Springer-Verlag: Berlin, Heidelberg and New York (in press).

Heap, R.B., Gwyn, M., Laing, J.A. and Walters, D.E. (1973). Pregnancy diagnosis in cows; changes in milk progesterone concentration during the oestrous cycle and pregnancy measured by a rapid radioimmunoassay. J. agric. Sci., Camb. 81, 151-157.

Huff, R.L. and Eik-Nes, K.B. (1966). Metabolism in vitro of acetate and certain steroids by six-day-old rabbit blastocysts. J. Reprod. Fert. 11, 57-63.

Marcal, J.M., Chew, N.J., Solomon, D.S. and Sherman, M.I. (1975). Δ^5-3β-hydroxysteroid dehydrogenase activities in rat trophoblast and ovary during pregnancy. Endocrinology 96, 1270-1279.

Masuda, H., Anderson, L.L., Henricks, D.M. and Melampy, R.M.(1967). Progesterone in ovarian venous plasma and corpora lutea of the pig. Endocrinology 80, 240-246.

McCracken, J.A. and Baird, D.T. (1969). The study of ovarian function by means of transplantation of the ovary in the neck. In: The Gonads. Ed. K.W. McKerns, pp.175-209. North Holland, Amsterdam.

Pack, B.A. and Brooks, S.C. (1974). Cyclic activity of estrogen sulfotransferase in the gilt uterus. Endocrinology 95, 1680-1690.

Perry, J.S., Heap, R.B. and Amoroso, E.C. (1973). Steroid hormone production by pig blastocysts. Nature, Lond. 245, 45-47.

Perry, J.S., Heap, R.B., Burton, R.D. and Gadsby, J.E. (1976). Endocrinology of the blastocyst and its role in the establishment of pregnancy. J. Reprod. Fert., Suppl. 25, 85-104.

Robertson, H.A. and King, G. J. (1974). Plasma concentration of progesterone, oestrone, oestradiol-17β and of oestrone sulphate in the pig at implantation, during pregnancy and at parturition. J. Reprod. Fert. 40, 133-141.

Seamark, R.F. and Lutwak-Mann, C. (1972). Progestins in rabbit blastocysts. J. Reprod. Fert. 29, 147-148.

Sherman, M.I. and Atienza, S.B. (1977). Production and metabolism of progesterone and androstenedione by cultured mouse blastocysts. Biol. Reprod. 16, 190-199.

DISCUSSION

Nalbandov: We have done the following experiments. Rabbits were mated or injected with LH, and at various time intervals the reproductive tract was taken out as well as pieces of the gut and the esophagus. The tissues were extracted and assayed for progesterone and estrogen. There was a large influx of progesterone and estrogen into the uterus. There was apparently no correlation between the changes which occur in the uterine lumen and the changes which occur in the plasma of the same animals. At the same time, the gut and the esophagus did not contain any of the steroids. What assurance is there that what you are assaying is not the estrogen that actually enters from the uterus into the blastocyst?

Gadsby: I consider that the evidence *in vitro* suggests the presence of active estrogen synthetic enzymes. I have shown also the data of Robertson and King that show very low levels of unconjugated estrogen within the maternal circulation. The data I presented suggest that estrogen cannot only be synthesized *in vitro* from added precursors, but is also present *in vivo*, since when blastocyst tissue is flushed out, estrogen can be measured in this tissue. The results obtained for the uterine flushings and the endometrium are not conclusive, but suggest the existence of a gradient across the uterine flushings to the endometrium. I would consider that strong evidence for the actual synthesis of estrogens rather than the uptake from the uterine flushings.

Schwartz: It seems to me that when you gave the measurements in the pig embryonic tissue that the orders of magnitude were essentially the same for the two estrogens and the progesterone. However, while in the uterine endometrium the estradiol and the estrone were in the same order of magnitude, the progesterone was much higher. The data that you showed later were just the estrogen; you dropped the progesterone. Are your data more convincing of a diffusion gradient from blastocyst to uterus with respect to estrogen than they are with respect to progesterone?

Gadsby: I didn't show the progesterone values because I cannot be absolutely sure that what I am measuring in the uterine flushings is progesterone. Uterine flushings taken 17 days p.c. appear to contain little progesterone but do contain a related steroid, possibly a metabolite of progesterone.

Schwartz: Let us make an assumption that it is progesterone. As you know, when one looks at blood measurements of estradiol and progesterone there is an order of magnitude difference. We have seen the same thing in ovarian follicular fluid. My suggestion is that when the biosynthetic pathways are utilized, there is really a difference in magnitude. The fact that you do not show it in the pig embryonic tissue, if I read the slide correctly, may be an argument that the estradiol synthesis is in the embryonic tissue but that the progesterone is really drifting in from the other direction. I do not think it hurts your hypothesis at all.

Gadsby: These data are very recent and I would not like to make any firm proposals regarding the "flow" of progesterone across the uterus at this stage, but it would be very convenient if it was maternal progesterone that was going across, being taken up, and utilized for the synthesis of estrogens by the blastocyst.

Ryan: I have two questions. Does the early embryo synthesize estrogen from cholesterol? And the second, could the gradient be explained in the opposite direction by having a greater estrogen binding capacity in the embryo than in the uterus.

Gadsby: I have not been able to obtain estrogen synthesis from cholesterol. I wondered whether that was because the tissue was unable to take up the cholesterol and get it into the cholesterol pool within the tissue. So I looked at the metabolism of cholesterol and I couldn't find any evidence of metabolism of cholesterol. This requires further examination. However, when blastocyst tissue is incubated with ^{14}C-acetate a compound is formed that runs chromatographically in a similar position to cholesterol. I haven't carried out crystallization on this compound yet, but this suggests that acetate may be incorporated into cholesterol. To answer your second question, I have to go back to the in vitro studies because I have shown the existence of an estrogen synthetic pathway in terms of in vitro synthesis. I believe that the estrogen found within the blastocyst tissue in vivo is in fact synthesized by the blastocyst. I haven't been able to determine whether the estrogen found in the blastocyst resides within the tissue or within the blastocoele fluid. There is evidence that almost all the progesterone found within the rabbit blastocyst is present within the blastocoele fluid. This may also be the case for the pig, I don't know. Certainly, there is the possibility that the blastocyst tissue may have a binding capacity for estrogens which may be related to a possible action of estrogens within the blastocyst itself. This would be very interesting to investigate.

Channing: It has been shown in the rabbit that there is gonadotropin-like activity in blastocyst fluid. We have confirmed his finding using a different bioassay; namely, adding rabbit blastocyst fluid to monkey granulosa cell cultures and observing stimulation of progesterone secretion and luteinization. Is it possible that there is also gonadotropin-like activity in the pig blastocyst that could induce these steroidogenic enzymes? Have you ever looked for a preimplantation blastocyst gonadotropin in pig embryos?

Gadsby: There is work in progress in our laboratory to look for gonadotropin-like activity in the pig blastocyst, but at the moment I have nothing to report.

PITUITARY AND GONADAL REGULATION

11

A Human Chorionic Gonadotropin-like Substance Distinct from Human Luteinizing Hormone

HAO-CHIA CHEN, SHUJI MATSUURA, G. D. HODGEN, and GRIFF T. ROSS

Although radioimmunoassays and receptor assays for gonadotropins have numerous laudable advantages, distinguishing hCG from hLH in urinary extracts with these assays is impractical. These difficulties are understandable on the basis of the extreme structural similarities of the two substances, which result in crossreactivity of antisera generated against either native hormone. Likewise, gonadal receptors for hLH bind hCG equally well; thus, receptor assays do not distinguish between the two gonadotropins. Vaitukaitis et al. (1972) developed a radioimmunoassay system based on an antiserum (Sb6) against a highly purified β-subunit of hCG and labelled hCG or hCGβ subunit that distinguished hCG from hLH in serum samples. The value of this system for basic and clinical investigations has been well-documented (Vaitukaitis et al., 1976). However, urinary extracts containing hLH and some pituitary extracts were antigenic in this system (Vaitukaitis et al., 1972).

Structural studies comparing hCG (Carlson et al., 1973; Morgan et al., 1975) and hLH (Shome and Parlow, 1973; Sairam and Li, 1975) have revealed that the beta subunit of hCG contains an extended carboxyl-terminal glycopeptide of 30 amino acid residues and four carbohydrate side chains. Accordingly, hCG is distinctive from hLH in its larger molecular size and weight, is more acidic, and has a longer biological half-life due to the higher sialic acid content of the carboxyl-terminal region of hCGβ

(Vaitukaitis et al., 1976).

We attempted to take advantage of this structural difference by producing antisera against this unique portion of the hCGβ carboxyl-terminal peptide in order to develop a specific radioimmunoassay for hCG (Chen et al., 1976). Antiserum H93 was generated and a suitable radioimmunoassay system developed. As we had hoped, this antiserum did not bind ^{125}I-hLH from highly purified pituitary preparations. However, we were surprised that both urinary extracts, previously thought not to contain hCG, and crude extracts from human pituitaries gave rise to dose-response curves indistinguishable from those of highly purified hCG (CR119) in this radioimmunoassay system. These observations prompted us to investigate further the possible existence of an hCG-like substance in urine and pituitary tissue of normal subjects.

Here, based on criteria derived from immunological, physicochemical and biological characterization, we present evidence to indicate the presence of an hCG-like gonadotropin in urinary and pituitary extracts from normal persons. Furthermore, we found a smaller peptide which was antigenically and physically similar to the carboxyl-terminal peptide of hCG. Preliminary studies suggest that this peptide can be produced by proteolytic hydrolysis occurring in urinary and pituitary extracts. These observations lead us to propose that the hCG-like substance might be a biosynthetic precursor of an LH within the human pituitary.

MATERIAL AND METHODS

Hormone Preparations. (A) Urinary extracts: (1) hCG-CR119 is a purified preparation contains 11,600 IU/mg (bioassay, second international standard hCG); (2) second international reference preparation human postmenopausal gonadotropin (2IRPHMG) with biological activity of 40 international units (IU) per ampoule or 1 IU = 0.2295 mg for both hLH and hFSH; (3) first international standard for human urinary FSH/LH for bioassay containing 46 IU of hLH and 54 IU hFSH per ampoule or 1 IU FSH = 0.11388 mg and 1 IU LH = 0.13369 mg of protein (according to the World Health Organization International Laboratory of Biological Standards); (4) Pergonal (Cutter Laboratories, Lot 2168), a commercial preparation of postmenopausal women's urine, manufacturer's potency estimate of 75 IU FSH and 75 IU LH per ampoule (bioassay, 2IRPHMG); (5) kaolin-acetone extracts prepared from 24 hr urine specimens by the methods of Albert (1956); (6) acetone precipitate of postmenopausal women's urine as described previously (Chen et al., 1976); (7) extract (PMBIII) of postmenopausal women's urine was provided by Dr. Alexander Albert of Mayo Clinic, Rochester, Minnesota.

(B) Pituitary Extracts: (1) first international reference preparation of human pituitary FSH and LH for bioassay (1IRP-pituitary) containing 25 IU LH and 10 IU FSH per ampoule or 1 IU LH = 0.0668 mg and 1 IU FSH = 0.1670 mg of protein (bioassay, 2IRPHMG) according to the World Health Organization International Laboratory of Biological Standards; (2) a purified preparation of hLH (LER960VI) distributed by the National Institutes of Health for use as tracer in radioimmunoassay for hLH, which contains 4620 IU LH per mg of protein (bioassay, 2IRPHMG); (3) a partially purified human pituitary fraction, LER1966 containing 415 IU FSH and 355 IU LH per mg protein (bioassay, 2IRPHMG) was provided by Dr. Leo E. Reichert, Jr. of Emory University, Atlanta, Georgia.

Synthetic Peptides. Peptide analogs from dipeptide to triacontapeptide were synthesized according to the sequence of Morgan et al. (1975) by the solid phase method. Synthesis and characterization of these peptides will be published elsewhere (Matsuura, Chen and Hodgen; to be published). Three peptides of middle segments, residues 125-130, 131-137 and 125-137 were a generous gift from Dr. V.C. Stevens of Ohio State University, Columbus, Ohio.

Radioimmunoassay. Antisera: (A) An anti-hCG serum B_1 (Odell et al., 1967) and H80 (Chen et al., 1976) for hCG/hLH; (B) an anti-hCGβ serum (Sb6) for hCG/hCGβ (Vaitukaitis et al., 1972); (C) an anti-hCGβ carboxyl-terminal peptide serum (H93) for hCG/hCGβ (Chen et al., 1976); (D) an anti-oFSH serum (H31) for FSH (Hodgen et al., 1976) were used as reported previously.

All radioimmunoassays were carried out by a modification of the double-antibody techniques of Odell et al. (1967) as described previously (Chen et al., 1976). Computation of potencies and calculations and comparisons of slopes of dose-response curves were carried out by a computer program developed by Rodbard (1974).

In Vitro Bioassay. The radioligand-receptor assay of Catt et al. (1972), and the rat interstitial cell testosterone production assay of Dufau et al. (1974) were employed.

Sulphopropyl-Sephadex Column Chromatography. A 6 x 110 cm column of SP-Sephadex C-25, preswollen and washed with 2 M solution of the Buffer (NH_4OAc-HOAc, pH4.75), was equilibrated with 0.01 M solution of the Buffer. A 100 g sample of PMBIII was dissolved in 1 liter of the Buffer (0.005 M). After centrifugation to remove the insoluble material in the sample, a clear solution was applied to the column, and followed by elution with 800 ml of the Buffer (0.01 M). The column was eluted with a gradient composed of 1 liter of each of the concentration of the Buffer from 0.01 to 0.2 M, followed by 0.5 M Buffer. Fractions showing H93 immunoreactivity were pooled, dialyzed and lyophilized.

Sephadex G-100 Gel Chromatography. Unless specified in the legends of figures or in the text, gel chromatography was performed according to the following procedure. A 1.48 x 190 cm column of Sephadex G-100 equilibrated with 0.05 M Tris-HCl+0.1 M NaCl, pH 7.8. The column was run at room temperature and samples were dissolved in 2 ml of the same buffer used to equilibrate the column, but containing a trace amount of ^{125}I-hCG as a marker.

Treatment of Urinary Extract with Protease Inhibitor. A crude urinary extract, 1.7 g derived from soluble fraction of PMBIII (5 g) was dissolved in 50 ml of 0.1 M NH_4OAc-ethylenediamine, pH 8.0. To this solution, a protease inhibitor, 2 ml of 1% phenylmethylsulfonyl fluoride (PMSF) in n-propanol, was added dropwise. Simultaneously, 2.0 M ethylenediamine solution was added to maintain pH at 8.0. After stirring for 2 hr at room temperature, 2 M acetic acid was added to bring down pH to 5.0 then applied to a pre-column containing 10 g of dry Sephadex G-25 powder, which was connected directly to a column (2.55 x 145 cm) of Sephadex G-100 previously equlibrated with 0.1 M NH_4OAc-HOAc, pH 5.0. The column was run at 4^oC. Samples taken from each fraction were lyophilized individually prior to the H93 radioimmunoassay and the receptor assay. In a parallel experiment, an equal amount of the same extract prepared at the same time was incubated without PMSF at pH 8.0 overnight in the cold. PMSF was added and pH adjusted just prior to application to the Sephadex G-100 column. The same column under the identical conditions was used for these two experiments. Both H93 radioimmunoassay and radio-ligand receptor assays for two experiments were performed simultaneously using identical sets of reagents.

RESULT AND DISCUSSION

HCG Specific Antiserum: Production and Characteristics

Bovine serum albumin (BSA) conjugates of the carboxyl-terminal glycopeptide (residues, hCGβ 123-145) isolated after tryptic digestion of an S-carboxymethylated desialylated hCGβ were employed as immunogen to produce an antiserum, designated as H93, in a rabbit (Louvet et al., 1974; Chen et al., 1976). Characterization of antibody recognition sites were conducted by studies of antigenic activity of series of synthetic peptides analogous to the unique carboxyl-terminal peptide of hCGβ in a radioimmunoassay using the H93 antiserum and ^{125}I-hCG as radioligand. As shown in Figure 1, the terminal dipeptide Pro-Gln (C2), but not Gln and Pro-Glu, was active. Activity increased with increasing chain length from the COOH-terminus, and reached a plateau after the pentadecapeptide (C15), which was equipotent on a molar basis with a highly purified hCG (CR119) (Figure 2). Furthermore, the rapid increase in activity when Pro, Leu and Arg or dinitro-Arg (residues 135, 134, 133) were added indicates a second locus of binding involving hydrophobic bonding. However, peptides which contain these amino acid residues, but no Pro-Gln were either inactive (residues 131-137) or significantly less active (residues 125-137) as compared with C13 or C14 synthetic peptide analogues. Four rabbits among ten injected with this immunogen generated antibodies of significant titer. All exhibited uniform binding characteristics as shown in Figure 2. These data indicate that peptides containing an amino acid sequence similar to the carboxyl-terminal pentadecapeptide of hCGβ were recognizable by the

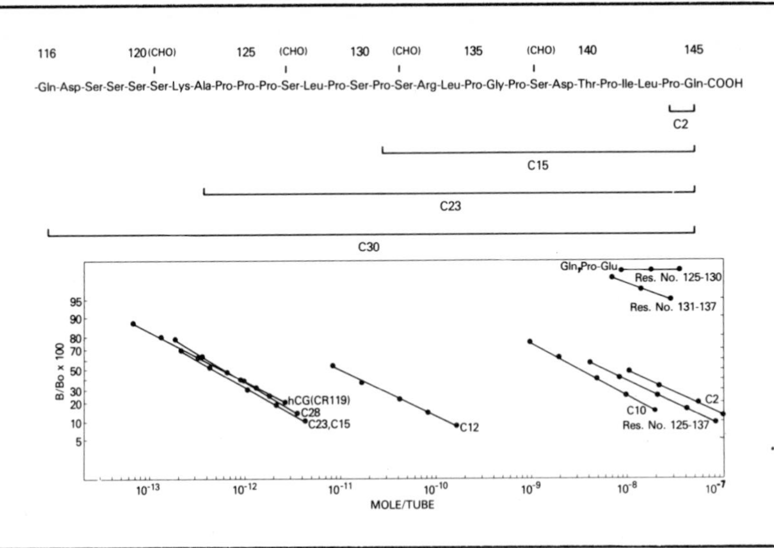

FIGURE 1. Dose-response curves of synthetic peptides analogous to the carboxyl-terminal sequence of hCGβ in a radioimmunoassay system based on an antiserum (H93) to a carboxyl-terminal hCG glycopeptide (residues 123-145 of hCGβ), diluted 1:8000. B=cpm bound in the presence of ^{125}I-hCG and unlabeled ligand, Bo=cpm bound in the presence of ^{125}I-hCG alone. Upper part shows the amino acid sequence of the carboxyl-terminal sequence of hCGβ (Morgan et al., 1975). CHO represents a polysaccaride moeity. In the synthetic peptides, CHO is not present. The number with prefix C indicates chain length elongated from the carboxyl-terminus as shown.

antibodies. Moreover, carbohydrate moieties occurring in native hCG were not required for antigenic recognition.

Specificity of the H93 antiserum was further demonstrated by its inability to bind either ^{125}I-hLH or ^{125}I-oLH. Several of the most highly purified glycoprotein hormones (at the levels per tube shown in parenthesis) also did not bind to the H93 antiserum: hLH-LER960VI (23 IU, 2IRPHMG), hFSH-LER1575 (40 IU, 2IRPHMG), NIH pregnant mare serum gonadotropin (100 IU, 2IRPHMG), and hTSH Pierce Fraction IV (4 μg). To further validate the specificity of this H93 radioimmunoassay system, the kaolin-acetone extracts of 248 randomly selected 24 hour urine samples, thought not to contain hCG, were analyzed in parallel by three radioimmunoassay systems for hCG. Only 2.8% of samples showed detectable

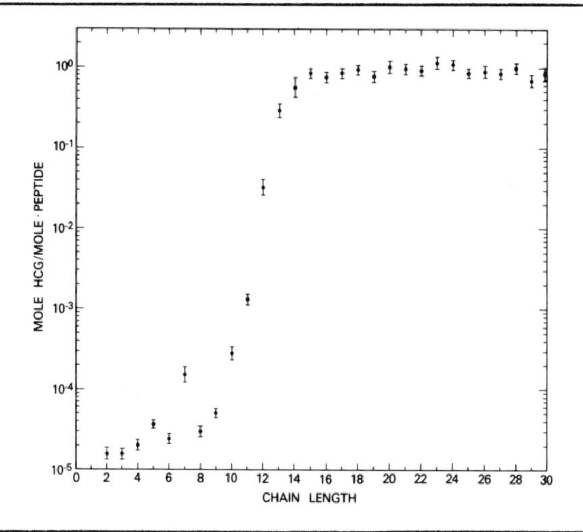

FIGURE 2. Competitive binding activities of synthetic peptides in H93 radioimmunoassay system. The ordinate indicates mole of unlabeled hCG (CR119) required to inhibit equally ^{125}I-hCG binding to H93 antiserum per mole of a peptide. The abscissa is the chain length of a peptide elongated from the carboxyl-terminus according to the sequence shown in Figure 1.

activity by H93 radioimmunoassay, whereas in assays with anti-hCGβ (Sb6) and anti-hCG (B1) sera 11.3 and 38.7% of these specimens showed activity above the minimum detectable level for H93 assay.

Collectively, these data suggest that the H93 radioimmunoassay system specifically measures hCG without crossreactivity with hLH or other glycoprotein hormones.

HCG-like Substance in Urinary and Pituitary Extracts

In spite of the high specificity of the H93-radioimmunoassay system for hCG, four preparations of partially purified urinary extracts from normal nonpregnant subjects and two pituitary extracts inhibited binding of ^{125}I-hCG in the H93 radioimmunoassay system, with dose-response curves parallel to that of highly purified hCG as shown in Figure 3. Potency estimates based upon these dose-response curves were not consistent with LH bio-potency of these preparations. Moreover, the purest, most potent LH preparation yielded the lowest activity. We interpreted these findings to mean that the H93 crossreacting substance

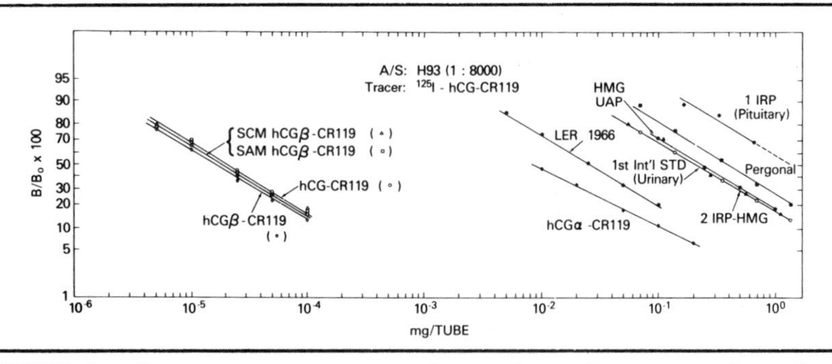

FIGURE 3. Dose-response curves of gonadotropin preparations in a radioimmunoassay system based on an antiserum of hCGβ-carboxyl-terminal peptide (H93) diluted at 1:8000. SCM is S-carboxyl-methyl; SAM, S-carboamidomethyl. Other abbreviations are described in "Material and Methods".

is distinct from hLH and contains a peptide structure similar to the unique carboxyl-terminal amino acid sequence of hCGβ. Furthermore, characterization of a postmenopausal urinary gonadotropin preparation (Pergonal) by Sephadex G-100 chromatography, revealed a single peak by H93 radioimmunoassay which is coincident with that of ^{125}I-hCG (Figure 4). The same peak was also monitored by two more hCG radioimmunoassay systems of different characteristics, namely Sb6 (anti-hCGβ serum) and H80 (anti-hCG serum) systems. In the same chromatographic system, both hLH and hFSH were retarded more than the first peak, indicating that the smaller molecular size of LH and FSH was associated with immunoreactivity in the H80 and H31 radioimmunoassays, respectively. In addition to antigenic activity in assays with the specific antiserum, fractions from the first peak showed proportional in vitro biological activity (Dufau et al., 1974). Similarly, gel filtration of the kaolin-acetone extract of a 24 hour urine specimen obtained from a boy with Klinefelter's syndrome showed an H93 and Sb6 immunoreactive peak which coeluted with ^{125}I-hCG and an immunoreactive hLH peak analogous to that seen following chromatography of Pergonal (Chen et al., 1976). In addition, we prepared an acetone precipitate of a pool of urine collected from postmenopausal women. This precipitate showed activity in the H93 assay. Although we cannot eliminate the possibility of inadvertent contamination of pooled specimens by pregnancy urine or urine from persons bearing neoplasms, the

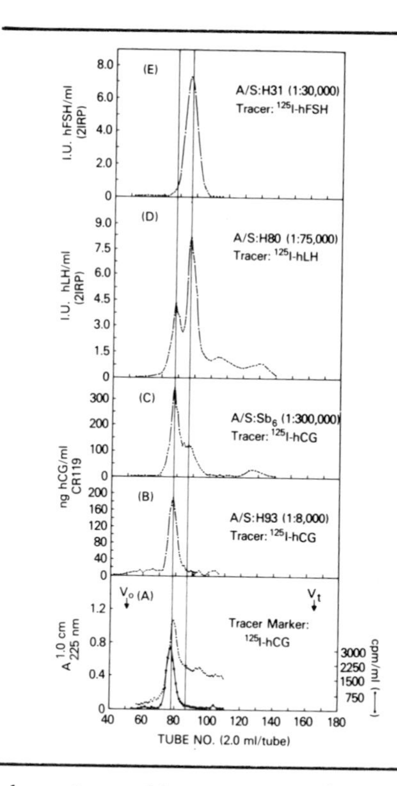

FIGURE 4. Sephadex G-100 gel chromatographic patterns of a human postmenopausal gonadotropin preparation (Pergonal).

possibility of such contamination seems extremely remote in all instances.

More recently, Dr. Alexander Albert of Mayo Clinic, Rochester, Minnesota, has provided us extracts of urine from postmenopausal women whose medical histories were examined thoroughly and revealed nothing abnormal. Among these extracts, a preparation designated as PMB III showed dose-response curves similar to hCG in the H93 radioimmunoassay system. This PMB III fraction was subjected to purification on sulphopropyl-Sephadex ion exchange chromatographic columns with gradient elutions using ammonium acetate-acetic acid, pH 4.75 as shown in Figure 5. H93 immunoreactive fractions were pooled, dialyzed, lyophilized and a portion was applied to the Sephadex G-100 column. Fractions were analyzed by H93 radioimmunoassay and the radioreceptor assay of Catt et al. (1972). The

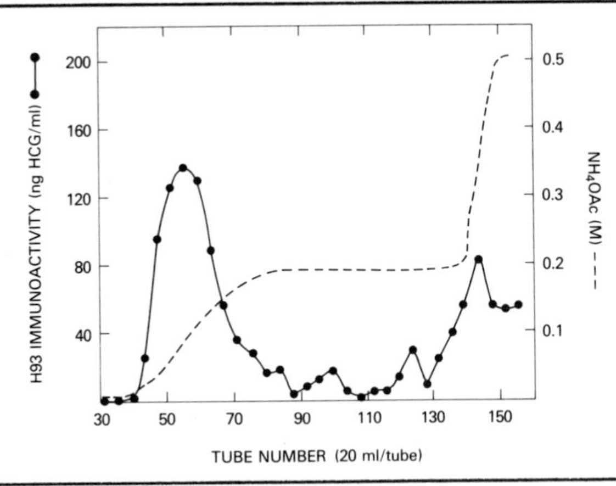

FIGURE 5. Sulphopropyl-Sephadex column chromatography of an extract (PMB III) obtained from urine of postmenopausal women.

results are shown in Figure 6. An H93 immunoreactive peak coincident with ^{125}I-hCG and exhibiting receptor activity was found. A later emerging peak, active in the receptor assay but not in the H93 assay, may represent LH. Though this material had been dialyzed exhaustively, large H93 immunoreactive peaks appeared very near the position of total column volume (Vt).

In addition, we have found that the first international reference preparation of human pituitary gonadotropin (1IRP-Pituitary) and a glycoprotein fraction of human pituitary extracts (LER1966) were active in the H93 radioimmunoassay system. The pituitary glycoprotein fraction LER1966 was derived from "Raben" glycoprotein fraction produced during extraction of growth hormone from human pituitary tissue. This fraction was subjected to DEAE-cellulose column chromatography and eluted at high salt concentration, conditions under which more acidic molecular species, such as FSH, are partially separated from hLH (Reichert, personal communication). Among all pituitary extracts we have tested, LER1966 showed the highest immunoreactivity in assays with the H93 antiserum (0.74 μg hCG-CR119/mg protein). Gel filtration of fraction LER-1966 on a Sephadex G-100 column as shown in Figure 7, revealed two peaks exhibiting H93 immunoreactivity: 1) a small one at the position of ^{125}I-hCG and 2) a large one in front of Vt. These patterns are very

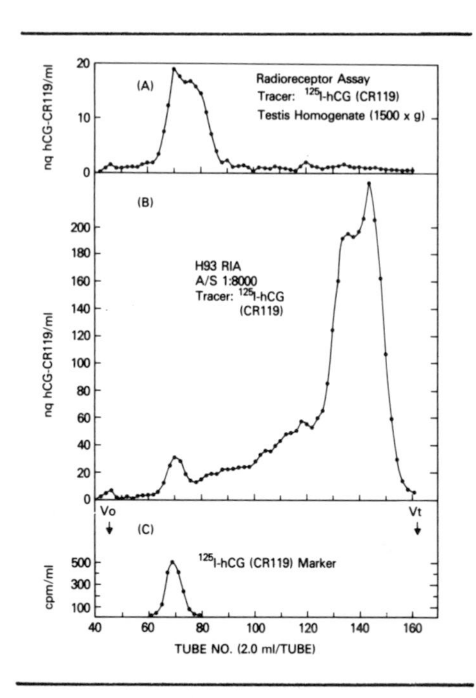

FIGURE 6. Sephadex G-100 gel chromatographic patterns of a fraction derived from postmenopausal women's urine (PMB III).

similar to those of the urinary extract shown in Figure 6.

Collectively, the presence of an hCG-like substance in pituitary and urinary extracts of normal subjects is based on the following criteria:

(1) immunoreactivity in a radioimmunoassay system specific for the unique carboxyl-terminal sequence in hCGβ,
(2) competition with hCG in gonadal receptor assays,
(3) separation from hLH by ion-exchange chromatography,
(4) coelution with ^{125}I-hCG in Sephadex G-100 gel chromatography.

The presence of a larger H93 immunoreactive peak eluted near the position of Vt as shown in Figure 6 and 7 is the most intriguing. During the isolation of an hCG-like substance from a urinary extract of postmenopausal women, it was also observed that an H93 immunoreactive substance appeared as nondialyzable solute at lower pH, but became dialyzable at high pH. Thus, it was recognized that some type of transformation via a proteolytic enzyme might have occurred. In this connec-

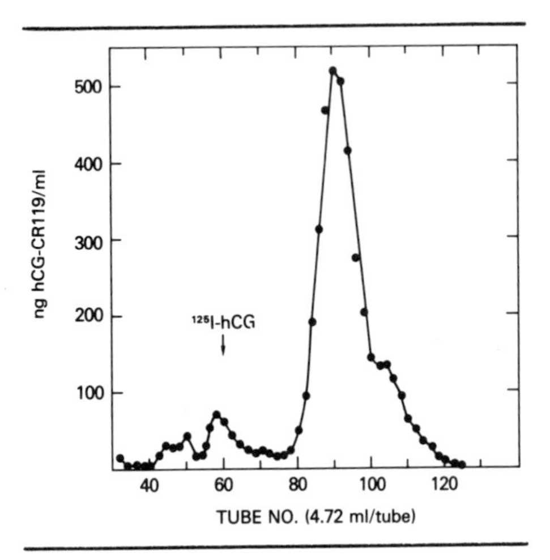

FIGURE 7. Sephadex G-100 column chromatography pattern of a human pituitary fraction (LER1966). Sample: 100 mg, column: 2.25 x 145 cm.

tion, a protease inhibitor, PMSF, was employed to determine if the transformation could be controlled.

Figure 8 shows Sephadex G-100 column chromatographic patterns of urinary extract incubated with or without PMSF. Table 1 describes results of immunological and biological activities as tabulated from each peak. Although H93 radioimmunoassay showed two peaks (I and II) in each case immunoreactivity of peak I in the presence of PMSF is higher than that in the absence of PMSF, whereas the reverse is true for peak II (Table 1). The increase in value for peak II was obviously derived at the expense of peak I. As shown in the upper part of Figure 8, the fractions corresponding to peak I have split into two components detected by receptor assay when incubation had been carried out without PMSF. Besides, the receptor activities were unchanged despite cleavage of an H93 immunoreactive portion. Peak II was biologically inactive. Therefore, it appears that the hCG-like substance in this urinary extract can be cleaved into two distinct components: 1) a peptide which antigenically resembles the carboxyl-terminal peptide of hCGβ, and 2) a protein having physical characteristics and possessing receptor activity

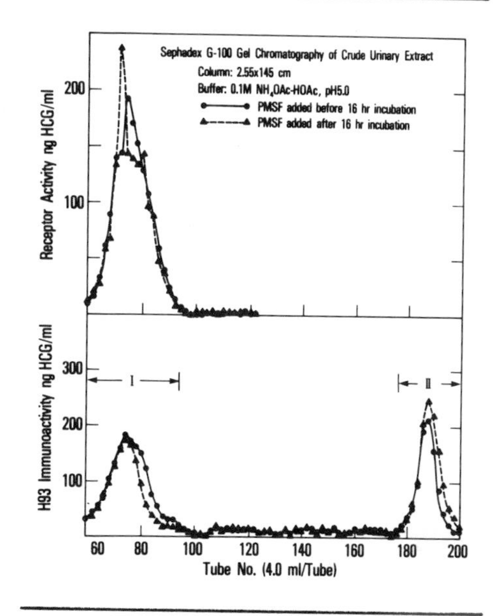

FIGURE 8. Sephadex G-100 gel chromatographic patterns of crude urinary extract.

Table I Distribution of H93 Immunoreactivity and Radio-receptor Activities in A Urinary Extract Incubated with and without PMSF

PMSF	H93 Immunoreactivity			Total
	Peak I	Peak II	Total	Receptor Activity
+	6.48	3.68	10.16	11.80
-	5.61	4.82	10.43	11.38

Unit of activities are expressed in μg HCG (CR119)/g

like LH. Furthermore, it was also found that the chimpanzee pituitary contained an enzyme capable of hydrolyzing hCG to release a peptide with physical and antigenic properties similar to the carboxyl-terminal peptide of hCGβ.

In addition to our findings, it has been reported that testicular tissue of normal men (Braunstein et al., 1975) and media from a microorganism isolated from urine of a patient with colon cancer (Livingston and Livingston 1974; Cohen and Strampp, 1976) contain a substance with physical and antigenic properties of hCG. Recently, Yoshimoto et al. (1977) reported that large amounts of human liver extract and of an extract of human colonic mucosa exhibited both immunoreactivity in an anti-hCGβ (Sb6) radioimmunoassay and radioreceptor activity using rat testicular homogenate. Thus, substances having biological, immunological and physical properties similar to hCG may occur more widely than previously recognized.

Possible Roles of HCG-like Substance

Traditionally, hCG from the synctiotrophoblast has been thought to extend the functional life-span of the corpus luteum during the fertile menstrual cycle and has been regarded exclusively as a hormone of pregnancy in normal subjects. Later, it was recognized that neoplastic cells of many varieties also secrete hCG. More recently, evidence was presented suggesting that development of fetal testis in man may require hCG during mid-gestation (Clement, et al., 1976). In addition to interstitial cell stimulating activity (LH), both crude and highly purified hCG preparations possess intrinsic, but modest, follicle stimulating (Louvet et al., 1976) and thyroid stimulating activities (Nisula et al., 1974). Also, it has a longer circulatory half-life in the body than any of the other gonadotropins (Vaitukaitis et al., 1976).

Although eutopic and ectopic secretions of hCG by tissue of neoplasms have been well-documented (Vaitukaitis et al., 1976), little is known of the meaning of its association with tumor tissue. Among the more credible hypotheses is that hCG may be a manifestation of tumor cell dedifferentiation to a primitive state characteristic of embryonic cell development, which results in the activation of repressed genes (Sherwood, 1976). On the basis of this hypothesis, the hCG-like substance found in normal subjects might represent "leakage" of incompletely repressed gene products from tissue. Further studies will be necessary to test this hypothesis.

Although the presence of hCG or an hCG-like substance may be more nearly universal than was thought in the past, the striking similarities between hCG and hLH in peptide sequence, as well as biological function, may not be merely coincidental occurrences. Current evidence indicates that hLH is secreted only by the pituitary. Our findings, as discussed in the previous section, indicate the relative abundance of an hCG-like substance in urinary extracts of persons with increased gonadotropin secretion including: postmenopausal women and a patient with Klinefelter's syndrome. Most importantly, human pituitary extracts contain a similar substance. These observations, coupled with the presence of the H93 immunoreactive peak appearing near Vt of gel chromatography as shown in Figure 6 and 7, and demonstration of the cleavage of an hCG-like substance into an LH-like component with release of the small H93 immunoreactive peptide, are consistent with the notion that the hCG-like substance might be converted into an hLH by cleavage of the carboxyl-terminal peptide, leaving a residual peptide like the unique "tail piece" on hCGβ not found in hLH.

To substantiate this hypothesis more concrete evidence will be required relating changes in occurrence of this phenomeneon to physiological processes affecting the synthesis and secretion of LH. Furthermore, this hypothesis is not intended to exclude other explanations for its occurrence. Further studies are in progress.

CONCLUDING REMARKS

The availability of a truly hCG specific antiserum (H93) and elucidation of its binding characteristics are the key to our current progress and findings that an hCG-like substance is present in the urinary and pituitary extracts of normal non-pregnant subjects. Distinction between hCG-like substance and hLH was determined after critical examination of immunochemical, physicochemical and biological properties. HGG and the hCG-like substances are reportedly found in many tissues. However, circumstantial evidence indicated that in normal subjects, the pituitary may be a major source where elevated gonadotropin secretion is associated with increased levels of the hCG-like material. Furthermore, our findings that generation of an hLH-like molecule from an hCG-like substance with concomitant release of a peptide antigenically similar to the unique extended carboxyl-terminal peptide of hCGβ, provide preliminary support for the proposal that an hCG-like substance may serve as a precursor of hLH. To be sure, more direct and concrete evidence is needed. Obviously, other roles for this substance cannot be ruled out at this time.

ACKNOWLEDGEMENT

We thank Drs. Alexander Albert of Mayo Clinic, Rochester, Minnesota, Leo E. Reichert, Jr. of Emory University, Atlanta, Georgia, and Vernon C. Stevens of Ohio State University, Columbum, Ohio for providing us with materials essential for these studies. We also thank Misses Patti Byrd and Ellen Herron for preparing this manuscript. One of us (S.M.) was supported in part by a fellowship from The Population Council, New York.

REFERENCES

Albert A. (1956). Human Urinary Gonadotropins. Recent. Prog. Horm. Res. 12, 227-296.

Braunstein, G.D., Rasor, J. and Wade, M.E. (1975). Presence in normal human testes of chorionic gonadotropin-like substance distinct from human luteinizing hormone. New England J. Med. 293, 1339-1343.

Carlson, R.B., Bahl, O.P. and Swaminathan, N. (1973). Human chorionic gonadotropin linear amino acid sequence of the β subunit. J. Biol. Chem. 248, 6810-6828.

Catt, K.J., Dufau, M.L. and Tsuruhara, T. (1972). Radioligand-receptor assay of luteinizing hormone and chorionic gonadotropin. J. Clin. Endocrinol. Metab. 34, 123-132.

Chen, H-C., Hodgen, G.D., Matsuura, S., Lin, L.J., Gross, E., Reichert, L.E. Jr., Birken, S., Canfield, R.E. and Ross, G.T. (1976). Evidence for a gonadotropin from nonpregnant subjects that has physical, immunological, and biological similarities to human chorionic gonadotropin. Proc. Natl. Acad. Sci. 73, 2885-2889.

Clements, J.A., Reyes, F.I., Winters, J.S.D., and Faiman, C. (1976).

Studies on human sexual development. III Fetal pituitary serum and aminiotic fluid concentrations of LH, CG, and FSH. J. Clin. Endocrinol. Metab. 42, 9-19.

Cohen, H. and Strampp, A. (1976). Bacterial synthesis of substance similar to human chorionic gonadotropin. Proc. Exp. Biol. Med. 152, 408-410.

Dufau, M.L., Mendelson, C.R. and Catt, K.J. (1974). A highly sensitive in vitro bioassay for luteinizing hormone and chorionic gonadotropin: testosterone production by dispersed Leydig cells. J. Clin. Endoclinol. Metab. 39, 610-613.

Hodgen, G.D., Wilks, J.W., Vaitukaitis, J.L., Chen, H.-C., Papkoff, H. and Ross, G.T. (1976). A new radioimmunoassay for follicle-stimulating hormone in Macaques: Ovulatory menstrual cycles. Endocrinology 99, 137-145.

Livingston, V.W.-C. and Livingston, A.M. (1974). Some cultural, immunological and biochemical properties of Progenitor cryptocides. Trans N.Y. Acad. Sci. 36, 569-582.

Louvet, J-P., Ross, G.T., Birken, S. and Canfield, R.E. (1974). Absence of neutralizing effect of antisera to the unique structural region of human chorionic gonadotropin. J. Clin. Endocrinol. Metab. 39 1155-1158.

Louvet, J-P., Harman, S.M., Nisula, B.C., Ross, G.T., Birken, S. and Canfield, R.E. (1976). Follicle stimulating activity of human chorionic gonadotropin: effective dissociation and recombination of subunits. Endocrinology 99, 1126-1128.

Morgan, F.J., Birken, S. and Canfield, R.E. (1975). The amino acid sequence of human chorionic gonadotropin: the α subunit and β subunit. J. Biol. Chem. 250, 5247-5258.

Nisula, B.C., Morgan, F.J. and Canfield, R.E. (1974). Evidence that chorionic gonadotropin has intrinsic thyrotropic activity. Biochem. Biophys. Res. Comm. 59, 86-91.

Rodbard, D. (1974). Statistical quality control and routine data processing for radioimmunoassays and immunoradiometric assays. Clin. Chem. 20, 1255-1270.

Sairam, M.R. and Li, C.H. (1975). Human pituitary lutropin. Isolation, properties, and the complete amino acid sequence of the beta-subunit Biochim. Biophys. Acta 412, 40-81.

Sherwood, L.M. (1976). Ectopic hormone syndromes. in "The Year in Endocrinology, 1975-1976" (Ingbar, S.H., Ed.), Plenum Press, New York, pp 249-276.

Shome, B. and Parlow, A.F. (1973). The primary structure of the hormone-specific β-subunit of hLH. J. Clin. Endocrinol. Metab. 36, 618-621.

Vaitukaitis, J.L., Braunstein, G.D. and Ross, G.T. (1972). A radioimmunoassay which specifically measures human chorionic gonadotropin in the presence of human luteinizing hormone. Am. J. Obstet. Gynecol. 113, 751-758.

Vaitukaitis, J.L., Ross, G.T., Braunstein, G.D. and Rayford, P.L. (1976). Gonadotropins and their subunits: basic and clinical studies. Recent Prog. Horm. Res. 32, 289-321.

Yoshimoto, Y., Wolfsen, A.R. and Odell, W.D. (1977). Human chorionic gonadotropin-like substance in non-endocrine tissues of normal subjects. Science 197, 575-577.

DISCUSSION

Ryan: Does the placenta have the capacity to split off the carboxy terminal peptide of hCG-β?

Chen: I don't believe so. HCG has been isolated from placental extracts, but no small molecular weight substances have been observed. The chromatographic profile gives a single peak.

Ryan: The reason for asking is that if this C-terminal tailpiece is a part of the precursor of the LH molecule, then the enzyme for cleaving, which is in the pituitary, should be lacking in the placenta.

Chen: We have never observed an LH-like substance in our chorionic gonadotropin preparations, suggesting that an enzyme is not present in placenta.

Peckham: Have you any idea about the size of the peptide that is cleaved from your urinary preparation?

Chen: If we run the dodecapeptide (C12) as a marker in Sephadex G-100, the peptide is larger; however, it may be a glycopeptide, we don't know. We have recently achieved greater purification of this peptide on Sephadex LH-20, and separated the peptide from most of the other non-reactive material.

Ward: Was that material pure enough to get a composition estimate?

Chen: We had a very rough analysis done with a crude material prior to purification by Sephadex LH-20, but the material contained histidine which should not be there. We need to do it again on a more purified preparation.

Peckham: What amounts of the cleaved peptide do you have?

Chen: We have looked at many preparations and the highest amounts were obtained from a pituitary preparation, LER-1966, which contained 0.7 μg hCG equivalents per mg. From urinary preparations the highest yield was 11.4 ng/mg protein.

Peckham: Are you talking about the amount of the hCG-like material or about the amount of cleaved peptide?

Chen: Both.

Ward: What are your buffer conditions with pH 5 vs. pH 8 where you get this big absorption difference?

Chen: I cannot state that the absorption difference was due simply to pH. Theoretically, if we run at low pH it should prevent absorption to Sephadex. However, at pH 5 more than half the carboxy groups in Sephadex are in ionized form, so you may get absorption due to charge.

Folkers: Are your results on the pentadecapeptide in harmony with those of Vernon Stevens?

Chen: We can't say with certainty, but our two antibodies have very similar characteristics.

Bullock: Is there any immunohistochemical evidence using H93 in the pituitary gland?

Chen: We have not been able to do those studies yet, because the H93 antiserum has a very low titer and we have only limited quantities.

Channing: Does this antiserum neutralize the biological activity of hCG?

Chen: No, it doesn't. We don't know if that is due to the complex *per se* or due to dissociation in the body. However, the dissociation constant of the H93 antiserum is two orders of magnitude lower than for the other antisera, *ie*. anti-hCG-β (Sb6) or anti-hCG (H80).

Channing: Using an antiserum made by Talwar against hCG-β assayed in monkey granulosa cell cultures, we found that antiserum to hCG-β can neutralize the activity of hCG but much less of LH.

Chen: We also have evidence that the Sb6 antiserum crossreacts very little with pituitary LH preparations, but to a much greater extent with urinary preparations.

Peckham: I wonder if you can quantitate the radioreceptor assay and the *in vitro* bioassay data sufficiently to tell whether they are commensurate with what you would expect on the basis of RIA?

Chen: On a standard Sephadex column, the biological activity and the immunological activity compare quite well. However, that is not the case with the material from the affinity columns where it is subjected to denaturing substances. Its immunoreactivity is always higher than its receptor activity.

Peckham: Is the peak of your hCG-like material well separated from any LH that's present?

Chen: Yes.

Ward: What type of affinity column did you use?

Chen: It was made with H100 antiserum which is similar to H93.

Ward: Did you bind the antiserum?

Chen: Yes, a 40% ammonium sulfate precipitate of the antiserum.

Ward: Your antibody binding constants are going to be fairly high. Have you tried ConA instead of that?

Chen: Yes, we did.

Ryan: Have you tried adding EDTA to remove calcium, because that will lower the affinity of the ConA for the ligand.

Chen: In ConA experiments we put EDTA in the system.

12

Some Properties of an LH-like Substance in Rhesus Monkey Sera

WILLIAM D. PECKHAM

Some time ago, shortly after an homologous radioimmunoassay (RIA) for rhesus monkey LH (rhLH) employing radioiodinated rhLH and an antiserum to rhLH was developed in this laboratory (Monroe et al., 1970), we had occasion to measure the LH in sequential daily bleedings of two hypophysectomized monkeys immediately following the operation. A comparison of the results obtained with the rhesus:anti-rhesus RIA and the ovine:anti-ovine rhLH RIA (Niswender et al., 1971), which was also operational in our laboratory at that time, revealed significant and changing levels of LH by the ovine:anti-ovine assay, whereas LH could not be detected in these same bleedings by the rhesus:anti-rhesus system. Several years later, Dr. Douglas Foster at the University of Michigan, using the ovine:anti-ovine RIA, found high and variable levels of LH in the circulation of a number of infant monkeys which he was studying. These infant monkey sera were subsequently shipped to our laboratory where they were tested in a newer version of our rhLH RIA, the rhesus:anti-hCG system (Karsch et al., 1973b), and found to contain no detectable LH. In the meantime, preliminary experiments by Dr. James Lemons of the Michigan group suggested that the molecular size of the LH-like substance (LHS) in infant monkey sera was similar to that of monkey pituitary LH.

At this point, we decided that the problem was of sufficient interest to warrant further investigation, and that the best way to implement this was in collaboration with the Michigan group. So far, these collaborative studies have yielded the information outlined below.

Ubiquity of serum LHS. Although at first it appeared that LHS could be found only in the circulation of infant monkeys and an occasional hypophysectomized animal, it soon became apparent that this was not the case. Because demonstration of its presence depends on a strong response in the ovine:anti-ovine RIA coupled with a negligible response in the rhesus:anti-hCG (or rhesus:anti-rhesus) RIA, it is difficult to know how much, if any, LHS is present in samples which contain large quantities of rhLH, such as sera from gonadectomized monkeys or from females during the preovulatory surge. However, examination of sera from a number of adult female and male monkeys with low circulating levels of rhLH showed that all contained some LHS. Similar results were obtained with sera from hypophysectomized (completeness unverified histologically) or estrogen-treated ovariectomized monkeys. As was the case with infant monkeys, the levels of LHS in the serum of adult monkeys varied a great deal from animal to animal (Table 1).

Table 1. LH activity in selected adult female monkey sera measured by two radioimmunoassays**

Animal Number	Rhesus:anti-hCG RIA	Ovine:anti-ovine RIA
831	2.2*	5*
705	3.4	7
736	5.2	11
606	<2	14
830	8.3	20
611	<2	22
571	<2	27
982	3.3	33
1543	2.5	35
741	4.6	41
573	2.0	46
801	3.2	54

**These sera were among those taken for LH assay from over 100 adult female rhesus monkeys during the follicular phase of the menstrual cycle, and were selected to illustrate the varied levels of LHS in different animals. All of the sera examined showed a greater response, relative to the standard, in the ovine:anti-ovine RIA than in the rhesus:anti-hCG assay. This same kind of variability was also apparent in infant monkeys of both sexes.
*ng rhLH standard (WP-X-47-BC) per ml.

Chromatographic behavior of serum LHS. When sera containing rather large quantities of LHS and negligible amounts of rhLH were chromatographed on Sephadex G-100, the LHS was eluted in a discrete peak at a position very similar to that of pituitary rhLH and that of the LH in sera from ovariectomized monkeys or females during the preovulatory surge (Fig. 1).

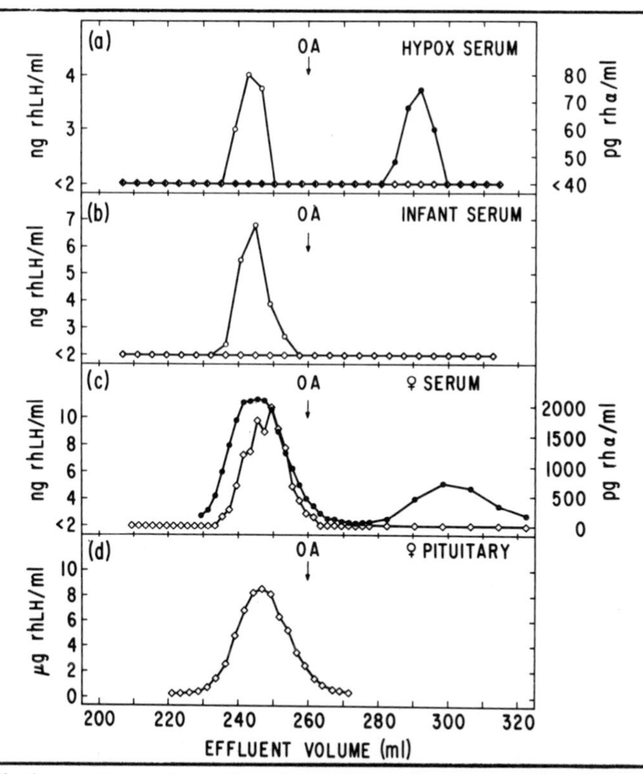

Fig. 1. Elution patterns from the chromatography on superfine Sephadex G-100 (2.5 x 100 cm columns) of (a) hypophysectomized monkey serum, (b) infant monkey serum, (c) adult female monkey serum taken during the preovulatory surge, and (d) adult female monkey pituitary extract. The vertical arrow represents the position of the peak of the internal marker, ovalbumin (OA). ○—○, LH measured by the ovine:anti-ovine RIA; ◇—◇, LH measured by the rhesus: anti-hCG RIA; ●—●, activity in the rhα-subunit RIA.

Immunological characteristics of serum LHS. Sephadex G-100 column fractions containing LHS, like the sera from which they were derived, produced log dose-response curves in the ovine:anti-ovine RIA which were not parallel to the curves generated by pituitary or serum rhLH (Fig. 2). These LHS fractions showed little activity in LH radioimmunoassays which employ an rhLH "tracer" and in an alpha subunit RIA which detects not only free rhesus alpha subunits but also the alpha component of undissociated rhLH, rhFSH, and rhTSH (Peckham et al., 1977).

Dissociation of serum LHS. Like rhLH, serum LHS after treatment with 4M guanidine HCl yielded a substance of significantly smaller molecular size

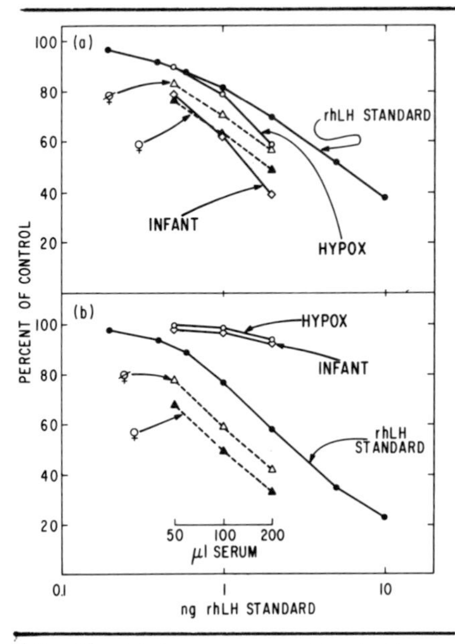

Fig. 2. Dose-response curves of rhesus monkey sera in (a) ovine:anti-ovine and (b) rhesus:anti-hCG RIAs for rhLH. Female serum was taken during the preovulatory surge. The rhLH standard is WP-X-47-BC, LH biopotency = 1.9 U/mg. HYPOX: hypophysectomized.

(Fig. 3) which was also active in the ovine:anti-ovine RIA. Insufficient material was available to characterize this product further.

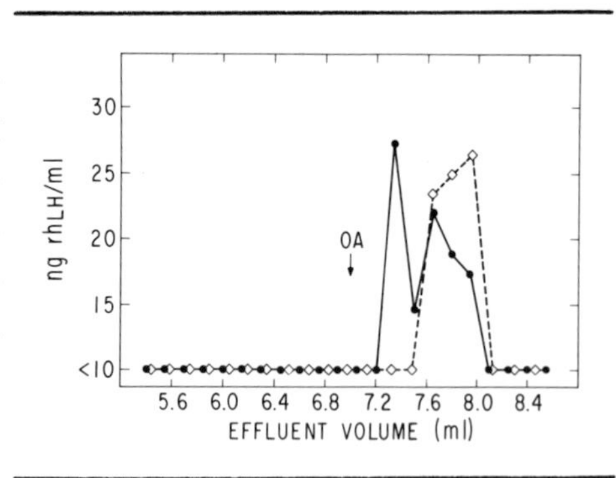

Fig. 3. Elution patterns from the chromatography of guanidine-treated serum LHS (●——●) and LH (◇- -◇) on small columns (12 ml) of superfine Sephadex G-100. The position of the peak of the internal marker ovalbumin is represented by the vertical arrow.

Control of secretion of LHS. Four adult female monkeys were selected for their high levels of serum LHS. Each animal was implanted with 6 estradiol-containing Silastic capsules (Karsch et al., 1973a) in order to produce circulating levels of estradiol of 200 to 600 pg/ml and suppress serum LH to undetectable levels. No reduction in LHS was seen in any of these animals (Fig. 4), demonstrating that LHS is not subject to negative feedback control by estrogen, at least under conditions where negative feedback control of LH (and FSH) can be readily demonstrated. Chronic estrogen treatment of four adult male monkeys also failed to evoke any significant reduction in serum LHS.

After serum LH in the estrogen-implanted females had remained undetectable for 3 to 4 months, three of these animals were placed in primate

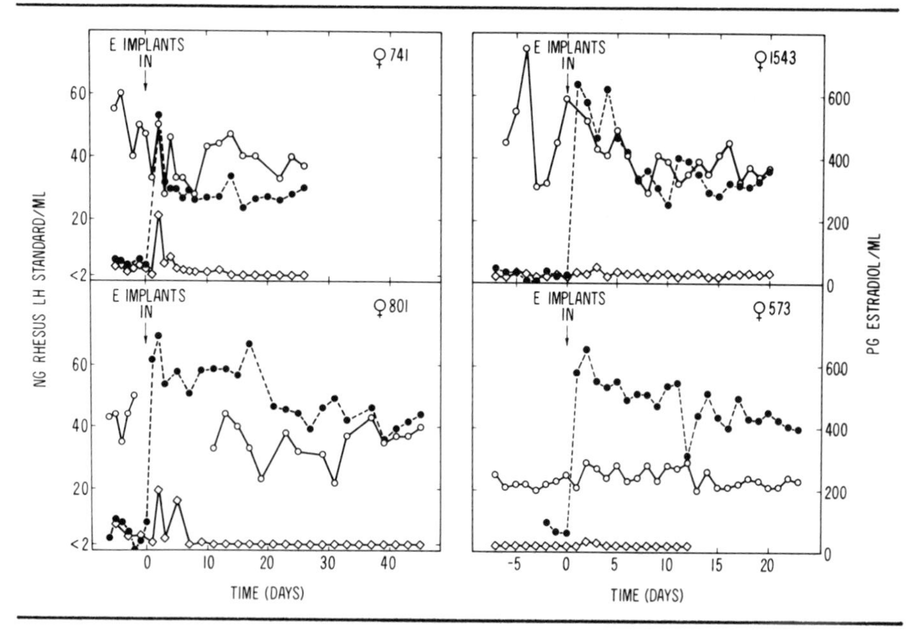

Fig. 4. Patterns of circulating estrogen (●- -●) produced by sc implantation of Silastic capsules containing estradiol-17β into female rhesus monkeys, and the serum LH immunoactivity in these animals, measured by the ovine:anti-ovine RIA (O—O) and by the rhesus:anti-hCG RIA (◇—◇).

chairs, fitted with saphenous vein catheters, and given an injection of 100 μg of GnRH. Blood samples were taken at 20 minute intervals beginning one hour before and ending three hours after injection. The samples were assayed by the ovine:anti-ovine and the rhesus:anti-hCG radioimmunoassays for LH, in order to determine: 1) if the pituitary reserves of LH in these animals were exhausted by the chronic estrogen treatment, and 2) if so, if a release of LHS was effected by the GnRH. The results (Fig. 5) indicate that no LH was released as a consequence of the GnRH injection, and that GnRH also failed to elicit a change in the already high concentrations of LHS in the circulation of these animals, thus suggesting that LHS secretion is independent of yet another mechanism involved in the control of LH secretion.

Source of serum LHS. The fact that LHS is found in the circulation of hypophysectomized monkeys leads to one of two conclusions: 1) this serum LHS does not come from the pituitary gland, or 2) the pituitary is the source of serum LHS, and some LHS is found in the circulation of "hypophysectomized" animals because of incomplete removal of the gland. At present, it is not

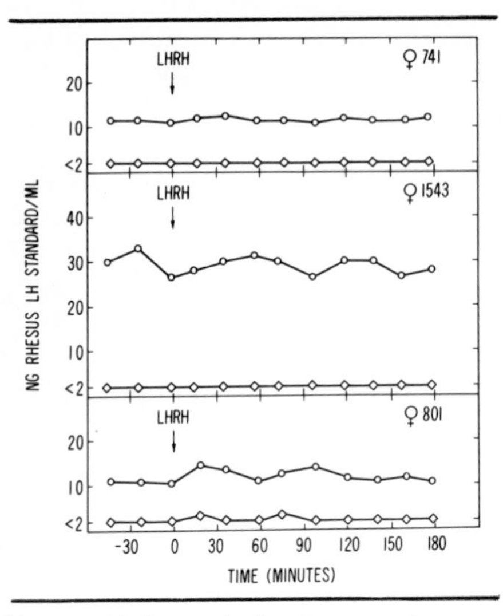

Fig. 5. LH immunoactivity, measured by the ovine:anti-ovine RIA (○—○) and by the rhesus:anti-hCG RIA (◇—◇), in the circulation of estrogen-suppressed female monkeys before and after iv injection of 100 μg GnRH. Circulating estrogen levels of 200 to 400 pg/ml had been maintained in these animals for 3 to 4 months by sc implanted Silastic capsules containing estradiol-17β .

known which conclusion is correct.

Biological activity of serum LHS. Concentrated fractions from Sephadex G-100 chromatography of serum from infant monkeys, from estrogen-treated ovariectomized monkeys, and from untreated ovariectomized monkeys were sent to Dr. Foster for assessment of biological activity using an *in vitro* system which involves the stimulation of progesterone secretion by dissociated rat ovarian cells (Bajpai et al., 1974). The fraction from ovariectomized monkey serum of course contained rhLH and was included for comparison with the other two serum fractions which contain LHS. Although all three concentrated fractions effected an increase in progesterone production by the ovarian cells which was considerably greater than that resulting from high doses of ovine pituitary LH, this "hypersecretion" appeared to be due to the presence of very large amounts of serum albumin in these fractions. It was thus apparent that a more satisfactory means of purifying serum LHS was required before its biological activity could be determined in this *in vitro* system. A simple three-step purification procedure was therefore devised to separate LHS from the bulk of the extraneous serum proteins. This consists of: 1) slow addition of an equal volume of cold ethanol to the well-stirred serum, and, after discarding the resulting precipitate, precipitation of the LHS-containing fraction by addition of a second volume of cold ethanol; 2) passage of the solution containing this fraction through a column of Concanavalin A-Sepharose (Pharmacia Fine Chemicals) and subsequent elution of the bound LHS-containing fraction by 0.2 M α-methyl-D-mannopyranoside (Fig. 6); and 3) exclusion chromatography on Sephadex G-100 (Fig. 7). Application of this procedure to several liters of serum which had been collected from a group of adult monkeys chronically treated with estrogen to suppress circulating rhLH to undetectable levels resulted in overall yields of approximately 25% and reduction of the total protein to but a few milligrams.

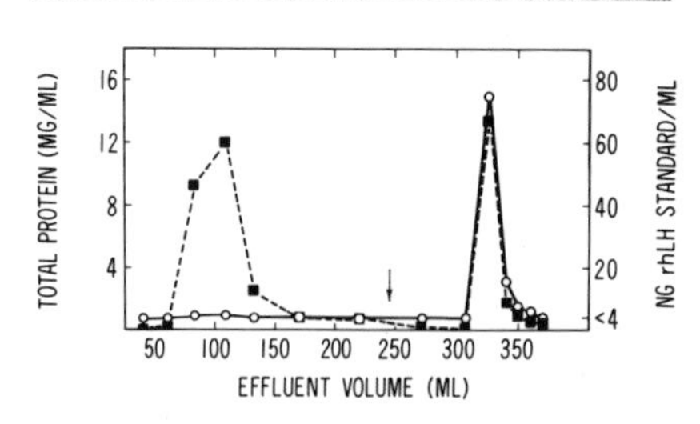

Fig. 6. Elution patterns from the chromatography on a Sepharose-linked Concanavalin A column (1.1 x 95 cm) of a crude LHS preparation obtained by ethanol fractionation of serum (see text). Volume of the solution applied to the column was 54 ml. ■— —■ , protein concentration (Lowry); ○—○, immunoactivity in the ovine:anti-ovine LH RIA. The vertical arrow indicates the point at which elution with 0.2 M α-methyl-D-mannopyranoside was begun.

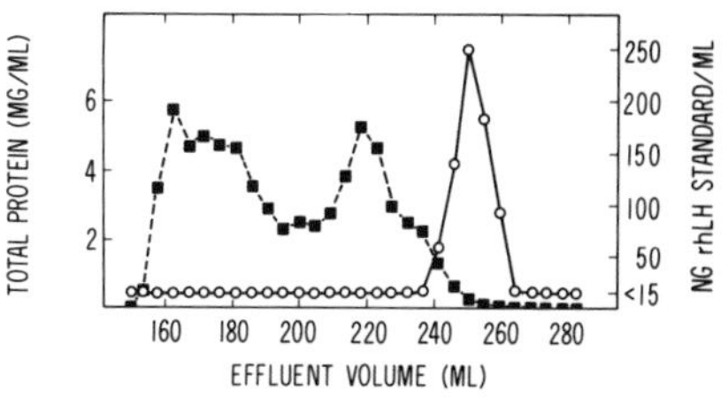

Fig. 7. Elution patterns from the chromatography on a superfine Sephadex G-100 column (2.5 x 100 cm) of serum LHS which had been previously purified by ethanol fractionation and affinity chromatography on Sepharose-linked Concanavalin A. ■— —■ , protein concentration (Lowry); ○——○, immunoactivity in the ovine:anti-ovine LH RIA.

After concentration by ultrafiltration, the activity of this serum LHS preparation in the ovine:anti-ovine RIA was equivalent to approximately 600 ng of rhesus LH standard (WP-X-47-BC) per ml, while its activity in the rhesus: anti-hCG RIA was only about 10 ng of rhLH standard per ml. Since the serum LHS solution contained 3 mg protein per ml, its potency, relative to the rhLH standard, was $\sim 200 \times 10^{-6}$ in the ovine:anti-ovine RIA, and $\sim 3 \times 10^{-6}$ in the rhesus:anti-hCG RIA.

When this preparation was tested by Dr. Foster in the rat ovarian cell-progesterone bioassay, it was found to have a potency of $\sim 12 \times 10^{-6}$ relative to the ovine LH standard NIH-LH-S16. Its activity in two radioligand-receptor assays was considerably greater; in a rat Leydig cell radioligand-receptor assay (Catt et al., 1972), its potency relative to NIH-LH-S16 was $\sim 90 \times 10^{-6}$, and in a rat luteal cell RRA (Midgley et al., 1974), the relative potency was $\sim 270 \times 10^{-6}$.

The same preparation was subsequently tested by Dr. J.D. Neill in his version (Neill et al., 1977) of the Dufau (Dufau et al., 1974) rat interstitial cell-testosterone (RICT) bioassay. In this system, the serum LHS preparation showed no activity, even at the maximum dose of 0.3 mg. Thus, its potency in the RICT bioassay was less than 0.5×10^{-6} relative to the rhLH standard WP-X-47-BC. Dr. Foster has recently tested this material in the RICT assay, and he too has found it to be without discernible activity in this system.

The underlying reasons for such variation in the assessments of potency of the serum LHS preparation by different methods are not readily apparent. These findings, nevertheless, are not inconsistent with our view (Peckham et al., 1977) that LHS may represent an atypical form of rhesus LH.

These studies were carried out in collaboration with Drs. E. Knobil and T.M. Plant of the Department of Physiology at the University of Pittsburgh School of Medicine, and with Dr. D. L. Foster, Department of Obstetrics and Gynecology, University of Michigan, and Dr. J.D. Neill, Department of Physiology, Emory University.

References

Bajpai, P.K., Dash, R.J., Midgley, A.R., Jr., and Reichert, L.E., Jr. (1974). Progesterone production by dissociated rat ovarian cells: a sensitive method for quantitation of hormones with luteinizing activity. J. Clin. Endocrinol. Metab. 38, 721-724.

Catt, K.J., Dufau, M.L., and Tsuruhara, T. (1972). Radioligand-receptor assay of luteinizing hormone and chorionic gonadotropin. J. Clin. Endocrinol. Metab. 34, 123-132.

Dufau, M.L., Mendelson, C.R., and Catt, K.J. (1974). A highly sensitive in vitro bioassay for luteinizing hormone and chorionic gonadotropin: testosterone production by dispersed Leydig cells. J. Clin. Endocrinol. Metab. 39, 610-613.

Karsch, F.J., Dierschke, D.J., Weick, R.F., Yamaji, T., Hotchkiss, J., and Knobil, E. (1973a). Positive and negative feedback control by estrogen of luteinizing hormone secretion in the rhesus monkey. Endocrinology 92, 799-804.

Karsch, F.J., Weick, R.F., Butler, W.R., Dierschke, D.J., Krey, L.C., Weiss, G., Hotchkiss, J., Yamaji, T., and Knobil, E. (1973b). Induced LH surges in the rhesus monkey: strength-duration characteristics of the estrogen stimulus. Endocrinology 92, 1740-1747.

Midgley, A.R., Jr., Zeleznik, A.J., Rajaniemi, H., Richards, J.S., and Reichert, L.E., Jr. (1974). Gonadotropin-receptor activity and granulosa luteal cell differentiation. In "Gonadotropins and Gonadal Function." (N.R. Moudgal, ed.), pp. 416-429, Academic Press, New York.

Monroe, S.E., Peckham, W.D., Neill, J.D., and Knobil, E. (1970). A radioimmunoassay for rhesus monkey luteinizing hormone (rhLH). Endocrinology 86, 1012-1018.

Neill, J.D., Dailey, R.A., Tsou, R.C., and Reichert, L.E., Jr. (1977). Immunoreactive LH-like substances in serum of hypophysectomized and prepubertal monkeys: inactive in an in vitro LH bioassay. Endocrinology 100, 856-861.

Niswender, G.D., Monroe, S.E., Peckham, W.D., Midgley, A.R., Jr., Knobil, E., and Reichert, L.E., Jr. (1971). Radioimmunoassay for rhesus monkey luteinizing hormone (LH) with anti-ovine LH serum and ovine LH-^{131}I. Endocrinology 88, 1327-1331.

Peckham, W.D., Foster, D.L., and Knobil, E. (1977). A new substance resembling luteinizing hormone in the blood of rhesus monkeys. Endocrinology 100, 826-834.

DISCUSSION

Wilks: Have you looked for LH-like substance in male monkeys, and have you compared the two radioimmunoassays for the male monkey?

Peckham: We have looked for the LH-like substance in male monkeys; and, as expected, we found it. It's very easy to measure because in the intact male monkey we don't detect LH with our rhesus:anti-hCG system; therefore, what we detect by the ovine:anti-ovine assay must be something other than rhLH, which indeed is the LH-like material. We have selected four male monkeys, because of their relatively high serum levels of this activity, as sources for material for preparative work and have implanted a number of estrogen-containing Silastic capsules into these males to try to ensure an absence of rhLH.

Ryan: I wonder if the ovine:ovine antiserum might recognize the CAGY sequence of LH-β, because that particular sequence happens to be in a whole variety of proteins, for example, cholera toxin and all of the serine proteases?

Peckham: It could be, yes. It does seem that this anti-ovine LH serum "sees" the β-subunit of LH exactly as it "sees" the whole molecule. I am not sure what sequence it does recognize, but I am reasonably sure that it isn't the so-called biologically active site of LH. I mention this because I think that it is not widely known that hCG and human LH are not recognized by this antiserum.

13

The Follicle Stimulating Hormone Releasing Hormone

KARL FOLKERS, S. FUCHS, J. HUMPHRIES, Y. P. WAN, and C. Y. BOWERS

As a chemist, I learn from physiology, and as a chemist, I hope to impart some of the aspects which chemistry allows for my subject, the putative follicle stimulating hormone releasing hormone (FSHRH). The adjective, putative, has crept into the semantics of research on some of the hypothalamic releasing and inhibiting hormones which have as yet eluded chemical characterization. If usage of this adjective has either a good or a bad connotation, I shall try to be very objective. Webster states that putative means: commonly thought of--or supposed.

Table I provides a list of the peptide releasing and inhibiting hormones which we have observed or perhaps discovered and purified since 1969 in cooperation with Cyril Bowers of Tulane University. In principle, we do the chemistry in Austin and Dr. Bowers does the endocrinology in New Orleans. These results came from our "scientific marriage" since 1969 in spite of our geographical separation.

We have found and purified an entity which behaves like FSHRH. Four other biologically active entities were found and purified. Both Factors A and B act like growth hormone releasing hormones. Factor C is an entity which behaves like a luteinizing hormone inhibiting hormone. Factor D behaves as a growth hormone inhibiting hormone, and is not somatostatin. Other investigators have pioneered on recognizing active fractions which show the activity of a prolactin inhibiting factor or hormone, and on other fractions which showed the activity of a corticotropin releasing factor or hormone. We have purified both PIF and CRF.

TABLE I.

CHEMICALLY UNKNOWN

PEPTIDE RELEASING AND INHIBITING HORMONES

Observed and Purified at U.T.A.

- FSH-RH
- Factor A-GHRH
- Factor B-GHRH
- Factor C-LHIH
- Factor D-GHIH

Purified at U.T.A.

- PIH
- CRF

PAST AND PRESENT POST-DOCTORATES AT THE UNIVERSITY OF TEXAS AT AUSTIN WHO CONTRIBUTED TO THIS AND RELATED PROGRESS:

YIEH-PING WAN
RONALD D. KNUDSEN
ISOMARO YAMAGUCHI
FRANZ ENZMANN
NILS GUNNAR JOHANSSON
BRUCE L. CURRIE
HANS STRAHSER
STEFAN FUCHS
JOHN HUMPHRIES
YIU-KUEN LAM
GERHARD RACKUR
CONNY BOGENTOFT
JAN JØRN HANSEN
JAW-KANG CHANG
EKKEHARD KUNZ
DING CHANG
WILLI FRICK
JAN BØLER
HANS SIEVERTSSON
TYGE GREIBROKK
JØRGEN WARBERG
GEORGE FISHER

As acknowledgement, I have listed in Table I those past and present postdoctorates who have contributed to this and related progress.

Table II is a partial list of the peptide hormones which are related to the hypothalamus and other tissues. It is no longer correct to classify all these peptides as hypothalamic hormones, because some of them have now been identified in tissues elsewhere in the brain and in the body, particularly, in the pancreas and the gastrointestinal tract.

I have classified these hormones as releasing hormones, inhibiting hormones, and other hormones. Since the chemistry of only two of at least seven releasing hormones, TRH and LHRH, is known, the chemistry of the releasing hormones in general is yet to be achieved. If somato-

TABLE II.

PARTIAL LIST OF HORMONES

RELATED TO THE HYPOTHALAMUS AND OTHER TISSUES

RELEASING HORMONES	CHEMISTRY
- Thyrotropin releasing hormone	known
- Luteinizing hormone releasing hormone	known
- Follicle stimulating hormone releasing hormone	unknown
- Corticotropin releasing hormone	unknown
- Growth hormone releasing hormone	
- Facotr A-GHRH	
- Factor B-GHRH	
- Prolactin releasing hormone	unknown
- Melanocyte stimulating hormone releasing hormone	uncertain
INHIBITING HORMONES	
- Growth hormone inhibiting hormone	unknown
- Somatostatin	known
- Hormone D-GHIH	unknown
- Prolactin inhibiting hormone	unknown
- Melanocyte inhibiting hormone	uncertain
- Factor C-LHIH	unknown
OTHER HORMONES	
- Substance P	known
- Somatostatin (see above)	known
- Neurotensin	known

statin is classified as a growth hormone inhibiting hormone, then it is the only inhibiting hormone of known chemistry. The chemistry of substance P and neurotensin is known, and although both have been isolated from hypothalamic tissue, their physiological roles are seemingly found outside of the pituitary gland.

FOLLICLE STIMULATING HORMONE RELEASING HORMONE

Now, I turn to an overview and some new information on the follicle stimulating hormone releasing hormone, and because its chemistry, assuming it exists, is unknown my presentation is somewhat of a "critique".

It was in 1970 that White of Abbott Laboratories reported that research on the isolation of LHRH led to the observation that the final product showed activity for the release of both LH and FSH, and at the level of 10-25 ng there was no separation of the two activities (White, 1970).

In 1971, Schally and White jointly published (Schally *et al.*, 1971a) the isolation and properties of the "FSH and LH-releasing hormone". They stated -- "this polypeptide appears to represent the hypothalamic hormone which controls the secretion of both LH and FSH from the pituitary".

Later in *Science*, Schally and White again elaborated on: "one polypeptide regulates secretion of luteinizing and follicle stimulating hormones" (Schally *et al.*, 1971b). They introduced the designation "gonadotropin releasing hormone" or "GnRH". This publication in 1971 is particularly interesting, because it reveals both their disbelief and belief that one peptide regulates both FSH and LH. The key points at the time were as follows: 1) during purification of extracts, the location of FSHRH activity always coincided with that of LHRH; 2) after the peptide was isolated in an essentially pure state by combinations of 12 different steps, or by counter current distribution, it also stimulated the release of FSH *in vivo* and *in vitro*; 3) further fractionation in 10 different solvent systems did not separate the LHRH activity from the FSHRH activity. When the synthetic peptide became available and was tested, it released both LH and FSH exactly like the isolated peptide. So, Schally and White and their co-workers proposed that the isolated and synthesized decapeptide is the gonadotropin releasing hormone which controls both LH and FSH.

Of course, Schally and co-workers proved that this decapeptide (GnRH) does have two activities, namely the release of LH and FSH, but did they really prove that there is no FSHRH which is a different hormone from the isolated decapeptide? If not, their concept and designation of the decapeptide as the intrinsic GnRH regulating both LH and FSH can be said to have been premature.

This decapeptide is not the only hormone of the hypothalamus and other tissues which has more than one activity. In fact, all five of those peptides in Table II which have known chemistry, and are associated with a hypothalamus, have more than one activity. There is no exception. TRH releases both TSH and PRL, even in man. Today we know that somatostatin suppresses the release of the growth hormone, insulin, glucagon, and gastric acid. This wide array of activities of somatostatin seemed to diminish the early enthusiasm for its therapeutic promise. Both substance P and neurotensin have multiple activities, including some effects on the pituitary hormones at elevated dosage levels. Even Leu-enkephalin and Met-enkephalin show some effects on pituitary hormones, such as the inhibition of the release of LH by LHRH.

The repeatedly confirmed finding that the decapeptide does release FSH does not prove that this decapeptide naturally regulates FSH as well as LH.

Our past fractionation efforts were scaled-up to hypothalamic fragments from over 1/2 million animals, and resulted in observations supporting the existence of an FSHRH which appears different from that of the decapeptide, <Glu-His-Trp-Ser-Tyr-Gly-Leu-Arg-Pro-Gly-NH_2. In 1973, we published three companion papers on FSHRH as summarized in Table III.

Johansson _et al_. (1973) conducted biosynthetic studies with ^{14}C-glutamic acid and glutamine in hypothalamic systems, followed by fractionations, including defatting, and steps of Bio-Gel P2, carboxymethylcellulose, Sephadex G-25, and partition chromatography. Certain fractions released 40,000- >128,000 ng/ml of FSH. The synthetic decapeptide, under comparable conditions, generally released about 18,000 and rarely up to 35,000 ng/ml of FSH. The possible presence of the decapeptide (GnRH) in the most purified fractions was considered negligible . The associated radioactivity also indicated that the entity releasing such high levels of FSH could have a <Glu moiety.

The second paper by Currie _et al_. (1973) stated that 13 fractions at stage P2 released FSH _in vitro_ at levels as high as 40,000- >128,000 ng/ml. In comparison with these fractions, the synthetic decapeptide released an average of about 16,700 units at the high dose of 500 ng/ml. These data seemed to support the existence of the unknown FSHRH by differentiation of its potency in release of FSH from that of the decapeptide. Such experimentation was not possible before the availability of the synthetic decapeptide.

TABLE III. FOLLICLE STIMULATING HORMONE RELEASING HORMONE (FSHRH OR FRH)

BIOSYNTHESIS AND EVIDENCE FOR EXISTENCE OF FSHRH - Associated radioactivity, high FSH release. Perhaps pGlu moiety.	Johansson, Currie Folkers, Bowers (1973)
EXISTENCE AND PURIFICATION OF FSHRH - Synthetic LHRH at 500 ng/ml up to 28,000 ng/ml FSH. - Partially purified FSHRH up to 128,000 ng/ml FSH.	Currie, Johansson, Folkers, Bowers (1973)
BIOLOGICAL EVIDENCE FOR SEPARATE LHRH AND FSHRH - The decapeptide is LHRH. - FSHRH is a separate hormone.	Bowers, Currie Johansson, Folkers (1973)

The third companion paper by Bowers *et al*. (1973) summarized the biological evidence which supported the conclusion that separate hypothalamic hormones release FSH and LH. It was found that a fraction from porcine hypothalmi, which was essentially free of the decapeptide, released both FSH and LH. Such a fraction *in vitro* released a greater amount of FSH than did the synthetic decapeptide. It was concluded that the decapeptide is actually LHRH, which releases predominently LH, and secondarily FSH, and that FSHRH is a separate hypothalamic releasing hormone. The interpretation of these data of 1972-73 is essentially the same today in our view.

Igarashi *et al*., in Japan, subsequently published evidence supporting the existence of an FSHRH which appeared different from the decapeptide, LHRH (Igarashi *et al*., 1974). Working with relatively crude rat hypothalamic extracts, they concluded that there was another substance in their extracts which was distinct from LHRH, as based on the properties of the synthetic decapeptide. This second entity showed a strong FSH releasing activity and a weak LH releasing activity, and this relationship coincided with the relative activities observed by Dr. Bowers for the fractions we provided him. We are still particularly impressed with the data showing that partially purified fractions could release up to and over 128,000 ng of FSH/ml in contrast to levels up to 12,000-28,000 ng of FSH by synthetic LHRH and at the high dosage of 500 ng/ml.

The data in Table IV are from Fawcett _et al_. (1975). They described chromatographic evidence for the existence of what they called "another species" of the luteinizing hormone releasing hormone. They extracted batches of 50-250 rat hypothalami, and after 5 extractions, tested fractions by two biological assays. One assay measured LHRH _in vitro_, and the other was a radioimmunoassay of LHRH. In their fractionations, two elution peaks were observed. The major peak coincided with LHRH. However, there was a minor peak which was completely separated from the major peak, and it was biologically and immunologically similar to LHRH. These data of Fawcett _et al_. also seem strongly supportive of the existence of FHSRH although these investigators designate the activity as "another species" of LHRH.

In 1976, Schally and Arimura and their co-workers (Schally _et al_., 1976) described their re-examination of stored porcine and bovine hypothalamic fractions in their search for an additional entity which would release LH and FSH. They re-examined 150 available fractions. Notably, they said that they expected that FSHRH would _not be destroyed_ by heating to 98°C. In any case, they could find no evidence _in vivo_ for an FSHRH, and they concluded that if FSHRH does exist, its level is lower than that of LHRH. One can say that their re-examination only demonstrated that they could not find the elusive FSHRH; possibly, it had been destroyed by heating to 98°C.

DISSOCIATED RELEASE OF LH AND FSH

In Table V are different but significant data, supporting the existence of an FSHRH, which pertain to the dissociation of the activities of synthetic analogs of LHRH for the release of LH and FSH. This table contains the data on the agonist activity of 2 analogs of LHRH for the

TABLE IV. CHROMATOGRAPHIC EVIDENCE FOR THE EXISTENCE OF ANOTHER SPECIES OF LUTEINIZING HORMONE RELEASING HORMONE

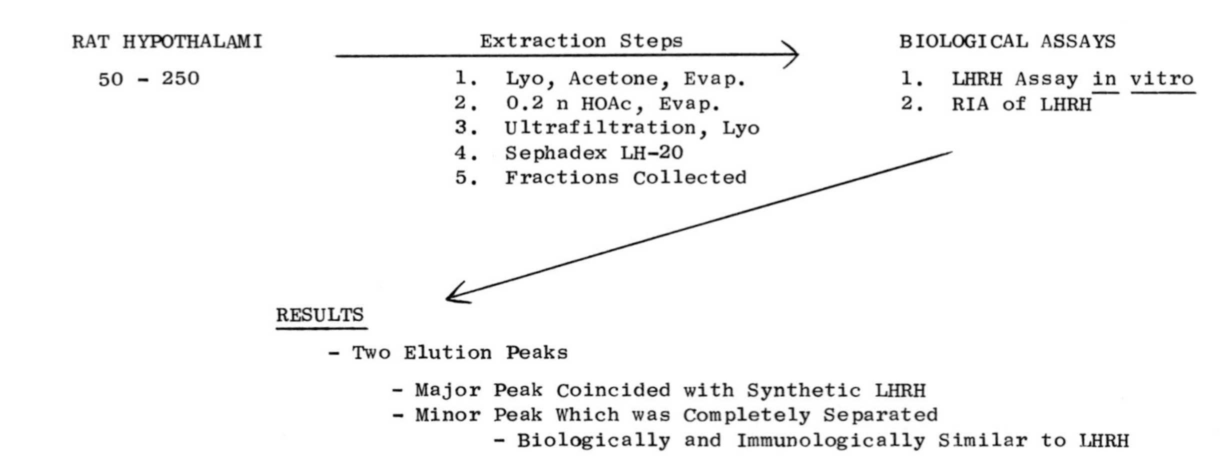

relative release of LH and FSH. Both [D-Phe^2, Phe^5, D-Phe^6]-LHRH and [D-Trp^2, Leu^3, D-Trp^6]-LHRH, at high levels, release LH, but little FSH. The control level of LH released by LHRH was about 200 and that of FSH was about 10,000. In contrast, analog number 3 gave a level of about 300 for LH and a level of only about 2,000 for FSH. Analog number 4 seems to show even greater dissociation, since the level of about 170 for LH was not significantly different from that of about 200 for LHRH, but analog number 4 resulted in a level of about 800 for FSH which may be compared to the level of about 10,000 for LHRH.

TABLE V. *IN VITRO* LH AND FSH AGONIST ACTIVITY OF LH-RH ANALOGS

Analogs	Dose µg/ml	LH Δmµg/ml medium	SEM	FSH Δmµg/ml medium	SEM
1. Saline	------	9	±10	-39	±225
2. LH-RH	0.0006	207	±34	10,213	±896
3. DPhe^2, Phe^5, DPhe^6-LH-RH	100	300	±36	1,994*	±246
4. DTrp^2, Leu^3, DTrp^6-LH-RH	100	172	±25	785	±404

*$P<0.001$ vs. saline.

We have not counted the number of analogs of LHRH which we have synthesized, but the number may be well over 200. Almost all of those analogs, which have agonist activity, do release LH and FSH in about the same ratios as does LHRH itself. However, there are notable exceptions as shown by the data in Table V. Since we do not have the structure of the apparent FSHRH, and because of the very large number of possible structures for it, it seems plausible that these analogs and others, which do show a dissociation of the release of LH and FSH, are more important than the many analogs which do not show such dissociation.

Figure I shows the comparative *in vitro* effects of a purified FSHRH fraction and that of synthetic LHRH on the release of LH and FSH. The first two bars on the left show that LHRH released an eight-fold greater amount of LH than did the fraction containing FSHRH. The third and fourth bars show the release of FSH was nearly the same for both. At about three-fold higher dosages, the fifth and sixth bars show that the LH response was nearly the same for both the fraction of FSHRH and LHRH. The seventh and eighth bars show that the FSH response of the fraction of FSHRH was considerably greater than that from LHRH. At the lower dosage, the fraction of FSHRH primarily released FSH and with little concommitant release of LH in comparison with LHRH. At the higher dosage, the fraction of FSHRH released considerable LH as does LHRH, and the fraction of FSHRH released more FSH than did LHRH.

FIGURE I.

COMPARATIVE EFFECT OF PARTIALLY PURIFIED "FSH-RH" AND SYNTHETIC LH-RH ON RELEASE OF LH AND FSH *IN VITRO*

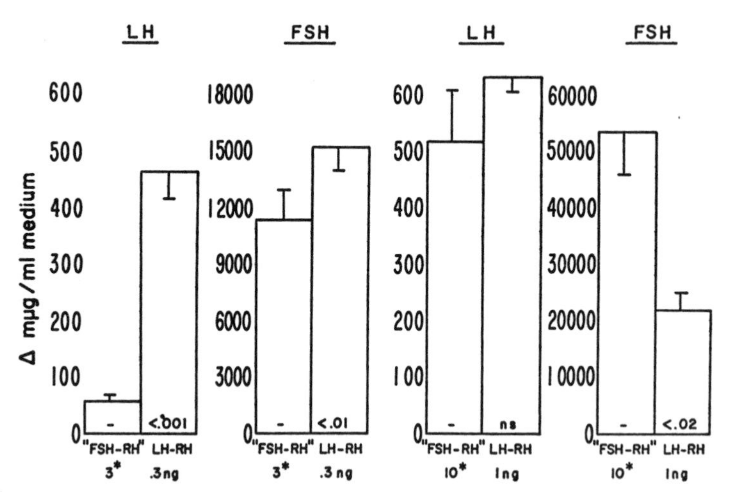

* Doses recorded as equivalents of porcine hypothalamic fragments/ml medium. Pituitaries of 20 day old female rats. Mean 3 ± SEM.

Figure II shows the results of the effect of a fraction of purified factor C - LIF (or LIH) on the release of LH and FSH by LHRH. This fraction (90 μg/ml) of factor C - LIF almost completely inhibited the release of LH, but the release of FSH was not inhibited. These data constitute another example of demonstrating a dissociation of the release of LH and FSH. If this apparent luteinizing hormone inhibiting factor is real and physiological, then it is important to note that it allows LHRH to release FSH since only FSH and little LH was released when both LIF and LHRH were present.

FIGURE II.

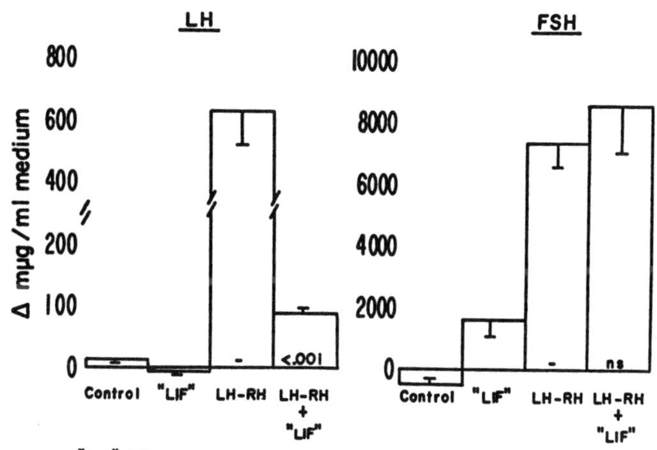

Figures I and II show data which may be interpreted on the basis that there is one entity which acts like FSHRH and another entity which acts like LIF (or LIH).

Figure III contains data on an analog which is an important exception to the general observation that most agonist analogs of LHRH release LH and FSH in the same ratio as does LHRH. This table shows that [D-Phe^2, Ala^4, D-Phe^6]-LHRH at the dosage of 100 μg/ml showed *in vitro* agonist activity, but this analog is most exceptional because it released LH with very little FSH. On the left part of this figure is the LH response; on the right part is the FSH response. The results are in the sequence: control, LHRH, the analog, and the analog plus LHRH. The analog released more LH than LHRH under these conditions, and more LH was released by LHRH plus the analog. The results on FSH are the opposite. LHRH released considerable FSH, but the analog released very little FSH. This unique analog rather completely inhibited the release of FSH by LHRH as shown by the result of testing them together.

FIGURE III.

ON THE DISSOCIATION OF LH AND FSH RELEASE BY DPHE^{2},ALA^{4},DPHE^{6}-LH-RH (ANALOG) IN VITRO

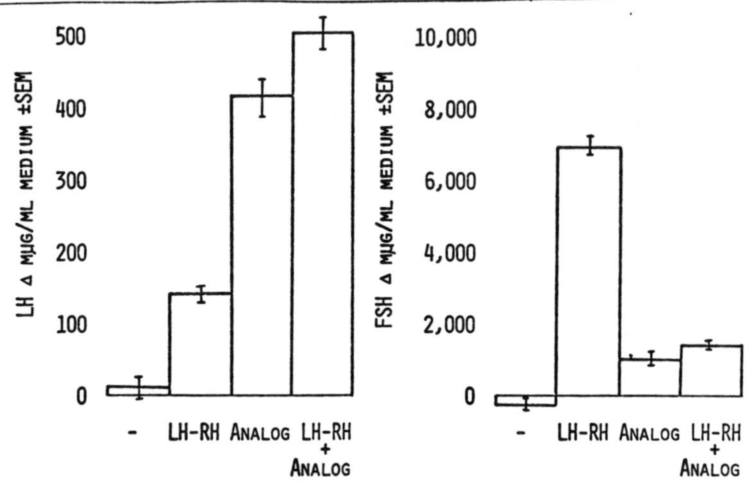

PITUITARIES OF 20 DAY OLD FEMALE RATS. MEAN (6). DOSE/ML: LH-RH 0.3 MμG, ANALOG 100 μG.

LHRH ANTAGONISTS

Alza's minipump (ALZET®) permits a continuous and subcutaneous release of a peptide in vivo over a few days of time. The minipump is based upon an osmotic driving agent which is the source of energy for the controlled delivery of about 1-2 μliters per hour during a period of 3-7 days. One of our recent effective inhibitors of LHRH is [D-Phe^{2}, Pro^{3}, D-Trp^{6}]-LHRH. This analog, in vitro, inhibits the LH and FSH response to LHRH at the low level of 30 nanograms, which represents a ratio of antagonist to LHRH of about 50:1. Figure IV shows the inhibition by this antagonist and the related [D-Phe^{2}, Leu^{3}, D-Trp^{6}]-LHRH on the LH levels of castrated male rats when it was infused over a 4-day period by the minipump at 375 μg/day. These analogs significantly lowered the elevated serum LH levels during the entire 4-day period of the infusion. The minipump is a distinct step forward in new pharmacological procedures for the delivery of metabolically labile peptide hormones.

FIGURE IV.

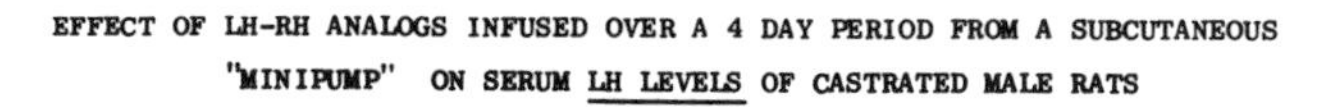

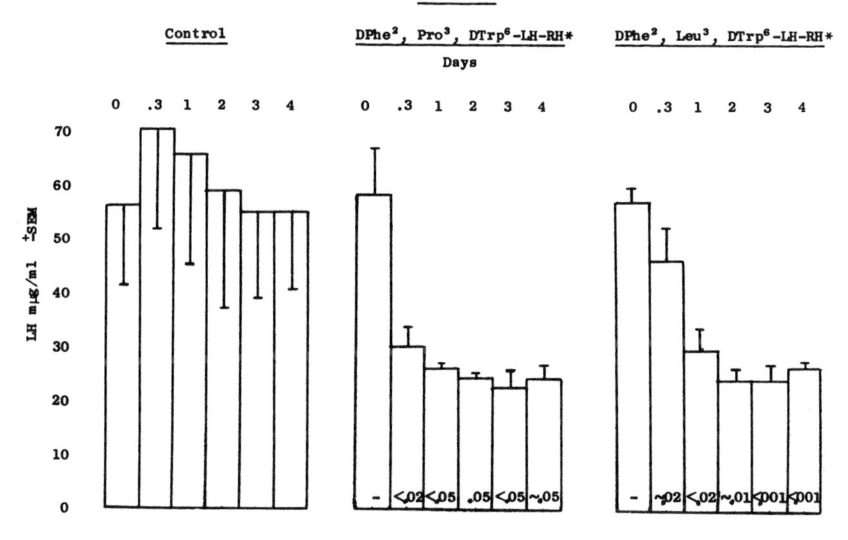

Δ Values = 0 time minus various times. p value = Δ Value of control versus experimental at respective time intervals. * 1.5 mg/minipump.

Table VI shows the effect of these antagonists on the ovulation of rats when they were infused over a 4-day period from a subcutaneous minipump. The number of ova in the ovulating control rats was 15.4 ± 1.8. There were no ova, or there was 100% inhibition, when the antagonist, [D-Phe^2, Pro^3, D-Trp^6]-LHRH was infused at the rate of 375 μg/day for 4 days. The closely related [D-Phe^2, Leu^3, D-Trp^6]-LHRH, when infused at 375 μg/day over 4 days resulted in only a 20% inhibition of ovulation. When the Pro^3-antagonist was assayed by a standard procedure, i.e., not using the minipump, a single injection of 750 μg caused 100% inhibition of ovulation, and 375 μg caused a 50% inhibition of ovulation.

LUTEINIZING HORMONE INHIBITING HORMONE

I wish to make the following additional comments on factor C which is an entity acting like a luteinizing hormone inhibiting hormone (Johansson et al., 1975). It was observed that factor C inhibited not

TABLE VI.

EFFECTS OF LH-RH ANALOGS CONSTANTLY INFUSED OVER A 4 DAY PERIOD FROM A SUBCUTANEOUS "MINIPUMP" ON OVULATION OF RATS

Analogs	Dose mg/rat s.c.	# rats	# rats ovulated	#ova/ ovulated rat $\pm$SEM	% inhibition of ovulat.
50% PG**	--	5	5	15.4 ± 1.8	0%
$DPhe^2$, Pro^3, $DTrp^6$-LHRH*	1.5mg+	4	0	0 ± 0	100%
$DPhe^2$, Leu^3, $DTrp^6$-LHRH	1.5mg	5	4	10.8 ± 2.7	20%

*kindly supplied by Alza Corporation. Palo Alto, California. **Propylene Glycol.

+ ca. 400 μg/day/4 days

STANDARD ASSAY ON [$DPhe^2$, Pro^3, $DTrp^6$]-LHRH (NO MINIPUMP)

Single injection at 750 μg, 100% Inhibition of Ovulation in Rats
Single Injection at 375 μg, 50% Inhibition of Ovulation in Rats

only the basal release of the lutinizing hormone, but it also inhibited the release of LH and FSH by synthetic LHRH. Factor C has been assayed in fractions which were and were not derived from biosynthesis. Factor C and FSHRH initially fractionated together, and FSHRH obscured factor C. During this obscurity, FSH assays guided fractionation. After partition chromatography on Sephadex G-25 in two systems, factor C and FSHRH were separated by Sephadex LH-20, purified by DEAE Sephadex and then by high pressure liquid chromatography to give fractions of factor C which inhibited LH release at levels of 100-500 nanograms.

FOLLICLE STIMULATING HORMONE INHIBITING HORMONE

From time to time, Dr. Bowers (unpublished data) has observed *in vitro* responses which could be interpreted on the basis of an entity behaving like an FSH inhibiting hormone, but we have less data available on an FSHIH than we have on a LHIH. In 1976, Steinberger and Steinberger (1976) investigated the ability of rat Sertoli cells to secrete an FSH-inhibiting substance *in vitro* by utilizing cultured Sertoli and pituitary cells. Sertoli cells were isolated from rat testes with a high degree of purity (>80%) and shown to retain in culture many morphologic characteristics and physiological responses. Rat anterior pituitary cells were also found to remain functional for prolonged culture periods. Their study provided data that isolated rat Sertoli cells secrete in culture a factor which selectively inhibits FSH release *in vitro* by acting directly upon pituitary cells.

EXTRA-HYPOTHALAMIC PEPTIDES

Of the diversified evidence which I have presented in support of the existence of FSHRH, and LHIH perhaps the most convincing evidence concerns those studies on the extraction of tissue and purification followed by the guidance of an appropriate bioassay. I turn now to additional evidence of this nature which has not yet been published, but will soon be submitted. We apparently observed the activity of the follicle stimulating hormone releasing hormone by bioassay, and also observed the activity of LHRH by both bioassay and radioimmunoassay in appropriate fractions from extracts of porcine caudate nucleus, cerebral cortex and cerebellum. Apparently, the activity of factor C-LHIH was again indicated. The assays appeared to show the presence of separate entities that released LH and FSH and inhibited the release of LH. We believed that the biologically active activities were probably identical to the corresponding activities which were previously recognized in extracts of hypothalamic tissue and which had been purified in an analogous manner. A gel filtration step served to reveal the overlap of the LH releasing and inhibiting entities that were together.

A direct radioimmunoassay for the decapeptide LHRH made possible the following observations: 1) the concentration of LHRH was the highest for the caudate nucleus, intermediate for the hypothalamus, and lowest for the cerebral cortex; 2) the concentration of LHRH in the cortex was small but significant; seemingly, the presence of factor C-LHIH was evident, since significant concentrations of the decapeptide were shown by radioimmunoassay, but there was diminished or negligible LH release; 3) the FSH which was released was not seemingly proportional to the amount of LHRH, which was present in the fraction.

It appears that these factors or hormones regulating gonadotropins are found in diverse areas of the brain. During the last several years, many investigators have reported diversified evidence for the presence of the now so-called hypothalamic hormones in tissues other than that of the hypothalamus.

CONCLUSION

In brief summary, there is evidence for the existence of an FSHRH, an LHIH and possibly an FSHIH, and the strength of the evidence of these three entities diminishes in the order they are named.

Why is the putative FSHRH so elusive? On the basis that it exists, some plausible explanations are:

1. It may be present in low concentration as Schally and others have considered.

2. It may be chemically unstable and be partly or entirely lost through months and years of storage of tissue and fracionations. It may be lost by either chemical or enzymic cleavage, or both.

3. It may be obscured by inhibiting hormones during assay of mixtures of hormones.

4. It may be simply missed by the nature of the horrendous fractionations.

We have just found two areas of eluates showing FSH release in a new fractionation scheme for hypothalami, and only one, but more potent area, for the thalamus and caudate nucleus. It could be a mistake to assume FSHRH does not exist, because it has not yet been isolated. We know about the difficult isolations of TRH and LHRH, and the isolation of FSHRH may also be very expensive and an exceptionally laborious task. The putative FSHRH is important, and deserves the best thinking and skill in the laboratories of all of us.

REFERENCES

Bowers, C.Y., B.L. Currie, K.N.G. Johansson, and K. Folkers (1973). Biological evidence that separate hypothalamic hormones release the follicle stimulating and luteinizing hormones. Biochem. Biophys. Res. Commun. 50:20-26.

Currie, B.L., K.N.G. Johansson, K. Folkers, and C.Y. Bowers (1973). On the chemical existence and partial purification of the hypothalamic follicle stimulating hormone releasing hormone. Biochem. Biophys. Res. Commun. 50:14-19.

Fawcett, C.P., A.E. Beexley, and J.E. Wheaton (1975). Chromatographic evidence for the existence of another species of luteinizing hormone releasing factor (LRF). Endocrinology 96:1311.

Igarashi, M., K. Taya, and J. Ishikawa (1974). In psychoneuro-endocrinology Workshop Conf. Int. Soc. Pshychoneuroendocrinology, Mieken, 1973, p. 178, Publ. Karger, Basal.

Johansson, K.N.G., B.L. Currie, K. Folkers, and C.Y. Bowers (1973). Biosynthesis and evidence for the existence of the follicle stimulating hormone releasing hormone. Biochem. Biophys. Res. Commun. 50:8-13.

Johansson, K.N.G., T. Greibrokk, R.L. Currie, J. Hansen, K. Folkers, and C.Y. Bowers (1975). Factor C-LHIH which inhibits the luteinizing hormone from basal release and from synthetic LHRH and studies on purification of FSHRH. Biochem. Biophys. Res. Commun. 63:62-68.

Schally, A.V., A. Arimura, Y. Baba, R.M.G. Nair, H. Matsuo, T.W. Redding, L. Debeljuk, and W.F. White (1971a). Isolation and properties of the FSH and LH-releasing hormone. Biochem. Biophys. Res. Commun. 43:393.

Schally, A.V., A. Arimura, A.J. Kastin, H. Matsuo, Y. Baba, T.W. Redding, R.M.G. Nair, L. Debeljuk, and W.F. White (1971b). Gonadotropin-releasing hormone: One polypeptide regulates secretion of luteinizing and follicle stimulating hormones. Science 173:1036-1037.

Schally, A.V., A. Arimura, T.W. Redding, L. Debeljuk, W. Carter, A. Dupont, and J.A. Vichez-Martinez (1976). Re-examination of porcine and bovine hypothalamic fractions for additional luteinizing hormone and follicle stimulating hormone releasing activities. Endocrinology 98:380.

Steinberger, A. and E. Steinberger (1976). Secretion of an FSH-inhibiting factor by cultured Sertoli cells. Endocrinology 99:918-921.

White, W.F. (1970.
In, Hypophysiotropic Hormones of the Hypothalamus, J. Meites, ed., p. 248. The Williams and Wilkins Co., Baltimore.

DISCUSSION

Peckham: Has anyone given antiserum against synthetic LH-RH and observed complete reduction of LH secretion in the face of continuing secretion of FSH?

Folkers: I do not recall the details of such studies.

Schwartz: Koch (Biochem. Biophys. Res. Comm. 55:623, 1973) used an antiserum to LH-RH in several different biological test systems. He observed a reduction in both serum FSH and LH. That, however, is not an argument against the existence of a FSH-RH.

Folkers: It is very difficult to make dogmatic interpretations about this work on the FSH-RH. The questions about FSH-RH may never be entirely answered until someone isolates it in pure form, sequences it, synthesizes it, and shows that the natural and synthetic products act identically. If we turn back time, there were a large number of doubts that TRH existed until the chemistry was done, and that is understandable. There are horrendous problems of purification and fractionation, particularly when you want to go to purity. One can extract 50 hypothalami and make some interesting and important observations, but if one wants to go to chemical purity, then one has a real big task.

Schwartz: It seems to me that the most potent argument in favor of a separate releasing factor for FSH is a physiological argument. There is evidence from a number of laboratories that there has to be a separate mechanism for controlling the secretion of FSH and LH. At the moment the burden of proof against the FSH releasing factor rests with people who can come up with a mechanism by which FSH can be released in the absence of LH release by the pituitary. They may evoke a direct gonadotroph mechanism, but it seems to me that the physiological evidence indicates that there must be some separate secretory control.

Folkers: I confess that if it weren't for all that physiological evidence, I don't think I would be doing what I am trying to do. The task is so long, so frustrating, and so expensive that few will attempt to do it unless it is important, and even then one may lose, because the pitfalls in this isolation are very, very many.

Schwartz: I will remind you of another piece of evidence. Fawcett and McCann some years ago did a three dimensional examination of the hypothalamus looking for FSH-RH and LH-RH. They were able to separate out locations in three dimensions which caused greater FSH release than other locations that caused LH release. They were using fresh frozen sections and not storing them; that may be an argument for the storage loss of FSH-RH.

Folkers: Chemical instability is possibly part of the trouble. I know of some peptides that are just inherently unstable, apart from any storage conditions. One can also speculate that the reasons TRH was synthesized first were at least two-fold. The assay was very good and that always facilitates the isolation of anything. Also, TRH is a remarkably stable tripeptide. TRH could adequately survive the "mauling and slugging" that isolation chemists gave it for several years. Chemically, LH-RH is a relatively stable peptide. So, for a hormone in hypothalamic tissue that is at all sensitive and delicate, it may be years before it is isolated in pure form.

Peckham: Do you have any basis for estimating how much of this FSH-RH might be present in the hypothalamus in comparison to the amount of LH-RH?

Folkers: This kind of fractionation produces many fractions and the number of assays are limited; and, therefore, it is a very qualitative business. One just hopes to keep enough hormone in the main stream to reach the goal. I think the only way to answer your question is for one to extract fresh tissue, and do some assays when the methods become available.

Ward: I think the early fractionation procedures, patterned after the early methods of Otto Kamm, were all designed to isolate small peptides. Alternative approaches using procedures that would initially isolate large molecules, and then take them apart or dissociate them, might be feasible and perhaps even advantageous. This seems to be one of the approaches you are suggesting.

Folkers: One must give an enormous amount of credit to Schally and Guillemin for the tremendous task that each of them performed; they were purifying something without knowing what it was. The idea that TRH was a peptide was both "on and off" as time progressed. Nobody had any idea what kind of peptide it was. So, the steps of fractionation were not geared to the exact chemical nature of what they were seeking, because they could not know what they were seeking. Today, knowing all about TRH, and LH-RH, and somatostatin, I think fractionation can be conducted more intelligently by the organic chemist or the biochemist.

14

The Suppression of Serum FSH by a Non-steroidal Factor from Porcine Follicles: Some Properties of the Inhibitor "Folliculostatin"

JANICE R. LORENZEN and NEENA B. SCHWARTZ

Several lines of evidence suggest that FSH and LH secretion rates are separately regulated in the female rat. (A) Following ovariectomy, FSH rises promptly, but LH does not. A decrease in serum estradiol is not necessary for this rise in FSH (Campbell, *et al*, 1977) since estradiol concentrations do not change during the time of FSH elevation. (B) In chronically ovariectomized rats, replacement therapy with high doses of estrogen can suppress LH levels to baseline; however, FSH levels cannot be reduced completely (Campbell and Schwartz, 1977). (C) Electrical stimulation of the dorsal anterior hypothalamic area, or injection of a bolus of LH, causes a prolonged secretion of FSH, in the absence of LH secretion (Chappel and Barraclough, 1977; Schwartz and Talley, *in press*). These observations may or may not militate against the current hypothesis of a single hypothalamic releasing factor controlling FSH and LH, but they do suggest the possibility of a second ovarian negative feedback factor which regulates separately the secretion of FSH rather than LH, or works in synergism with estrogen to control the secretion of FSH. This kind of hypothesis is not new - it has long been proposed that a polypeptide from the testis, inhibin, inhibits FSH secretion in the male (Baker, et al, 1976). The object of the following studies was to investigate the possibility that a similar substance may be found in the ovaries, particularly in the fluid from the follicles.

Experiment 1: Can follicular fluid block the secondary FSH surge?

Female rats treated with pentobarbital at 1330h and 1550h on proestrus will not show the usual preovulatory LH or FSH primary surges. If an artificial LH surge is superimposed on the pentobarbital block at 1530h, all the rats will have a synchronous FSH secondary surge which is independent of estradiol feedback (Schwartz and Tally, in press; Chappel and Barraclough, 1977). If there is a substance in follicular fluid capable of suppressing FSH release it should be able to block the rise of FSH seen at 0400h estrus in this model. This was tested in the following experiment (Schwartz and Channing, in press).

Porcine follicular fluid, collected from 3-10 mm follicles of ovaries acquired from a meat packing plant, was charcoal extracted to remove steroids. After extraction, the levels of estradiol, progesterone, and testosterone were less than 1% of the original values (Marder, et al, in press). Each pentobarbital blocked, LH injected rat received two intraperitoneal injections (at 1545h and 1830h) of 0.5 ml of either a) undiluted follicular fluid, b) follicular fluid diluted 1:5, 1:25, or 1:125, or c) porcine serum (also charcoal extracted). The results are shown in Table 1. Not only could the follicular fluid inhibit the FSH rise at 0400h in this model, but the response could be seen to diminish as the dose declined ($p<0.01$). Thus, follicular fluid contains a substance(s) which is non-steroidal and can inhibit the secondary FSH surge in a pentobarbital blocked, LH injected rat.

Experiment 2: Can follicular fluid block the post-ovariectomy rise in FSH?

If there is a substance in follicular fluid which normally maintains the tonic levels of FSH, then replacement of this substance after removal of the ovaries should inhibit the rise in FSH seen by nine hrs post-castration, since estradiol does not (Campbell, *et al*, 1977; Chappel and Barraclough, 1977). Female rats were gonadectomized at 0800h on metestrus and injected with 250 µl porcine follicular fluid or porcine serum at 1130h. They were autopsied at 1700h the same day. Serum FSH levels were suppressed to near sham-operated control levels in the follicular fluid treated rats (see Figure 5 from Marder, *et al*, 1977 in the Channing, *et al*, paper in this symposium). Porcine serum treated rats demonstrated the typical post-castration increase in FSH. The suppression by follicular fluid was dose-dependent (Marder, *et al*, 1977). Thus, replacement of the factor from follicular fluid (folliculostatin) can prevent the rise of FSH after ovariectomy.

Table 1. Effect of follicular fluid on an artificially induced secondary FSH surge[1]

FOLLICULAR FLUID	FSH (ng/ml)
undiluted	217.7 ± 52.1[2]
1:5 dilution	489.7 ± 66.1
1:25	674.3 ± 12.7
1:125	573.2 ± 78.7
porcine serum	728.2 ± 31.7

[1]Female rats exhibiting 2 consecutive 4 day cycles were injected on proestrus with 3.15 mg pentobarbital/100g body weight at 1330h and 0.75 mg pentobarbital/100g BW at 1500h, both given intraperitoneally (IP). At 1530h, an intravenous injection of 8 μg NIH-LH-S16 (in 0.5 ml) was given. At 1545h and 1830h, each rat received an IP injection of 0.5 ml of one of the above dilutions. Animals were sacrificed at 0400h.

[2]Values represent the mean ± standard error (N = 3)

(Schwartz and Channing, *in press*)

Experiment 3: Is folliculostatin present in all follicles?

To assess the physiologic function of this substance found in follicular fluid in its role as an FSH modulator, it was of interest to know if it is present in follicles of all sizes. Is it being formed and released in a steady state with constant storage levels in follicles; is it diluted out as the follicle increases in size; or does its concentration increase (and release perhaps decrease?) as the fol-

Table 2. The ability of follicular fluid from small, medium, and large follicles to suppress the post-ovariectomy rise of FSH in rats.

	FSH (ng/ml)	LH (ng/ml)
OVARIECTOMIZED[1]		
Small[2]	180, 150[3]	1.6, 2.1
Medium	165.8 ± 51.2[4]	1.5 ± 0.3
Large	205.3 ± 62.6	1.4 ± 0.3
Porcine serum control	503.3 ± 37.6	1.3 ± 0.2
SHAM		
Small	137, 100	<1.0, 1.0
Medium	104.3 ± 23.2	1.0 ± 0.0
Large	138.3 ± 22.1	1.0 ± 0.0
Porcine serum control	253.5 ± 16.0	1.2 ± 0.2

[1]All rats were ovariectomized or sham-operated at 0800h on the day of metestrus. At 0830h and 1130h each rat received an intraperitoneal injection of 250 μl of the indicated fluids. Autopsies were done at 1700h.

[2]Small follicles are those 1-2 mm in size, medium follicles are 3-5 mm, and large follicles are 6-12 mm.

[3]Only two values were acquired with follicular fluid from small follicles

[4]Represents mean ± standard error (N = 4).

(Marder, et al, in press)

Table 3. Failure of substances <10,000 daltons MW in follicular fluid to suppress serum FSH.

	FSH (ng/ml)	LH (ng/ml)
OVARIECTOMIZED[1]		
Retentate[2]	235.0 ± 19.6[3]*	1.8 ± 0.2
Filtrate	561.8 ± 43.1	1.9 ± 0.3
Porcine serum control	582.5 ± 26.5	3.6 ± 1.2
SHAM		
Retentate	167.0 ± 7.3*	1.2 ± 0.1*
Filtrate	282.5 ± 32.9	1.5 ± 0.1
Porcine serum control	262.8 ± 18.0	1.6 ± 0.1

[1]All rats were ovariectomized or sham-operated at 0800h on the day of metestrus. At 1130h each rat received an intraperitoneal injection of the indicated fluid. Autopsies were done at 1700h.

[2]Follicular fluid was filtered through an Amicon PM-10 membrane. The fluid remaining (retentate) has substances >10,000 daltons molecular weight. The filtrate has substances $\leq$ 10,000 daltons.

[3]Each value represents the mean ± standard error (N = 4).

*Significantly ($p<0.01$) different than serum controls.

licle prepares to ovulate? Using the metestrous ovariectomized or sham operated rat as the bioassay models, porcine follicular fluid from small (1-2 mm), medium (3-5 mm), and large (6-12 mm) follicles was tested for presence of folliculostatin. All size follicles showed evidence of FSH inhibiting activity (Table 2). However, from a single dose it is impossible to assess possible concentration differences in follicles.

Experiment 4: What is the molecular nature of folliculostatin?

Since all the follicular fluid used in these studies had been charcoal extracted to remove steroids, the effective agent must be nonsteroidal. The first estimate of its molecular weight was obtained by passing the follicular fluid through an Amicon PM-10 membrane which has a molecular weight cut-off of 10,000 daltons. Again, using the ovariectomized or sham operated metestrous rat models, the FSH suppressing activity was found to be retained by the membrane (Table 3). Thus, folliculostatin appears to have a molecular weight of greater than 10,000 daltons.

Experiment 5: Can follicular fluid inhibit the postcastration rise of FSH in the male?

This experiment tested whether or not this follicular fluid substance could act in a manner similar to the elusive testicular "inhibin", giving us an opportunity to see if FSH could be suppressed without simultaneous suppression of LH. Male rats were castrated at 2000h; injected with 250 μl follicular fluid or pig serum at 0800h and 1130h the next day; given a third injection of 500 μl at 1500h; and autopsied at 2000h, 24 hours post-castration (Table 4). The follicular

Table 4. Effect of porcine follicular fluid on the post-castration gonadotrophin rise in the male rat.

	FSH (ng/ml)	LH (ng/ml)
CASTRATES[1]		
Follicular fluid	869.0 ± 119.5[2]*	23.2 ± 6.7
Serum control	1574.0 ± 226.9	33.0 ± 18.9
SHAM		
Follicular fluid	389.0 ± 110.0	1.5 ± 0.1
Serum control	439.0 ± 21.2	1.3 ± 0.1

[1]All animals were castrated or sham-castrated at 2000h. Intraperitoneal (IP) injections of 250 μl of either porcine follicular fluid or porcine serum were given at 0800h and 1130h the next day. This was followed by a third IP injection of 500 μl at 1500h. Autopsies were done at 2000h; 24 hours after surgery.

[2]All values represent the mean ± standard error (N = 4).

*Significantly ($P<0.05$) different from serum controls.

fluid treated rats had much less of a rise in FSH than the pig serum treated rats, although the serum FSH was not suppressed completely to the level seen in the sham-castrate controls. The elevated post-castration LH values were not affected by the follicular fluid. Thus, folliculostatin inhibits the post-castration rise of FSH in both males and females.

SUMMARY

The multiplicity of roles of follicular fluid in reproductive events is becoming increasingly better defined (Edwards, 1974; Eiler and Nalbandov, 1977). Many of these roles, such as facilitating oocyte transport and protecting and nourishing the oocyte, are directly related to maintaining and delivering a fertilizable egg to the oviduct. Others deal indirectly with ovum delivery; for example, follicular fluid may contain proteolytic enzymes which result in rupture of the follicular wall during ovulation and which aid in spermatazoan capacitation. Recently several factors have been discovered in the follicular fluid, which are capable of inhibiting certain functions. Tsafriri, *et al*, (1976) and Channing, *et al*, (1977) have reported an oocyte maturation inhibitor found in follicular fluid which delays the completion of the first meiotic division in the soon-to-be ovulated oocyte. Ledwitz-Rigby, *et al* (*in press*) have demonstrated the presence of a luteinization inhibitor in fluid from small and medium (but not large) follicles which inhibits luteinization, as well as secretion of progesterone, by cultured granulosa cells. The results presented in this paper, as well as other reports (Schwartz and Channing, *in press*; Marder, *et al*, *in press*; DeJong and Sharpe, 1976; Welschen, *et al*, 1977) suggest the presence of another inhibiting factor in follicular fluid, folliculostatin, which inhibits the rise in FSH secretion seen in an artificially or naturally induced secondary FSH surge (Schwartz and Channing, *in press*), as well as the post-castration rise in FSH seen in females (Marder, *et al*, *in press*;

Welschen, et al, 1977) and males (Table 4; DeJong and Sharpe, 1976). It is present in follicles of all sizes (Table 2) and is more than 10,000 daltons in molecular weight (Table 3).

These three inhibitory factors in follicular fluid may serve the function of suppressing oocyte maturation, granulosa cell luteinization, and FSH secretion for recruitment of the next crop of follicles, until the preovulatory LH surge takes place. This surge in some manner "turns off" these three inhibitory functions, permitting ovulation of a prepared ovum, luteinization of the granulosa cells in preparation for pregnancy, and increased secretion of FSH which recruits a new follicular cohort.

Acknowledgements: We wish to thank Drs. C.P. Channing and S.H. Pomerantz for generous supplies of the follicular fluid, William L. Talley, Brigette G. Mann, Julian A. Alvarez, and John Gesme for excellent technical assistance; The National Institutes of Health, Endocrinology Study Section, for generous supplies of gonadotropic hormones used as standards for assays and for the FSH kit; Dr. L.E. Reichert, Jr., for the LH used for radioiodination; Dr. G.D. Niswender for contributing the antiserum used for the radioimmunoassays. Supported in part by U.S. Public Health Service grant HD 07504.

BIBLIOGRAPHY

Baker, H.W.G., W.J. Bremmer, H.G. Burger, D.M. de Kretser, A. Dulmanis, L.W. Eddie, B. Hudson, E.J. Keogh, V.W.K. Lee, and G.C. Rennie.

Testicular control of follicle-stimulating hormone secretion. Rec. Prog. Hor. Res. 32: 429-476, 1976.

Campbell, C.S. and N.B. Schwartz. Steroid feedback regulation of LH and FSH secretion rates in male and female rats. J. Toxic. Envir. Health 3: 61-95, 1977.

Campbell, C.S., N.B. Schwartz, and M.G. Firlit. The role of adrenal and ovarian steroids in the control of serum LH and FSH. Endocrinology 101: 162-172, 1977.

Channing, C.P., S.L. Stone, A.S. Kripner, S.H. Pomerantz. Studies on an oocyte maturation inhibitor present in porcine follicular fluid. Brook Lodge Symposium, present volume.

Chappel, S.C. and C.A. Barraclough. Further studies on the regulation of FSH secretion. Endocrinology 101: 24-31, 1977.

deJong, F.H. and R.M. Sharpe. Evidence for inhibin-like activity in bovine follicular fluid. Nature 263: 71-72, 1976.

Edwards, R.G. Follicular fluid. J. Reprod. Fert. 37: 189-219, 1974.

Eiler, H. and A.V. Nalbandov. Sex steroids in follicular fluid and blood plasma during the estrous cycle of pigs. Endocrinology 100: 331-338, 1977.

Ledwitz-Rigby, F., B.W. Rigby, V.L. Gay, M. Stetson, J. Young, and C.P. Channing. Inhibitory action of porcine follicular fluid upon granulosa cell luteinization in vitro; assay and influence of follicular maturation. J. Endocrinol., in press.

Marder, M.L., C.P. Channing and N.B. Schwartz. Suppression of serum follicle stimulating hormone in intact and acutely ovariectomized rats by porcine follicular fluid. Endocrinology, in press.

Schwartz, N.B. and C.P. Channing. Evidence for ovarian "inhibin": suppression of the secondary rise in serum follicle stimulating hormone levels in proestrus rats by injection of porcine follicular fluid. Proc. Nat. Acad. Sci., in press.

Schwartz, N.B. and W.L. Talley. Effects of exogenous LH or FSH on endogenous FSH, progesterone and estradiol secretion. Biol. Reprod., in press.

Tsafriri, A., S.H. Pomerantz, and C.P. Channing. Inhibition of oocyte maturation by porcine follicular fluid: partial characterization of the inhibitor. Biol. Reprod. 14: 511-516, 1976.

Welschen, R., W.P. Hermans, J. Dullaart, F.H. deJong. Effects of an inhibin-like factor present in bovine and porcine follicular fluid on gonadotrophin levels in ovariectomized rats. J. Reprod. Fertil. 50: 129-131, 1977.

DISCUSSION

Ryan: Does LH increase the amount of folliculostatin in follicular fluid?

Schwartz: We do not know yet. We have not found a difference, in preliminary studies, in concentration between large, medium, and small size follicles. We need a more careful search of that before we can really say for sure.

Peckham: Has anyone looked for neuraminidase activity in this pig follicular fluid?

Ryan: It's been very difficult to demonstrate extracellular neuraminadase activity. I think people have been unable to identify it in serum. Whether it's in follicular fluid or not I don't know.

Peckham: If neuraminadase activity is present in follicular fluid, that might possibly explain your observations. Neuraminadase cleavage of sialic acid residues from FSH would result in more rapid clearance of this hormone from the circulation and produce marked reductions in serum FSH levels. Since LH has relatively little sialic acid to begin with, the effect on serum LH levels would be minor.

Schwartz: Are you suggesting that the neuraminidase is circulating in the blood and cleaving the sialic acid from the FSH as it circulates? We are injecting the PFF intraperitoneally, and we now know that it takes more than 3 hrs to see the FSH drop. Do you really think that you can explain this effect by desialation of the already circulating FSH? That would seem to me rather difficult.

Baker: You could presumably check that out by injecting the animal with hCG and seeing if this was subsequently desialated.

Schwartz: We have not done that.

Channing: We have done a number of studies in which follicular fluid was added to the hCG in cultures. It doesn't seem to alter the hCG effect on cyclic AMP levels. We have shown in other studies that removal of sialic acid and other sugars does diminish cyclic AMP stimulatable activity.

Peckham: I don't think that's right. I don't think desialated hCG has diminished activity *in vitro*.

Channing: Further removal of sugars does diminish the activity.

Peckham: Yes, but I'm not talking about that, I'm just talking about desialation.

Schwartz: My major argument against you would be the time and the dose at which we are operating. We are giving the PFF intraperitoneally, and we do not see the first effects until at least 3 hrs later in a chronically ovariectomized preparation. I wonder where the neuramindase could be stored in that 3 hr period before it starts to cleave the FSH.

Peckham: I haven't any idea. Let me say that I don't know if neuraminidase is involved at all, but I think that it's something which should be checked out.

Bullock: I was wondering how specific this factor is for the follicular fluid. You had a pig serum control; but if this substance is any use in the pig, why can't you find it in the serum?

Schwartz: I suppose it is concentrated in the fluid. There is certainly no evidence that there is an effect of the pig serum that we get from commercial houses. But that could be a concentration problem.

Baker: Are you suggesting that it acts on the hypothalamus or on the pituitary? In either case the route seems to be rather long.

Schwartz: It is no longer than for estradiol to inhibit LH or FSH. As a signal between organs, not as an intraovarian signal, it seems perfectly reasonable to me that there would be a peptide signal as well as a steroid signal.

Baker: I was thinking of its route after injection into the peritoneal cavity. Some of the substance would probably be metabolised in the liver while the remainder would enter most of the organs in the abdomen, including the ovary itself where there might be a local effect. The dose level in blood will be much lower than that originally injected. It's a matter of at what level the substance is going to act and whether there could be sufficient amounts of the substance left to have an effect.

Schwartz: There are a couple of levels in which to answer your question. Since it works in animals without ovaries it does not need the ovary to manifest the effect. I hear your argument that "very little would get into the animal" as an argument in our favor, rather than against us. We are already getting an effect from about 100 μl of the fluid; if only 50 μl are showing up in the blood volume of a 250 g rat, then I do not see that as an argument against what we are saying. Yes, a remarkably small amount of this must be getting to the pituitary-hypothalamic axis. I do not know at what level it works though.

Lobl: What is the state of purification of this material?

Schwartz: All we know at this point is that it has a molecular weight greater than 10,000 daltons and it appears to be a protein because it is trypsin sensitive.

Feigelson: Have you looked at homologous follicular fluid, i.e. rat follicular fluid?

Schwartz: We hope to. We have looked at human follicular fluid with Jay Holt, and we have not been able to demonstrate an effect in the same dose range. I am inclined to think that those negative results are due to two things. One is that the dose range may be different.

It has been shown by others that bovine follicular fluid may have a greater FSH suppressing activity than porcine, so there may be a species difference there. The other thing is that most human follicular fluid comes from ovaries from older, perimenopausal women, and my guess is that that may be the wrong ovary in which to be looking for our follicular factor.

Baker: It's interesting that the very follicular fluid you are using as an inhibitor probably contains fairly large amounts of FSH, LH, estradiol, progesterone, androstenediol, and prolactin. You are presumably taking this into account.

Schwartz: We are extracting the steroids. The steroid content of the extracted fluid is lower than the steroid levels in the animals into which we are injecting it, which I think is a pretty potent argument against the "steroid" hypothesis. We have measured the FSH in the rat assay, which was parallel, and found the FSH content was fairly low.

Baker: I was referring specifically to the human follicular fluid.

Schwartz: The human follicular fluid was extracted by the same means, but we have not measured FSH and LH.

15

Some Properties of Small Molecular Weight Tissue and Serum Inhibitors of Follitropin Binding to Receptor

LEO E. REICHERT, JR.

1. INTRODUCTION:

Recent studies have described a variety of relatively low molecular weight affectors of hormonally related processes. Bailly *et al*. (1977) have reported an inhibitor of glucocorticoid receptor activation in cytosols prepared from a variety of rat tissues that is dialyzable and retarded by Sephadex G-25. Kent (1975) has sequenced a tetrapeptide (threonyl-prolyl-arginyl-lysine) which inhibits ovulation in hamsters. Tsafriri *et al*. (1976) have described an inhibitor of oocyte maturation in porcine follicular fluid with an estimated molecular mass of about 2000 daltons. Jagiello *et al*. (1977) reported a factor with similar properties in ovine and bovine follicular fluid, possibly derived from a large molecular weight (greater than 15000 daltons) precursor molecule present in serum. Leidenberger *et al*. (1976) reported a "mini LH" obtained from normal human serum by dialysis and gel filtration having a molecular weight of about 2000 daltons which inhibited binding of LH to testicular receptor, stimulated testosterone synthesis by mouse Leydig cells *in vitro* and had immunologic activity as determined by radioimmunoassay. Yang *et al*. (1976) have described a low (3800 daltons) molecular weight tissue-specific inhibitor of luteinizing hormone receptor binding which can be extracted from pregnant and pseudopregnant rat ovaries. Although with the exception of the Kent peptide (Kent, 1975), the chemical nature of these substances is not known, their potential theoretical and practical significance can be readily appreciated. In this report, we wish to summarize our observations on small molecular weight tissue and serum factors which inhibit the interaction of ^{125}I-hFSH with gonadal receptors.

2. EXPERIMENTAL:

The system utilized in the hormone binding and dissociation studies described herein consisted of radioiodinated, biologically active human follitropin (^{125}I-hFSH) as the ligand and testes from mature rats as the receptor source. The procedure utilized for preparation of ^{125}I-hFSH has been described in detail previously (Reichert and Bhalla, 1974). The receptor source was either homogenates of whole testis (Reichert and Bhalla, 1974) or membranes prepared from homogenates of tubules through use of sucrose density gradient centrifugation (Abou-Issa and Reichert, 1976). Our previous experience has indicated that the latter type of preparation is essential in studies of inhibitor effects on follitropin-receptor complex formation and dissociation since these may be masked when tissue homogenates are utilized as the receptor source. The procedure for preparation and dialysis of small molecular weight fractions from testes and other tissues for use in studies of FSH binding inhibition has been described in detail elsewhere (Reichert and Abou-Issa, 1977). Serum fractions were prepared using the identical dialysis procedures described for testes. The ratio of serum to dialysate (water) was 1:20, and after lyophilization, the dialysate was reconstituted with water or buffer as required.

Binding inhibition was expressed either in terms of ^{125}I-hFSH specific uptake in the presence or absence of inhibitor, or in terms of binding inhibition by intact hFSH. In the latter case, highly purified hFSH (as the reference preparation) and FSH-BI fractions were run at several increments of dose to obtain linear response curves, which were then compared by standard parallel line assay statistics. The inhibitory potency of the test fractions were then expressed in terms of ng hFSH. That is, if a fraction had a FSH-BI potency of 5 ng/mcg, that meant that one mcg of test substance inhibited binding of ^{125}I-hFSH to receptor to the same degree as 5 ng of unlabeled native hFSH run simultaneously as the standard in the assay. Often, non-parallelism of test substance with hFSH made quantitative analysis by the latter procedure statistically invalid. When this occured, an estimate of hFSH equivalence was made at 50% binding inhibition.

3. FOLLITROPIN BINDING INHIBITORS IN RAT TESTES:

Our interest in follitropin binding inhibitors (FSH-BI) of rat testes stemmed from the observation that ^{125}I-hFSH unbound to testicular receptors and remaining in the tissue supernatant after a specific period of incubation did not bind to fresh receptor as well as ^{125}I-hFSH incubated in buffer alone, suggesting the presence of a binding inhibitor (Abou-Issa and Reichert, 1976). Studies with high speed supernatants (150,000 xg) of rat testes homogenates confirmed the presence of such an inhibitor. The FSH-BI activity of such testes supernatants could be markedly reduced, by about 60%, by simple dialysis against distilled water using Spectrapor #1 membrane (Spectrapor Industries) as the dialyzing barrier. This membrane passes molecules of 8000 daltons or less. It was, however, not possible to completely eliminate FSH-BI activity from the testicular supernatant, even after exhaustive and vigorous dialysis against water or 0.05M TRIS-HCl buffer pH 7.5. This result could be explained by suboptimal conditions of dialysis or by binding of the putative inhibitor to non-dialyzable carrier molecules. Another possibility would be the existence of two components in 150,000 xg rat testis supernatants with FSH-BI activity, one dialyzable and of relatively small molecular size (less than 8000 daltons), the other non-dialyzable and of larger size or possibly an aggregate of the small molecular weight component. The possibility that non-dialyzable FSH-BI activity could be due to endogenous hormone released from testicular receptors by homogenization and so forth, could not be ruled out in these experiments.

The positive detection of FSH-BI activity in dialysates of 150,000 xg rat testes supernatants supported earlier impressions concerning its small molecular size. The experiment was performed by subjecting testes supernatants to vigorous dialysis for three days against deionized, distilled water in the cold, using a ratio of supernatant to dialysate of 1:20. The dialysates were lyophilized to dryness and reconstituted with 0.05M TRIS-HCl buffer pH 7.5 to 1/3 the original volume of supernatant. 400 μl of such buffer reconstituted lyophilized dialysates (LD) inhibited binding of 5 ng ^{125}I-hFSH to testicular receptor by 80%, with smaller volumes inhibiting binding in a dose related fashion.

Although direct conductivity measurements rendered the possibility unlikely, we were concerned that the observed FSH-BI activity might be artifactual, resulting from high ionic strength properties of the concentrated rat testes LD. To check this point, rat testes LD were equilibrated with 0.05M TRIS-HCL buffer pH 7.5 by dialysis ultrafiltration (6-fold sample volume turnover) utilizing an ultrafiltration system fitted with an Amicon UM-05 membrane. The latter has an exclusion limit of 500 daltons. Testicular LD fractions equilibrated with TRIS-HCl buffer in this fashion and not passing the membrane barrier retained FSH-BI activity. This affirmed that the binding inhibition was not due to system perturbation caused by ionic strength effects and also more closely defined the molecular weight (MW) range of the inhibitor as falling between 500 and 8000 daltons.

To obtain additional information on the molecular mass of the FSH-BI, rat testis LD fractions were sieved through a column of Sephadex G-50 equilibrated and developed with distilled water. Tubes of the column eluate were scanned for FSH-BI activity using the standard FSH receptor assay (Reichert and Abou-Issa, 1977) (Reichert and Bhalla, 1974). Inhibitory activity was located in a fraction not retarded by the gel (MW 30,000 or greater) and in a fraction eluting in a position approximately equivalent to that of bacitracin

(MW about 1500). The appearance of FSH-BI activity in the large MW fraction was surprising since the preparation applied to the column had been obtained by dialysis through Spectrapor #1 membrane reported to retain molecules of 8000 or greater. This phenomenon is similar to what was subsequently observed in studies of a dialyzable serum FSH-BI and will be discussed in greater detail later in this report (Section 5). The MW estimate for the smaller sized component having FSH-BI activity was consistent with that derived from dialysis and ultrafiltration experiments (vida supra). However, this conclusion is subject to caveats associated with gel filtration of an unknown substance through Sephadex G-50 in a medium of low ionic-strength and additional studies are required to verify the figure.

FSH-BI activity in buffer reconstituted rat testicular dialysates was thermolabile, loosing approximately 50% of inhibitory activity after heating for 10 minutes at neutral pH in a boiling water bath, whereas that present in the initial tissue supernatant (not dialyzed) was quite stable loosing only 20% FSH-BI activity after 80 minutes of heating under similar conditions.

4. FOLLITROPIN BINDING INHIBITORS IN NON-GONADAL RAT TISSUE:

In order to determine whether FSH-BI activity seen in rat testis supernatant fractions was tissue specific, 150,000 xg supernatants were prepared from 0.05M TRIS-HCl pH 7.5 homogenates of rat non-gonadal tissue such as liver, kidney and brain in exactly the same fashion as had been utilized for testis. Care was taken to remove blood from all tissues to the extent possible to eliminate complications arising from the presence of serum (Reichert and Leidenberger, 1976) (Reichert and Abou-Issa, 1976) (Reichert et al., 1975) (Section 5). FSH-BI activity was detected in all tissue extracts examined and could be reduced, but not eliminated, by prolonged (3 days) dialysis utilizing Spectrapor #1 membrane as the dialyzing barrier. Whereas in the case of testes extracts, such retained activity might be explained by the presence of released endogenously bound FSH, this is not a likely explanation for the presence of large (greater than 8000 MW) FSH-BI activity in retentates of the various non-gonadal tissues studied. FSH-BI activity present in non-gonadal tissue supernatants could be detected in the dialysate and was retained after dialysis-ultrafiltration through an Amicon UM-05 membrane (vida supra).

Since it is not known whether the FSH-BI activity present in the various tissue extracts studied is due to the same or different, single or multiple tissue components, the test of total binding inhibition utilized in these studies cannot be considered specific for a unique inhibitor substance. Attempts to measure relative concentrations of the inhibitor(s) in these tissues, therefore, seems premature at this time.

5. LOW MOLECULAR WEIGHT FOLLITROPIN BINDING INHIBITORS IN HUMAN SERUM:

In previous studies we have utilized a rat testes tissue receptor assay, developed for measurement of pituitary follitropin (FSH) activity during its purification (Reichert and Bhalla, 1974), for measurement of serum levels of that hormone (Reichert and Leidenberger, 1976) (Reichert *et al*., 1975). The amount of FSH-like activity measured in human serum by the tissue receptor assay seemed spuriously high and the dose response curves obtained were non-parallel to those of pituitary FSH. In addition, serum from hypophysectomized patients, having no detectable FSH activity by radioimmunoassay, showed high levels of FSH-like activity when measured in the tissue receptor assay (Reichert *et al*., 1975). The response seen with hypophysectomized serum could be virtually eliminated, the very high levels of FSH-like activity present in normal serum reduced and non-parallelism with binding inhibition curves obtained with highly purified hFSH corrected for, by simple dialysis of the serum prior to performance of the tissue receptor assay, in a manner similar to that employed previously in studies of testes FSH-BI (Section 3). We have recently demonstrated FSH-BI activity in concentrated dialysates of serum (Reichert *et al*., 1977) and observed that such activity would not pass Amicon UM-05 membrane on dialysis-ultrafiltration. When sieved through a column of Sephadex G-50, buffer reconstituted lyophilized dialysates of serum were resolved into two broad areas having FSH-BI activity, one emerging about the void volume of the column, indicating a molecular weight of 30,000 or greater, the other strongly retained, eluting in a region suggesting a molecular weight of about 2000 daltons. In addition to FSH-BI activity as measured in the tissue receptor assay, these serum components were tested for immunologic activity, utilizing ^{125}I-hFSH as the ligand and antiserum to hFSH (NPA-hFSH antiserum batch #3). The procedure for radioimmunoassay was essentially that described by Midgley (1967). Immunologic activity was found associated with both the large and small molecular weight fraction obtained by gel filtration. As discussed in our studies with testicular FSH binding inhibitors (Section 3), the presence of an apparently large molecular weight serum component active in both tissue receptor and immunologic assays was unexpected in view of the dialyzable nature of the FSH-BI. The explanation for this phenomenon is not certain at this time. It would not seem related to pituitary FSH, since serum from presumably hypophysectomized patients also exhibited FSH-BI activity. The large MW component derived from the G-50 filtration was refractionated by filtration through Ultragel ACA-44 (molecular weight exclusion about 130,000 daltons). Several components were resolved having ultraviolet absorbancy peaks (280 nm) varying in apparent molecular size between that excluded from the gel to about 2000 daltons. Scanning assays localized FSH-BI activity (tissue receptor assay) and immunologic activity (radioimmunoassay) to a broad area of the elution profile with a fastigium of activity appearing in a region of low UV absorbancy (280 nm) corresponding to an approximate molecular weight of about 26,000. The nature of this serum component is currently under study.

The small molecular weight component derived from Sephadex G-50 gel filtration was further fractionated by filtration first through Sephadex G-25, then through Sephadex G-10. Although it was not possible to obtain a reliable molecular weight estimate in the latter case, fractions showing FSH-BI activity as measured by receptor assay and RIA were retarded by Sephadex G-10, emerging with a Ve/Vo ratio suggesting a molecular mass of about 700 daltons. Thus, it appears that serum contains a relatively small molecular weight component having the property of inhibiting ^{125}I-hFSH binding to hormone specific testicular receptors and also having immunologic activity as determined by radioimmunoassay.

6. SOME PROPERTIES OF THE SMALL MOLECULAR WEIGHT SERUM FOLLITROPIN BINDING INHIBITOR:

Preliminary studies were performed in order to obtain information on the chemical nature of the small molecular weight serum FSH-BI. The sample used in studies to be described below was derived from normal human serum by dialysis, lyophilization of the dialysate and gel filtration through Sephadex G-50. The strongly retarded fraction from the latter separation was utilized for characterization studies. The binding inhibitor was not adsorbed to charcoal or to concanavalin A-Sepharose and was quite heat stable, with no loss of FSH-BI activity noted after 60 minutes of exposure to 80°C at neutral pH. Extraction of an aqueous solution of the small molecular weight binding inhibitor with diethyl ether did not result in loss of FSH-BI activity. Activity was lost, however, after treatment of the FSH-BI with N-bromosuccinimide.

The chemical nature of the inhibitor is not yet clear. It would not appear to be a steroid or nucleotide, based on lack of adsorption to charcoal, nor does it appear to be a carbohydrate containing C3, C4 or C6 hydroxyl groups of the D-mannopyranose or D-glucopyranose types, since these would be expected to bind to concanavalin A-Sepharose. A positive halogen reagent such as N-bromosuccinimide reacts mostly with tryptophan but also with tyrosine and histidine and various sulfur species. The results tend to suggest the FSH-BI may be peptide in nature, although the specificity of a potent oxidizing agent such as N-bromosuccinimide for amino acids in the likely presence of a variety of non-peptide substances is not certain.

The small molecular weight inhibitor fraction used in these characterization studies was quite crude in nature. The influence of impurities on the results obtained in the above experiments is not known. Therefore, unequivocal elucidation of the properties of the FSH-BI must await its final purification.

7. STUDIES ON THE MECHANISM OF ACTION OF FOLLITROPIN BINDING INHIBITORS:

The system utilized in the previous experiments for detection and measurement of FSH-BI activity in small molecular weight serum and tissue fractions requires a mixing of the putative FSH-BI fraction with ^{125}I-hFSH and testis receptor preparation in the cold, after which incubation is allowed to proceed for 3 hours at 37°C. Under such conditions, decreases in ^{125}I-hFSH binding could be attributed to at least three possible mechanisms: a) an effect of FSH-BI on the ^{125}I-hFSH itself, b) an effect of FSH-BI on or near the membrane bound receptor, thereby interfering with efficient formation of the hormone-receptor complex and c) FSH-BI induced dissociation of the hormone-receptor complex once formed.

Possible direct effects of the FSH-BI on ^{125}I-hFSH were studied by incubating the two together for two hours at 37°C followed by removal of the FSH-BI by dialysis through Visking Membrane tubing and gel filtration of the retentate through a column of Ultragel ACA-44. ^{125}I-hFSH is retarded by this filtration medium, emerging with a Ve/Vo ratio of about 1.65. There was no detectable difference in elution profile (as determined by counting of radioactivity) after incubation of ^{125}I-hFSH with the serum or tissue FSH-BI compared to that observed when incubation was carried out in the absence of the binding inhibitor. This suggests that a gross fragmentation of ^{125}I-hFSH had not occured, although removal of a small unlabeled peptide essential for hFSH binding could not be ruled out by this experimental design.

Effects of serum and tissue FSH-BI fractions on formation and dissociation of the FSH-testicular receptor complex were studied utilizing highly purified tubule membranes, prepared by sucrose gradient centrifugation, as the receptor source and inhibitor fractions derived from human and rat serum and testis, hypophysectomized human serum and rat liver and kidney. The inhibitor fraction from each of these sources was prepared by dialysis through Spectrapor #1 membrane, lyophilization of the dialysate and dialysis-ultrafiltration of buffer reconstituted lyophilized dialysates through Amicon UM-05 membrane. The fraction not passing the membrane was used in these studies.

Studies on the effect of FSH-BI on <u>dissociation</u> of the bound hormone were done by incubation of ^{125}I-hFSH and receptor under conditions favoring formation of the hormone-receptor complex, removal of unbound radioligand by centrifugation and washing of the pellet, followed by incubation of buffer reconstituted hormone-receptor complex with various FSH-BI fractions. The amount of radioactivity remaining bound after that incubation was compared to that remaining bound following incubation of the preformed hormone-receptor complex with buffer alone. By this calculation a value of 50%, for example, indicates that incubation of the hormone-receptor complex with the inhibitor resulted in 50% less ^{125}I-hFSH remaining bound to the membrane fraction compared to that remaining bound when incubation was with buffer alone. The extent of dissociation seen with comparable volumes of various dialyzed and Amicon UM-05 retained serum and tissue fractions were: rat testes, 30%, rat kidney, 44%, rat liver, 38%, rat serum, 32% human serum, 47% and hypophysectomized human serum, 70%. It should be emphasized that for reasons discussed in Section 4, these values (as well as those given below) are qualitative in nature and cannot as yet be interpreted in a quantitative sense.

For studies of inhibitor effects of <u>formation</u> of the hormone receptor complex, receptor membrane pellets were preincubated with the same small molecular weight tissue fractions used in the dissociation studies described above. After appropriate washing, the amount of subsequently added ^{125}I-hFSH specifically bound to receptor was compared to that bound after preincubation with buffer alone. By this calculation, a value of 50% would indicate that 50% fewer counts of ^{125}I-hFSH were bound to membranes preincubated with the tissue fraction, compared to that bound when preincubation was with buffer alone. The results obtained with comparable volumes of sample were: rat testes, 89%, rat liver, 77%, rat kidney, 80%, rat serum, 67%, human serum, 60%, and human hypophysectomized serum, 90%.

It appears that the various tissue and serum factors studied may be more effective in interfering with formation of ^{125}I-hFSH receptor complex than in promoting dissociation of the complex once formed. Such an inference, however, must be drawn with extreme caution. Although effects on dissociation of ^{125}I-hFSH could be attributable to direct interaction of inhibitor with the bound hormone this would not seem a likely explanation for the observed inhibition of formation of the hormone-receptor complex, since in the latter experiment excess tissue or serum FSH-BI fractions were removed by washing of the pellet prior to addition of ^{125}I-hFSH for receptor binding. This result would be consistent with the interpretation placed on the gel filtration studies described earlier, wherein direct effects of inhibitor on radioligand could not be demonstrated.

CONCLUSIONS:

The significance of small molecular weight inhibitors of FSH binding described in this report is uncertain. The FSH-BI fraction derived from gonadal tissue does not appear tissue specific, being also identified in similar preparations from a variety of non-gonadal tissue. In this sense, the FSH-BI is different from the inhibitor of luteinizing hormone-receptor site binding reported by Yang et al. (1976) since the latter was found only in extracts of pregnant and pseudopregnant rat ovaries and not in testes or liver extracts. However, similarity of action does not prove identity and it should be recalled that such non-gonadal tissue as liver and kidney does not represent target tissue for FSH specific binding as does testis (Abou-Issa and Reichert, 1976). The presence of a binding inhibitor in the latter, therefore, may have relevance with regard to the mechanism of hormone action and its control, despite a ubiquitous presence. It has not yet been possible to test for anti-gonadotropic effects of FSH-BI fractions in vivo, due primarily to logistical considerations. An effective anti-hFSH, however, would presumably be of interest regardless of tissue of origin or involvement in normal physiologic processes.

FSH-BI in fractions from human serum appear to have several properties in common with those derived from testes. The apparent immunologic activity of serum FSH-BI fractions may have implications with regard to the significance and correct interpretation of serum FSH levels as measured by radioimmunoassay. The source of serum FSH-BI is uncertain but it does not appear related to the pituitary gland, since sera from hypophysectomized patients have FSH-BI activity. Perhaps the serum FSH-BI is derived from various tissue membranes as a result of metabolism or degradation. Also, it seems likely that the low molecular weight serum inhibitor would not be retained by the kidney, and therefore may be present in the urine. This would provide a ready source for purification and chemical identification. Results of N-bromosuccinimide treatment suggest that the FSH-BI may be peptide in nature. If so, and in view of its apparent small size, chemical synthesis should be feasible.

Although clearly at an early stage in development, studies of serum and tissue inhibitors of gonadotropin binding offers sufficient theoretical and practical potential to warrant continued efforts at purification and identification. Studies along such lines are currently underway in our laboratory.

Acknowledgement

It is a pleasure to acknowledge the able technical assistance of Nancy L. Shih, Bettie G. Wright, Rosemary B. Ramsey and Eloise B. Carter in various phases of this work supported by UPHS Grant HD-08228.

REFERENCES:

Abou-Issa, H. and Reichert, L.E., Jr. (1976). Properties of the follitropin-receptor interaction. Characterization of the interaction of follitropin with receptors in purified membranes isolated from mature rat testes tubules. J. Biol. Chem. 251, 3326-3337.

Bailly, A., Sallas, N. and Milgron, E. (1977). A low molecular weight inhibitor of steroid receptor activation. J. Biol. Chem. 252, 858-863.

Jagiello, G., Graffeo, J., Ducayen, M. and Prosser, R. (1977). Further studies of inhibitors of in vitro mammaliam oocyte maturation. Fertility and Sterility 28, 476-481.

Kent, H.A., Jr. (1975). Contraceptive polypeptide from hamster embryo: Sequence of amino acids in the compound. Biology of Reproduction 12, 504-507.

Leidenberger, F.A., Graesslin, D., Scheel, H.J., Hess, N., Lichtenberg, V. and Bettendorf, G. (1976). A low molecular weight substance obtained from serum which has luteinizing hormone like activity ("mini LH"). J. Clin. Endocrinol. Metab. 43, 1410-1413.

Midgley, A.R., Jr. (1967). Radioimmunoassay for human follicle stimulating hormone. J. Clin. Endocrin. Metab. 27, 295-299.

Reichert, L.E., Jr. and Abou-Issa, H. (1976). Inhibitors of FSH binding to tubule receptors in Regulatory Mechanisms of Male Reproductive Physiology (C.H. Spilman, T.J. Lobl and K.T. Kirton eds). American Elsevier:New York pp. 71-78.

Reichert, L.E., Jr., Abou-Issa, H., Carter, E.B., and Shih, N.L. (1977). Inhibition of FSH binding to testes by small molecular weight components of human serum. Endocrinology 100, A-222.

Reichert, L.E. and Abou-Issa, H. (1977). Studies on a low molecular weight testicular factor which inhibits binding of FSH to receptor. Biology of Reproduction 17, 300-308.

Reichert, L.E., Jr., and Bhalla, V.K. (1974). Development of a radioligand-receptor assay for human follicle stimulating hormone. Endocrinology 94, 483-491.

Reichert, L.E., Jr., and Leidenberger (1976). Measurement of human serum follitropin (FSH) and lutropin (LH) by tissue receptor assay in Ovulation in the Human (P.G. Crosignani and D.W. Mishell eds) Academic Press:London pp. 155-166).

Reichert, L.E., Jr., Ramsey, R.B. and Carter, E.B. (1975). Application of a tissue receptor assay to measurement of serum follitropin (FSH). J. Clin. Endocrinol. Metab. 41, 634-637.

Tsafriri, A., Pomerantz, S.H., and Channing, C.P. (1976). Inhibition of oocyte maturation by porcine follicular fluid: Characterization of the inhibitor. Biology of Reproduction 14, 511-516.

Yang, K.P., Samaan, N.A., and Ward, D.N. (1976). Characterization of an inhibitor for luteinizing hormone receptor site binding. Endocrinology 98, 233-241.

DISCUSSION

Ward: How much serum equivalent gives you approximately 60% inhibition?

Reichert: If the serum lyophilized dialysate is reconstituted to ½ the original volume of serum used for dialysis, then 75 μl will give approximately 50% inhibition of binding. This would, then, be equivalent to 150 μl of serum.

Peckham: What is the significance of the heat inactivation of the high molecular weight inhibitor?

Reichert: The instability of the high molecular weight fraction upon heating is a reproducible phenomenon, but its significance is not yet known.

Channing: Did you correlate this with the physiological state of the serum in any experiments?

Reichert: None other than to utilize serum from hypophysectomized patients and to note the presence of FSH binding inhibition in the lyophilized dialysate.

16

Pineal-Mediated Reproductive Events

RUSSELL J. REITER

Even though the mammalian pineal gland was described before most other classical organs of internal secretion, the function of this midline azygous structure remained a mystery until very recent times. Its discovery predates that of the pituitary and adrenal glands and can be found in the earliest anatomical descriptions of Claudius Galen (130-201 A.D.), a medical practitioner of great renown who frequently is recognized for giving the pineal its name. However, probably the greatest attention was drawn to the pineal by Rene Descartes (1596-1650 A.D.) who considered the body a machine governed by a sensitive soul residing in the pineal gland. Because of his eminence as a philosopher-physiologist, Descartes' idea that the pineal is the "seat of the soul" persisted for hundreds of years. His _De Homine_ is credited with being one of the first physiological textbooks; in it the pineal has a very central role in many of the illustrations as well as in the accompanying narrative. Descartes in fact prophetically related the eyes to the function of the pineal gland. Only in the last two decades has it been unequivocally established that the biochemical and secretory activities of the pineal gland are inextricably linked to photic information detected by the lateral eyes

(Kappers, 1976).

The modern era of pinealology has witnessed the elucidation of the relationships between the pineal gland and reproductive physiology and the isolation of a number of potential pineal hormones. Few investigators currently doubt that the major influence of the pineal gland on reproduction is one of inhibition. In spite of this, it is inappropriate to expect that, under conditions of long daily photoperiods, surgical pinealectomy will lead to a stimulation of the reproductive system. Under long photoperiods and with proper nutrition etc., the hypothalamo-pituitary-gonadal axis of long-day breeding animals is likely functioning at its maximal level and simple removal of the pineal gland will not hyperstimulate the system. What in fact pineal removal does is prevent short days from suppressing various facets of reproduction. Reducing the length (the critical length is probably different for different species) of the daily photoperiod to which photosensitive species are exposed stimulates the pineal gland to produce a factor or factors which inhibit, by mechanisms not completely understood, sexual physiology. During long photoperiods the consequences of pineal removal are negligible since an already nonfunctional (because long days are inhibitory to the pineal) organ is removed. Hence, pinealectomy does not induce gonadal hypertrophy but rather prevents regression of the sexual organs which is often a consequence of exposure of the animals to restricted photoperiods.

Manipulations of the photoperiod to demonstrate the antigonadotrophic activity of the pineal gland have raised some objections because the critics claim that reduced photoperiods (usually less than 12 hours light per 24 hour period) represent an unnatural or artificial situation for experimental animals. This criticism is completely unwarranted considering that the species (e.g., the Syrian hamster) which most frequently have been utilized to illustrate the potential of the pineal gland are nocturnal animals. Under field conditions, the warm temperatures of the summer days essentially restrict the animals to the darkness of their subterranean burrows; when they do emerge (at night) they are exposed either to darkness or to low intensity lunar light. Additionally, during the winter months hamsters hibernate which relegates the species to 5-6 months of total or near total darkness per year. Considering this, the long carefully regulated photoperiods of the laboratory setting must be considered to be unnatural while short days are seemingly more reminiscent of the environment in which these animals evolved.

The nature of the pineal secretory product which mediates gonadal regression in animals exposed to short photoperiods has not been satisfactorily established. Two general categories of potential hormones, i.e., indoleamines and polypeptides, have been thoroughly investigated (Reiter and Vaughan, 1977) but no consensus has been reached as to which of these categories is more important in terms of the ability of the pineal gland to inhibit reproduction. In general, the indoleamines, and especially N-acetyl-5-methoxytryptamine (melatonin), frequently have been designated pineal antigonadotrophins (Wurtman, 1973; Cardinali, 1974). Indeed, it is difficult to exclude melatonin from the antigonadotrophic category of substances in view of some very compelling recent findings (Tamarkin et al., 1977; Reiter, 1977) which revealed that its daily administration during a very restricted portion of the light:dark cycle induced reproductive collapse in Syrian hamsters. On the other hand, polypeptidic factors from the pineal also exhibit potent antigonadotrophic activity (Reiter and Vaughan, 1977; Ebels and Benson, 1978). The only pineal peptide with gonad-inhibiting capability that has been structurally identified is arginine vasotocin (AVT) (Vaughan et al., 1976). Pavel (1973) has speculated that the release of AVT from the pineal gland may be under the control of melatonin, a proposition also tentatively put forth by Reiter and colleagues (1976a). The release of a peptide, AVT, by an amine, melatonin, within the pineal is attractive since in the hypothalamus the release of another group of polypeptides, the hypothalamic releasing hormones, are seemingly also governed by amines (monoamines) (Fuxe et al., 1976).

Gonads and Gonadotrophins

Although the rat has been widely used in the study of reproductive physiology, because of its photoinsensitivity it is far from an ideal animal model in which to investigate the actions of the pineal gland. For example, pinealectomy in rats has minimal influence on the gonads while light deprivation, which is known to stimulate the biochemical activity of the pineal gland, has a very modestly suppressive effect on reproduction (Kinson, 1976). Although total light restriction may lead to a slight regression in the weights of the testes, seminal vesicles and prostate glands in males and the ovaries and uteri in females, the embarrassment to the reproductive system is usually insufficient to prohibit the animals from successfully mating. Reproducibly detectable changes in circulating gonadotrophin levels also are not characteristic of rats that have been merely pinealectomized or deprived of light, e.g., by blinding.

There are certain manipulations, however, which render the neuroendocrine-reproductive axis of the rat highly sensitive to inhibition by the pineal gland. Thus, light restriction combined with any of several procedures including rendering the animals anosmic, early neonatal steroid treatment, or underfeeding, cause marked regressive changes in both the reproductive organs and the gonadotrophin levels of rats. These procedures have been referred to as potentiating factors in that they increase the sensitivity of the pituitary-gonadal axis to the pineal gland which has been stimulated by light restriction (Reiter, 1974a). Since the potentiating factors are not known to change the biochemical or secretory activity of the pineal it has been theorized that they act at the level of the hypothalamus where they increase the sensitivity (decrease the threshold) of the inhibited site making it more susceptible to inhibition by pineal antigonadotrophic factors. The effects of the combined treatments are prevented, for the most part, by pineal removal proving that this organ is responsible for the suppressed status of the reproductive organs. In these situations the pineal may act by restricting the release of the gonadotrophin releasing hormone (Blask and Reiter, 1975).

Commercially available pineal substances have been tested for their antigonadotrophic capabilities in rats. Usually the changes induced by these compounds are similar in magnitude to those described above as sequelae of pineal manipulations. Thus, at one time or another serotonin, N-acetylserotonin, melatonin, 5-hydroxytryptophol and 5-methoxytryptophol have been noted to suppress the pituitary-gonadal axis of rats (Cardinali, 1974; Kappers, 1976; Reiter and Vaughan, 1977). In these experiments the degree of inhibition varied markedly with the usual findings including slight to modest suppression of sexual physiology. Characteristically, unless excessively large dosages of indole were administered, the alterations induced probably would not have jeopardized the reproductive potential of the animals. The potentiating factors described above do not seem to increase the sensitivity of the neuroendocrine axis to inhibition by melatonin (unpublished observations).

AVT is the only pineal polypeptide which has been examined to any extent for its gonad-inhibiting ability. This compound has been tentatively identified in the rat, bovine, porcine and human fetal pineal gland as well as in the cerebrospinal fluid of cats and humans. When synthetic AVT is exogenously administered (in microgram quantities) to experimental animals, most frequently to mice, it causes changes indicating that it suppresses the release or action of pituitary gonadotrophins (Pavel et al., 1973) while stimulating prolactin release (Blask et al., 1976). Unidentified presumptive pineal polypeptides, which are probably structurally distinct from AVT, produce effects in rats and mice similar to those of other pineal constituents. The studies of Benson et al. (1972) and Orts and colleagues (1974) have clearly shown that peptides which they isolated from the pineal could depress circulating LH levels and inhibit compensatory growth of the remaining ovary after unilateral ovariectomy. Unlike with melatonin, there is no convincing evidence that any of the polypeptides are normally discharged from the pineal gland into bodily fluids.

In contrast to the rat, the reproductive system of the Syrian hamster is acutely sensitive to inhibition by darkness acting via the pineal gland. Merely limiting the amount of light hamsters witness (less than 12 hrs per 24 hr period) renders the animals completely incapable of reproducing (Hoffman and Reiter, 1965). The gonadal atrophy requires 6-10 weeks to become manifested and is associated with changes in gonadotrophin and testosterone levels. The testes and accessory sexual organs of light restricted hamsters regress in size (Fig. 1) and exhibit spermatogenic arrest (Reiter and Johnson,

FIG. 1. Testicular and accessory organ weights and pituitary prolactin levels in hamsters kept under naturally short winter photoperiods. The accessory sex organs included the seminal vesicles and coagulating glands; animals were necropsied in mid-January. Pituitary prolactin stores are expressed in reference to a pool of standard hamster anterior pituitaries (SHAP). PINX = pinealectomy. All INTACT values differ from PINX values with a $p<0.001$. From Reiter et al. (1976a).

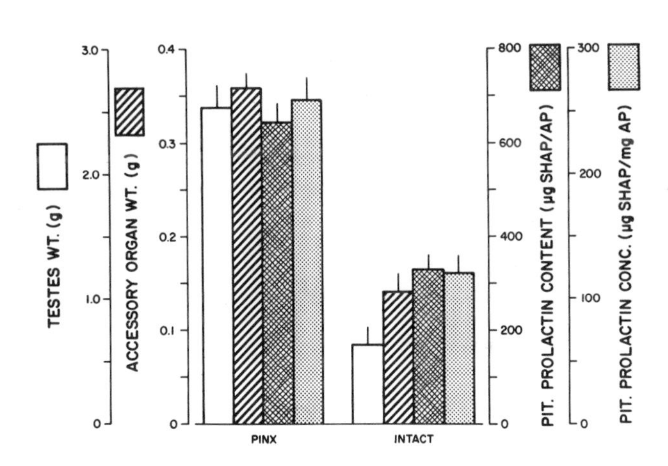

1974a). These changes are accompanied by decreases in both pituitary LH and prolactin stores while plasma titers of these hormones respond in a variable manner. According to Berndtson and Desjardins (1974), plasma concentrations of both LH and FSH drop steadily as the testes regress; however, these changes have not been documented by all investigators studying this problem. Bartke and colleagues (1975) feel that the deficiency in prolactin may be largely responsible for the gonadal atrophy which is a consequence of light restriction.

The effects of light restriction in female hamsters are similar to those in males. The uteri become infantile and the ovaries undergo functional involution; ovulation ceases as indicated by the lack of graffian follicles and corpora lutea (Reiter and Johnson, 1974b). Predictably, the estrous cycles of light restricted hamsters are interrupted with the vaginal smears containing cells typical of both metestrous and diestrous smears (Sorrentino and Reiter, 1970). In contrast to males, female hamsters that have involuted reproductive organs due to prolonged darkness have elevated pituitary stores of LH; pituitary prolactin concentrations are, however, greatly decreased (Reiter and Johnson, 1974b). Plasma titers of LH exhibit a peculiar change under these conditions. Rather than the hormone exhibiting a surge on the afternoon of every fourth day (which is the case in reproductively normal females), the elevation occurs every afternoon; this is accompanied by a mid-afternoon rise in FSH as well (Seegal and Goldman, 1975).

The reproductive sequelae of short daily photoperiods described in the previous paragraphs undoubtedly involve the pineal gland as an intermediary in that surgical removal of this organ prevents darkness from interferring with the normal functioning of the neuroendocrine-reproductive axis (Reiter, 1973). Likewise, any procedure which interferes with the sympathetic innervation to the pineal gland also renders it incapable of inhibiting sexual physiology. Hence, transection of the nervi conarii (post ganglionic sympathetic fibers which enter the pineal gland), removal of the superior cervical ganglia, interruption of the preganglionic sympathetic fibers inferior to the superior cervical ganglia, or cutting various hypothalamic nerve bundles inactivate the pineal in terms of its gonad-inhibiting ability. In each case, the effects of these manipulations are essentailly identical to those which follow pinealectomy. Finally, the suprachiasmatic nuclei in the hypothalamus must be intact before darkness will induce involution of the reproductive organs in hamsters (Stetson and Watson-Whitmyre, 1976).

Findings concerning the role of pineal indoleamines in the control of sexual physiology in hamsters have been extremely variable and difficult to interpret. When the utility of the hamster as a model in which to test for pineal-gonadal interactions was initially described, it was anticipated that it would be an equally valuable species in which to examine indole-gonadal relationships. The rationale for this was as follows. If melatonin, for example, is the pineal constituent responsible for the gonad inhibiting ability of the activated pineal gland, then the administration of this substance on a regular basis should induce the same degree of atrophy of the neuroendocrine-reproductive axis as does exposure of hamsters to short daily photoperiods. However, a number of early experiments in hamsters failed to show that melatonin had any influence on reproduction in this species (Reiter, 1974a). In fact, the repeated failures in these experiments led to the provisional conclusion that melatonin may not be the pineal antigonadotrophin in the hamster.

In 1974 two publications appeared which revolutionized the thinking of the scientific community concerning the role of melatonin in controlling sexual functions in hamsters. Almost simultaneously, Hoffmann (1974) and Reiter et al. (1974) published papers dealing with what has come to be known as the counter antigonadotrophic or progonadal action of melatonin. Hoffmann, working with the dwarf hamster (<u>Phodopus</u> <u>sungorus</u>) (another photosensitive species whose reproductive organs undergo regression in response to reduced photoperiodic length), found that if these animals received subcutaneous deposits of melatonin (in beeswax) their gonads did not regress in response to short daily periods of light. In other words, rather than inhibiting reproduction chronically available melatonin prevented the degenerative response of the testes to the activated pineal gland. In these experiments it was assumed that melatonin was released continuously from the subcutaneous deposits. The findings of Reiter and co-workers (1974) using the Syrian hamster corroborate the observations of Hoffmann (1974). They reported that the exposure of male hamsters to 1 hour light and 23 hours darkness daily for 9 weeks led to total involution of the testes and accessory sex organs and to 60% drop in pituitary prolactin values. The inhibitory effects of darkness were prevented in hamsters that received subcutaneously each week 1 mg melatonin (in beeswax) and in animals that were pinealectomized. Thus, chronic melatonin availability duplicated the effects of pineal removal, i.e., it caused a "functional pinealectomy".

It could be argued that the counter antigonadotrophic actions of melatonin are a consequence of the continuous availability of pharmacological amounts of melatonin. However, subsequent studies have shown that only small amounts of melatonin are required for it to act in a counter antigonadotrophic manner. When light-restricted hamsters received bi-weekly subcutaneous implants of melatonin-beeswax pellets (containing between 1 μg and 1 mg melatonin), all but the smallest dosage overcame the inhibitory effects of the pineal gland on reproduction (Reiter et al., 1975). Thus, 50 μg melatonin was equally effective as 1 mg per two week period in preventing the gonadal and hormonal changes associated with prolonged darkness. If we make the unlikely assumption that the entire 50 μg melatonin was absorbed into the tissue fluid during each two week interval, the daily dosage of melatonin would have been 3.6 μg. However, since it is likely that at least a portion of the melatonin remained in the beeswax pellet, it seems safe to conclude that the daily effective dosage of melatonin required to combat the antigonadotrophic acitivity of the pineal gland in the Syrian hamster is somewhat less than 3.6 μg daily. It must be re-emphasized that for melatonin to act in a counter antigonadotrophic manner it must be continuously available.

Having made these discoveries relative to the counter antigonadotrophic actions of continuously available melatonin, the indoleamine temporarily was overlooked as a potential antigonadotrophic agent in the hamster. However, at a scientific meeting in late 1975, Tamarkin and colleagues reported that indeed melatonin could inhibit reproductive physiology in Syrian hamsters provided it was acutely administered during a very restricted portion of the light: dark cycle. Hence, injecting melatonin subcutaneously just prior to lights out on a daily basis into hamsters maintained in long daily photoperiods (LD 14:10) was followed by collapse of the gonads, equivalent in magnitude to that which follows exposure of hamsters to restricted photoperiods. When hamsters are kept in LD cycles of 14:10 the so-called sensitive period for melatonin injections falls between 6.5 and 13.75 hours after lights on with the sensitivity apparently increasing with increased length of exposure to light (Tamarkin et al., 1976). The effective dosages of melatonin tested in these experiments varied from 2.5 to 25 μg daily.

Confirmation of the findings of Tamarkin and co-workers (1975, 1976) was forthcoming soon after they described the phenomenon of the antigonadotrophic nature of acutely administered melatonin in male and female hamsters. Reiter et al. (1976b) followed the experimental paradigm of Tamarkin and housed hamsters in LD cycles of 14:10; the animals were injected daily with 25 μg melatonin (13 hours after lights on) for 50 consecutive days. This treatment was accompanied by atrophy of the testes and adnexa and with alterations in gonadotrophin levels. Similar injections of melatonin given 3 hours after lights on failed to have any measurable influence on the reproductive organs (Fig. 2) or associated hormone levels. Furthermore, the late afternoon injections of melatonin were effective only if the animal's pineal gland was intact and if the pineal's sympathetic innervation was not disturbed. Thus, in pinealectomized hamsters and in animals in which the innervation of the pineal gland had been interrupted anywhere between the suprachiasmatic nuclei and the pineal, single injections of melatonin lost their antigonadotrophic capability (Fig. 2). The failure of melatonin to be effective as a gonad-inhibiting agent in pinealectomized hamsters or in animals in which the pineal is sympathetically innervated could be interpreted to mean that the injected melatonin acts on the pineal to induce the release of possibly a different pineal antigonadotrophin. Conversely, possibly the afternoon injections of exogenous melatonin must be supplemented by a mid-dark rise in endogenous melatonin before gonadal regression will result. Judging from data obtained in other species it would be expected that the nighttime rise in plasma melatonin would not occur in animals lacking their pineal; however, this finding still requires confirmation in the hamster.

The findings described in the preceding paragraphs seem almost contradictory. Thus, when melatonin is injected daily during a critical interval it inhibits reproduction while its continuous availability negates the suppressive influence of the pineal gland on sexual functions. What is even more perplexing is that the actions of daily acute melatonin injections can be prevented by subcutaneous deposits of the indoleamine, i.e., melatonin can prevent its own

FIG. 2. Absolute (clear bars) and relative (cross-hatched bars) testicular weights of hamsters that received 25 μg melatonin at either 9 A.M. (AM-Mel) or at 7 P.M. (PM-Mel) for 50 consecutive days. Animals were maintained under LD cycles of 14:10; lights were automatically regulated and were turned on at 0600 hours. INT = intact hamsters; PINX = pinealectomy; SCGX = superior cervical ganglionectomy; deSCG = decentralization of the superior cervical ganglia; DEAF = deafferentiation of the anterior hypothalamus. *p<0.001 vs absolute weights of all other groups; **p<0.001 vs relative weights of all other groups. From Reiter et al. (1976b).

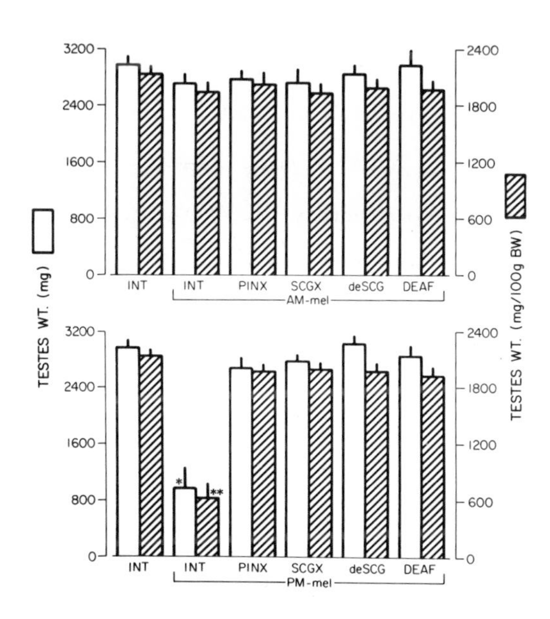

action (Reiter et al., 1977). Finally, whereas the results of the majority of investigators agree on what action melatonin will have under a given experimental situation, there are some data which currently defy explanation (Turek et al., 1975).

Seasonal Reproduction

Since the annual reproductive cycle of many mammals relies on seasonal changes in photoperiodic length, it is not surprising that investigators began to suspect that the pineal gland served as an intermediary between the changing status of reproduction and the annual variations in day length. Again, the most useful animal for examining these interactions has been the Syrian hamster. In the wild, this species is normally a seasonal breeder and hibernator. Males and females breed during the spring when day lengths are on the increase and the pups are born 16 days later. Through the summer months animals probably remain reproductively competent. With decreasing day lengths in the fall of the year hamsters spend increasingly more time in their underground burrows in preparation for hibernation. The exposure to less light activates the pineal gland which in turn secretes a hormone which inhibits reproduction. During the winter months (the hibernatory state) the gonads are held in the atrophic condition by the stimulated pineal gland. As spring approaches the gonads spontaneously regenerate and, as the hamsters emerge from hibernation, mating ensues. The annual reproductive cycle as summarized has been divided into four distinctive phases as follows: the sexually active phase (spring and summer), the inhibition phase (the fall), the sexually quiescent phase (early and mid-winter), and the restoration phase (late winter) (Reiter, 1975).

The pineal gland has been shown to be involved in the annual reproductive cycle as illustrated by the fact that removal of the gland prevents short winter days from inducing involution of the reproductive organs. Indeed, it has been demonstrated that, unlike intact animals, pinealectomized hamsters are quite capable of successful mating throughout the winter months (Reiter, 1975). The late winter restoration of sexual function seems to be independent of the increasing day lengths and, therefore, it has been referred to as a spontaneous or endogenous regeneration. This ensures that the animals are sexually competent when they emerge from hibernation. During the summer months the neuroendocrine-reproductive axis seems to be refractory to the antigonadotrophic influence of the pineal gland. However, with decreasing day lengths after the autumnal equinox the pineal again induces reproductive involution. These theoretical relationships have been summarized in two recent reviews (Reiter, 1974b; 1975).

Concluding Remarks

Pinealology has rapidly developed as a subdiscipline of endocrine physiology within the last decade. Several active pineal factors have been identified although it is not yet known which of these substances is the normal envoy of the pineal which modifies reproduction. The influence of the pineal gland on sexual functions has now been clearly and convincingly demonstrated and the possible role of the pineal in the control of annual reproductive cycles of some mammals, especially the long day breeders, has been partially defined. Now that appropriate and dependable models are available for testing pineal-gonadal interactions the amount of knowledge in the field should increase expotentially in the years to come.

Acknowledgments

Work by the author was supported by NSF grant PCM-77-05734.

References

Bartke, A., Croft, B. T., and Dalterio, S. (1975) Prolactin restores plasma testosterone levels and stimulates testicular growth in hamsters exposed to short day-length. Endocrinology, 97, 1601-1604.

Benson, B., Matthews, M. J., and Rodin, A. E. (1972) Studies on a non-melatonin pineal antigonadotrophin. Acta Endocr., 69, 257-266.

Berndtson, W. E., and Desjardins, C. (1974) Circulating LH and FSH levels and testicular function in hamsters during light deprivation and subsequent photoperiodic stimulation. Endocrinology, 95, 195-205.

Blask, D. E., and Reiter, R. J. (1975) The pineal gland of the blind-anosmic female rat: Its influence on medial basal hypothalamic LRH, PIF and/or PRF activity *in vivo*. Neuroendocrinology, 17, 362-374.

Blask, D. E., Vaughan, M. K., Reiter, R. J., Johnson, L. Y., and Vaughan, G. M. (1976) Prolactin-releasing and release-inhibiting factor activities in the bovine, rat, and human pineal gland: *In vitro* and *in vivo* studies. Endocrinology, 99, 152-161.

Cardinali, D. P. (1974) Melatonin and the endocrine role of the pineal gland. In: Current Topics in Experimental Endocrinology, Vol. 2, pp. 107-128. V. H. T. James and L. Martini, eds. Academic Press, N.Y.

Ebels, I. and Benson, B. (1978) A survey of the evidence that unidentified pineal substances affect the reproductive system in mammals. In: The Pineal and Reproduction, in press. R. J. Reiter, ed. Karger, Basel.

Fuxe, K., Hökfelt, T., Löfström, Johansson, O., Agnati, L., Everitt, B., Goldstein, M., Jeffcoate, S., White, N., Eneroth, P., Gustafsson, J.-A., and Skett, P. (1976) On the role of neurotransmitters and hypothalamic hormones and their interactions in hypothalamic and extrahypothalamic control of pituitary function and sexual behavior. In: Subcellular Mechanisms of Reproductive Neuroendocrinology, pp. 193-246. F. Naftolin, K. J. Ryan and J. Davies, eds. Elsevier, Amsterdam.

Hoffman, R. A., and Reiter, R. J. (1965) Pineal gland: Influence on gonads of male hamsters. Science, 148, 1609-1611.

Hoffmann, K. (1974) Testicular involution in short photoperiods inhibited by melatonin. Naturwissenschaften 61, 364-365.

Kappers, J. A. (1976) The mammalian pineal gland, a survey. Acta Neurochir., 34, 109-149.

Kinson, G. A. (1976) Pineal factors in the control of testicular function. Adv. Sex Horm. Res. 2, 87-139.

Orts, R. J., Benson, B., and Cook, B. F. (1974) Some antigonadotrophic effects of melatonin-free bovine pineal extracts in rats. Acta Endocr., 76, 438-448.

Pavel, S. (1973) Arginine vasotocin release into cerebrospinal fluid of cats induced by melatonin. Nature 246, 183-184.

Pavel, S., Petrescu, M., Vicoleanu, N. (1973) Evidence of central gonadotrophin inhibiting activity of arginine vasotocin in the female mouse. Neuroendocrinology, 11, 370-374.

Reiter, R. J. (1973) Comparative physiology: Pineal gland. Annu. Rev. Physiol., 35, 305-328.

Reiter, R. J. (1974a) Pineal regulation of hypothalamicopituitary axis: Gonadotrophins. In: Handbook of Physiology, Endocrinology IV, Part 2, pp. 519-550. E. Knobil and C. H. Sawyer, eds. American Physiological Society, Washington.

Reiter R. J. (1974b) Circannual reproductive rhythms in mammals related to photoperiod and pineal function: A review. Chronobiologia, 1, 365-395.

Reiter, R. J. (1975) The pineal and seasonal reproductive adjustments. Int. J. Biometeor., 19, 282-288.

Reiter R. J. (1977) The Pineal - 1977, pp. 96-106. Eden Press, Montreal.

Reiter, R. J., and Johnson, L. Y. (1974a) Depressant action of the pineal gland on pituitary luteinizing hormone and prolactin in male hamsters. Horm. Res., 5, 311-320.

Reiter, R. J., and Johnson, L. Y. (1974b). Elevated pituitary LH and depressed pituitary prolactin levels in female hamsters with pineal-induced gonadal atrophy and the effects of chronic treatment with synthetic LRF. Neuroendocrinology, 14, 310-320.

Reiter, R. J. and Vaughan, M. K. (1977) Pineal antigonadotrophic substances: Polypeptides and indoles. Life Sci. 21, 159-172.

Reiter, R. J., Vaughan, M. K., Blask, D. E., and Johnson, L. Y. (1974) Melatonin: Its inhibition of pineal antigonadotrophic activity in male hamsters. Science, 185, 1169-1171.

Reiter, R. J., Vaughan, M. K., and Waring, P. J. (1975) Studies on the minimal dosage of melatonin required to inhibit pineal antigonadotrophic activity in male golden hamsters. Horm. Res., 6, 258-267.

Reiter, R. J., Lukaszyk, A. J., Vaughan, M. K., and Blask, D. E. (1976a) New horizons of pineal research. Amer. Zool. 16, 93-101.

Reiter, R. J., Blask, D. E., Johnson, L. Y., Rudeen, P. K., Vaughan, M. K., and Waring P. J. (1976b) Melatonin inhibition of reproduction in the male hamster: Its dependency on time of day of administration and on an intact and sympathetically innervated pineal gland. Neuroendocrinology 22, 107-116.

Reiter, R. J., Rudeen, P. K., Sackman, J. W., Vaughan, M. K., Johnson, L. Y., and Little, J. C. (1977) Subcutaneous melatonin implants inhibit reproductive atrophy in male hamsters induced by daily melatonin injections. Endocr. Res. Commun., 4, 35-44.

Seegal, R. F. and Goldman, B. D. (1975) Effects of photoperiod on cyclicity and serum gonadotropins in the Syrian hamster. Biol. Reprod., 12, 223-231.

Sorrentino, S., Jr., and Reiter, R. J. (1970) Pineal-induced alteration of estrous cycles in blinded hamsters. Gen. Comp. Endocr., 15, 39-42.

Stetson, M. H., and Watson-Whitmyre, M. (1976) Nucleus suprachiasmaticus: The biological clock in the hamster? Science, 191, 197-199.

Tamarkin, L., Brown, S., and Goldman, B. (1975) Neuroendocrine regulation of seasonal reproductive cycles in the hamster. Abstr. 5th Ann. Mtg. Soc. Neuroscience, N.Y. p. 458.

Tamarkin, L., Westrom, W. K., Hamill, A. I., and Goldman, B. D. (1976) Effect of melatonin on the reproductive systems of male and female Syrian hamsters: A diurnal rhythm in sensitivity to melatonin. Endocrinology 99, 1534-1541.

Tamarkin, L., Lefebvre, N. G., Hollister, C. W. and Goldman, B. D. (1977) Effect of melatonin administered during the night on reproductive function in the Syrian hamster. Endocrinology 101, 631-634.

Turek, F. W., Desjardins, C., and Menaker, M. (1975) Melatonin: Antigonadal and progonadal effects in male golden hamsters. Science, 190, 280-282.

Vaughan, M. K., Vaughan, G. M., Blask, D. E., Barnett, M. P. and Reiter, R. J. (1976) Arginine vasotocin: Structure-activity relationships and influence on gonadal growth and function. Amer. Zool. 16, 25-34.

Wurtman, R. J. (1973) Introduction: Neurotransmitters and monoamines. Fed. Proc. 32, 1769-1772.

DISCUSSION

Lynch: One of the graduate students in my laboratory, Mr. Ozaki, has recently completed a study that is relevant to Dr. Reiter's presentation. Four-day cycling female rats were housed individually in metabolic cages, and urine specimens were collected during the daily dark period for 12 consecutive days. Each urine sample was assayed for melatonin and norepinephrine. The phase of the vaginal estrous cycle associated with each urine specimen was determined by preparation of daily vaginal smears. We found that proestrus consistently was associated with a depression in melatonin excretion. Also, there was a relationship between the level of norepinephrine in the urine and its melatonin content. The slope of the curve relating these two compounds in proestrus - estrus differed from that in metestrus - diestrus specimens, suggesting that factors other than the catecholamine were involved in the regulation of melatonin secretion. We suspected that gonadal steroids might be involved. Ovariectomized rats showed an increase in their serum melatonin content. Combined treatment with estrogen and progesterone depressed circulating melatonin levels, whereas neither of the steroids were effective alone. These data seem to suggest an antipineal effect of the gonadal steroids.

Ward: I noticed that by simply turning out the lights on the hamster you get a testicular cross-section that looks as if it is loaded with Sertoli cells. Are any of the people who are interested in Sertoli cell function working with the hamster?

Reiter: Yes. There is an individual in our department who is interested in Sertoli cells, and his method for getting a preparation with a large number of Sertoli cells is to place the animals in darkness.

Schwartz: Does he find the reflection of the "Sertoli cell-only syndrome" with respect to FSH and LH in the serum? Fred Turek looked at serum FSH and LH in short photoperiods and both of them are suppressed.

Reiter: To my knowledge, he has not measured circulating gonadotropins.

Nalbandov: Can the same kind of evidence be presented for laboratory rats?

Reiter: My feeling is that breeders have bred out of rats the ability to respond to external environmental cues. If rats are placed in darkness they undergo a very modest suppression of reproductive physiology which can be prevented by pinealectomy. It is necessary to do other things to rats to get them to respond to the pineal gland in a dramatic way, for example, give them estrogen shortly after birth. It is not a very highly photosensitive species, and is a poor animal choice in which to demonstrate the physiological consequences of the pineal gland.

Baker: Is there any evidence in primates, either the human or rhesus monkey, that blinding produces effects like this?

Reiter: It has been reported that 17-keto steroid excretion and fertility are depressed in blind individuals. However, there may be other effects of blindness, such as psychological effects, which may cause the depressed fertility. There are also studies from Lapland which show a definite seasonal change in conception rate. The higher conception rate occurred during the summer months, and the lower conception rate occurred during the short winter photoperiods. We are interested in the role of the pineal gland in the human, particularly in blind subjects and in individuals with pineal tumors. Photoperiod is not the only synchronizer of annual reproductive cycles. Some are synchronized by rainfall, some are synchronized by other factors. If it is a photoperiodic response, the pineal is probably involved.

van Tienhoven: Is anything known about sheep? In these animals a decrease in photoperiod is stimulatory rather than inhibitory.

Reiter: Animals have all kinds of ploys to ensure that they give birth in the spring. There is only one study that I am aware of in sheep. Animals were pinealectomized, but were only followed for one year. The animals came into estrus at the normal time the following fall, and it was concluded that pinealectomy had no effect. It is unfortunate that these animals were not studied longer because it is possible that the following year the onset of estrus may have been delayed. At this point we simply don't know if the pineal is involved in short-day breeders. If the sexual cycle is sensitive to changing photoperiod, then it is likely that the pineal acts as a mediator of the cyclic changes. It is natural that animals would evolve a mechanism that relies on photoperiod because it is so regular from year to year.

EPILOGUE

Epilogue

TERRY G. BAKER

At the start of this meeting I thought that the problems to be discussed were fairly clear-cut and understood. However, on reflection I now feel somewhat confused by the apparent complexity of the processes involved in the control of reproductive function. Perhaps we all need to spend some time considering the implications of these new concepts, and to review all of the data in the light of the earlier classical studies between 1920 and 1970, before attempting to revise the previously held concepts.

The first question that I would raise concerns the large number of regulatory agents that we have been told are involved in the control of the hypothalamus, pituitary gland, ovary, testis, and even the early embryo. A cytologist might question whether the cells within these organs are sufficiently complex to produce the additional substances involved, or to house the multitude of receptors which are seemingly required. It could be asked whether there really are so many specific receptors on cell membrane systems or merely a small number whose activity is modified by hormonal interaction. Are there really so many proteins (inhibitors, stimulators, regulators) being produced or are we merely looking at the way in which one substance can act (or interact), perhaps at different stages in a cycle of the cell's activity, or in respect to changing relationships between closely associated cell types?

It should be remembered that not so many years ago 'new' steroid hormones were being reported in almost every issue of the journals relating to endocrinology. However, it was soon realized that many of these substances were either intermediates in pathways of steroid biosynthesis, or were metabolic products associated with hormone excretion, and had very little biological activity in their own right. Might this not also be true of some of the substances which are now being involved in endocrine regulation?

Perhaps in about 10 years time The Upjohn Company will need to convene another Brook Lodge Workshop to critically reassess some of the ideas which have seemed so interesting during the past few days. There can be little doubt that some of the concepts (and the 'new' substances involved) will not stand the test of time. But perhaps the real value of the Seventh Brook Lodge Workshop will then become apparent; namely that the problems posed made us reconsider our views and carry out the definitive *in vivo* and *in vitro* experiments required to evaluate the present status of knowledge on the control of reproductive function.

INDEX